TABLES OF

INDEFINITE INTEGRALS

By G. PETIT BOIS

DOVER PUBLICATIONS, INC.
NEW YORK

Published in Canada by General Publishing Company, Ltd., 30 Lesmill Road, Don Mills, Toronto, Ontario.

Published in the United Kingdom by Constable and Company, Ltd., 10 Orange Street, London WC 2.

This new Dover edition, first published in 1961, is a new unabridged English translation of the work first published by B. G. Teubner in 1906 under the title *Tafeln der unbestimmten Integrale.*

Standard Book Number: 486-60225-7

Manufactured in the United States of America
Dover Publications, Inc.
180 Varick Street
New York, N.Y. 10014

Preface

The textbooks on infinitesimal calculus contain a great many solved indefinite integrals. If, however, one wishes to look up one or another of these integrals, one often loses much time in hunting for it in these books.

It can therefore be very useful to put together these formulas in methodically classified groups.

Such a collection seems to me to be of great use to students of the infinitesimal calculus. For this reason I have also included numerous applications of the given formulas and at the center of the work I have furnished the principal transformations of the integrals.

The following is a list of the books which I have found most useful in my work:

J. A. SCHUBERT, *Sammlung von Differential und Integral Formeln*

FD. MINDING, *Sammlung von Integraltafeln*

L. A. SOHNCKE'S, *Sammlung von Aufgaben aus der Integralrechnung*

F. FRENET, *Recueil d'exercices sur le Calcul infinitésimal*

J. GRAINDORGE, *Exercices de Calcul intégral*

ED. BRAHY, *Exercices méthodiques de Calcul intégral*

D. F. GREGORY, *Examples of the Processes of the Differential and Integral Calculus*

RALPH A. ROBERTS, *A Treatise on the Integral Calculus*

CARR, *Integral Calculus*

Notations

I have represented by X, Y polynomials in x, complete or incomplete, of any degree whatsoever, and by X^n, Y^n these same polynomials when their degree is n.

I have designated by U, V the functions of x, generally rational; their successive derivatives are represented respectively by U', U'', U''' . . . V', V'', V''' . . . and their successive integrals by U_1, U_2, U_3 . . . V_1, V_2, V_3. . . . I have often made use of the following abbreviation:

Instead of $$\int \frac{x}{(a+bx+cx^2)^2}\,dx = -\frac{1}{4c\,(a+bx+cx^2)^2}$$
$$-\frac{b}{2c}\int \frac{dx}{(a+bx+cx^2)^3} + C.$$

I have written $$\int \frac{x}{(a+bx+cx^2)^2}\,dx = -\frac{1}{4c\Phi^2} - \frac{b}{2c}\int \frac{dx}{\Phi^3} + C.$$

Instead of $$\int \frac{x}{\sqrt{x^2+x+1}}\,dx = \sqrt{x^2+x+1}$$
$$-\frac{1}{2}\log\left(1+2x+2\sqrt{x^2+x+1}\right) + C,$$

I have written $$\int \frac{x}{\sqrt{x^2+x+1}}\,dx = \sqrt{\Phi}$$
$$-\frac{1}{2}\log\left(1+2x+2\sqrt{\Phi}\right) + C.$$

It will therefore be necessary when comparing two formulas containing the letter Φ, first to replace this letter by its appropriate value.

For the sake of simplicity of classification, I have introduced among the binomial integrals, $\int x^m (a + bx^n)^p\, dx$, (pp. 74-97), the integrals in which p may be a whole number and m equal to zero. It is therefore in this class that one must look for

$$\int \frac{dx}{1+x^n}, \qquad \int \frac{x}{(a+bx^n)^2}\,dx, \qquad \text{etc.}$$

I have grouped under the title "Function x^m" the integrals characterized by this function and which could not be classified in other groups.

I have omitted the arbitrary constant which should accompany each integral.

The sign *log* designates a Napierian logarithm.

In the Index, the notation $\dfrac{1}{X^4\,(2)}$ has been used to indicate that the denominator of the fraction includes only two terms.

In the transcendental functions, I have designated the functions $\sin x$, $\cos x$, etc., in a general way by Cir x.

The letter $\mathfrak{D}$., used in the chapter which follows, signifies "derivative of." For example: $\mathfrak{D}.\ x^m = mx^{m-1}$. This chapter might have been abridged inasmuch as several of the relations which figure in it are directly deduced one from the other.

Transformation of Integral Expressions

1. If $\mathrm{U} = \mathrm{V}$, one has $\mathfrak{D}.\log \mathrm{U} = \mathfrak{D}.\log \mathrm{V}$.

Examples:

$$\mathfrak{D}.\log \frac{\sqrt{x^2 - a^2}}{x + a} = \mathfrak{D}.\log \frac{x - a}{\sqrt{x^2 - a^2}}. \qquad \mathfrak{D}.\log \frac{\sqrt{x^2 + 1}}{x\sqrt{2} - \sqrt{x^2 - 1}} = \mathfrak{D}.\log \frac{x\sqrt{2} + \sqrt{x^2 - 1}}{\sqrt{x^2 + 1}}$$

$$\mathfrak{D}.\log \frac{\mathrm{U} + \mathrm{V}}{\mathrm{U} - \mathrm{V}} = \mathfrak{D}.\log \left\{ \frac{\mathrm{U} + \mathrm{V}}{\sqrt{\mathrm{U}^2 - \mathrm{V}^2}} \right\}^2 = \mathfrak{D}.\, 2 \log \frac{\mathrm{U} + \mathrm{V}}{\sqrt{\mathrm{U}^2 - \mathrm{V}^2}}.$$

2. $\mathfrak{D}.\log k\mathrm{U} = \mathfrak{D}.\log \mathrm{U}. \quad \mathfrak{D}.\log \left(\frac{\mathrm{U}}{a} + \frac{\mathrm{V}}{b}\right) = \mathfrak{D}.\log \frac{b\mathrm{U} + a\mathrm{V}}{ab} = \mathfrak{D}.\log (b\mathrm{U} + a\mathrm{V}).$

$$\mathfrak{D}.\log \left\{ ax + \sqrt{a(ax^2 + b)} \right\} = \mathfrak{D}.\log \left\{ x\sqrt{a} + \sqrt{ax^2 + b} \right\}.$$

$$\mathfrak{D}.\log \frac{\sqrt{x - b} + \sqrt{x - a}}{\sqrt{x - b} - \sqrt{x - a}} = \mathfrak{D}.\log \frac{(\sqrt{x - b} + \sqrt{x - a})^2}{a - b} = \mathfrak{D}.\, 2 \log (\sqrt{x - b} + \sqrt{x - a}).$$

$$\mathfrak{D}.\log (x + 2 + \sqrt{x^2 + 4x}) = \mathfrak{D}.\log (2x + 4 + 2\sqrt{x^2 + 4x}) = \mathfrak{D}.\log (\sqrt{x} + \sqrt{x + 4})^2.$$

$$= \mathfrak{D}.\, 2 \log (\sqrt{x} + \sqrt{x + 4}).$$

3 $\mathfrak{D}.\log \mathrm{U} = \mathfrak{D}.\log (-\mathrm{U}). \quad \mathfrak{D}.\log \frac{a + x}{b - x} = \mathfrak{D}.\log \frac{a + x}{x - b}. \quad \mathfrak{D}.\log \frac{-kx}{\sqrt{ax^2 + b}} = \mathfrak{D}.\log \frac{x}{\sqrt{ax^2 + b}}.$

4. $\mathfrak{D}.\log \frac{\mathrm{U}}{\mathrm{V}} = -\mathfrak{D}.\log \frac{\mathrm{V}}{\mathrm{U}}. \quad \mathfrak{D}.\log \frac{bx - a}{bx + a} = -\mathfrak{D}.\log \frac{bx + a}{bx - a}. \quad \mathfrak{D}.\log \frac{1}{x} = -\mathfrak{D}.\log x.$

5. If $\mathrm{U}^2 - \mathrm{V}^2 = k$, one has $\mathfrak{D}.\log (\mathrm{U} + \mathrm{V}) = -\mathfrak{D}.\log (\mathrm{U} - \mathrm{V})$. $\quad \mathfrak{D}.\log (\sqrt{x^2 + 1} + x) = -\mathfrak{D}.\log (\sqrt{x^2 + 1} - x).$

$$-\mathfrak{D}.\, 2\log (\sqrt{x + 1} - \sqrt{x - 1}) = -\mathfrak{D}.\log (2x - 2\sqrt{x^2 - 1}) = -\mathfrak{D}.\log (x - \sqrt{x^2 - 1}) = \mathfrak{D}.\log (x + \sqrt{x^2 - 1}).$$

6 If $\mathrm{U}^2 - \mathrm{V}^2 = k$, one has $\mathfrak{D}.\log \frac{\mathrm{U} + \mathrm{V}}{\mathrm{U} - \mathrm{V}} = \mathfrak{D}.\, 2 \log (\mathrm{U} + \mathrm{V})$ et $\mathfrak{D}.\log \frac{\mathrm{U} - \mathrm{V}}{\mathrm{U} + \mathrm{V}} = \mathfrak{D}.\, 2 \log (\mathrm{U} - \mathrm{V}).$

$$\mathfrak{D}.\log \frac{\sqrt{ax + b} + \sqrt{ax}}{\sqrt{ax + b} - \sqrt{ax}} = \mathfrak{D}.\, 2 \log (\sqrt{ax + b} + \sqrt{ax}).$$

7. If $\mathrm{U}^2 - \mathrm{V}^2 = k\mathrm{W}^2$, one has $\mathfrak{D}.\log \frac{\mathrm{U} + \mathrm{V}}{\mathrm{W}} = -\mathfrak{D}.\log \frac{\mathrm{U} - \mathrm{V}}{\mathrm{W}}. \quad \mathfrak{D}.\log \frac{\sqrt{a^2 + b^2x^2} + a}{x} = -\mathfrak{D}.\log \frac{\sqrt{a^2 + b^2x^2} - a}{x}.$

$$\mathfrak{D}.\log \frac{2c + bx + 2\sqrt{c(ax^2 + bx + c)}}{x} = -\mathfrak{D}.\log \frac{2c + bx - 2\sqrt{c(ax^2 + bx + c)}}{x}.$$

8. $\mathfrak{D}.\log (\mathrm{U} - \mathrm{V})^{\frac{p}{q}} = \mathfrak{D}.\log (\mathrm{V} - \mathrm{U})^{\frac{p}{q}}. \quad \mathfrak{D}.\log (x - a)^m = \mathfrak{D}.\,\mathrm{og} (a - x)^m. \quad \mathfrak{D}.\log \sqrt{a^2 - x^2} = \mathfrak{D}.\log \sqrt{x^2 - a^2}$

$$\mathfrak{D}.\log\frac{a\sqrt{1-x^2}-x\sqrt{1-a^2}}{\sqrt{x^2-a^2}}=\mathfrak{D}.\log\frac{a\sqrt{1-x^2}-x\sqrt{1-a^2}}{\sqrt{a^2-x^2}}.$$

9. $\mathfrak{D}.\log\left\{\sqrt{ax+b}+\sqrt{mx+n}\right\}=\mathfrak{D}.\frac{1}{2}\log\left\{\frac{a+m}{2}x+\frac{b+n}{2}+\sqrt{(ax+b)(mx+n)}\right\}.$

$$\mathfrak{D}.\log\left(\sqrt{x+a}+\sqrt{x-a}\right)=\mathfrak{D}.\frac{1}{2}\log\left(x+\sqrt{x^2-a^2}\right).$$

10. $\mathfrak{D}.\log\frac{x\sqrt{a}-\sqrt{ax^2+bx+c}+\sqrt{c}}{x\sqrt{a}-\sqrt{ax^2+bx+c}-\sqrt{c}}=\mathfrak{D}.\log\frac{x}{2c+bx+2\sqrt{c(ax^2+bx+c)}}$

$$\mathfrak{D}.\log\frac{x-\sqrt{x^2+x+1}+1}{x-\sqrt{x^2+x+1}-1}=\mathfrak{D}.\log\frac{x}{2+x+2\sqrt{x^2+x+1}}.$$

11. $\mathfrak{D}$ arc sin U $= -\mathfrak{D}.$ arc sin $(-\text{U})$. (*) $\qquad \mathfrak{D}.$ arc sin $(2x-1) = -\mathfrak{D}.$ arc sin $(1-2x)$.

12. $\mathfrak{D}.$ arc sin U $= -\mathfrak{D}.$ arc sin $\sqrt{1-\text{U}^2}$ $\qquad \mathfrak{D}.$ arc sin $\sqrt{1-ax} = -\mathfrak{D}.$ arc sin $\sqrt{ax}$

$$\mathfrak{D}.\ \text{arc sin}\,\frac{x\sqrt{2}}{1+x^2}=-\mathfrak{D}.\ \text{arc sin}\,\frac{\sqrt{x^4+1}}{1+x^2}$$

13. $\mathfrak{D}.$ arc sin U $=\mathfrak{D}.\frac{1}{2}$ arc sin $2\text{U}\sqrt{1-\text{U}^2}$. $\qquad \mathfrak{D}.\ \text{arc sin}\,\frac{\sqrt{2x^2-a}}{x\sqrt{2}}=\mathfrak{D}.\frac{1}{2}\ \text{arc sin}\,\frac{x^2-a}{x^2}.$

14. $\mathfrak{D}.$ arc sin $(2a^2\text{U}^2-1)=\mathfrak{D}.$ 2 arc sin aU. $\qquad \mathfrak{D}.\ \text{arc sin}\,\frac{a^2x^2-b^2}{b^2}=\mathfrak{D}.\ 2\ \text{arc sin}\,\frac{ax}{b\sqrt{2}}.$

15. $\mathfrak{D}.$ arc sin $\sqrt{1-a\text{U}}=\mathfrak{D}.\frac{1}{2}$ arc sin $(1-2a\text{U})$. $\qquad \mathfrak{D}.\ \text{arc sin}\sqrt{\frac{1+x^m}{2}}=\mathfrak{D}.\frac{1}{2}\ \text{arc sin}\ x^m.$

16. $\mathfrak{D}.\ \text{arc sin}\sqrt{a\text{U}+\frac{1}{2}}=\mathfrak{D}.\frac{1}{2}$ arc sin $2a$U. $\qquad \mathfrak{D}.\ \text{arc sin}\sqrt{\frac{x+a}{2a}}=\mathfrak{D}.\frac{1}{2}\ \text{arc sin}\,\frac{x}{a}.$

17. $\mathfrak{D}.$ arc sin U $= -\mathfrak{D}.$ arc cos U.

18. $\mathfrak{D}.$ arc sin U $=\mathfrak{D}.$ arc cos $\sqrt{1-\text{U}}$. $\qquad \mathfrak{D}.\ \text{arc sin}\,\frac{x}{\sqrt{1+x^2}}=\mathfrak{D}.\ \text{arc cos}\,\frac{1}{\sqrt{1+x^2}}.$

19. $\mathfrak{D}.$ arc sin U $=\mathfrak{D}.\ \text{arc tg}\,\frac{\text{U}}{\sqrt{1-\text{U}^2}}$ $\qquad \mathfrak{D}.\ \text{arc sin}\ x\sqrt{\frac{a}{ax^2+b}}=\mathfrak{D}.\ \text{arc tg}\ x\sqrt{\frac{a}{b}}.$

$$\mathfrak{D}.\ \text{arc sin}\,\frac{bx-2c}{x\sqrt{b^2+4ac}}=\mathfrak{D}.\text{arc tg}\,\frac{bx-2c}{2\sqrt{c(ax^2+bx-c)}}$$

20 $\mathfrak{D}.$ arc sin U $=\mathfrak{D}.\frac{1}{2}\ \text{arc tg}\,\frac{2\text{U}\sqrt{1-\text{U}^2}}{1-2\text{U}^2}$ $\qquad \mathfrak{D}.\ \text{arc sin}\sqrt{1-x^2}=\mathfrak{D}.\frac{1}{2}\ \text{arc tg}\,\frac{2x\sqrt{1-x^2}}{2x^2-1}.$

(*) One has $(x\text{-}a)^2 = (a\text{-}x)^2$. It is important, in the derivation of certain functions, to choose one or the other of these forms. One mav ascertain which is appropriate by verifying the equality

$$\mathfrak{D}.\ \text{arc sin}\,\frac{a-x}{x-b}=\mathfrak{D}.\ \text{arc sin}\,\frac{x-a}{b-x}.$$

21. $\mathfrak{D}.\ \text{arc sin}\ U = \mathfrak{D}.\ 2\ \text{arc tg}\ \frac{1-\sqrt{1-U^2}}{U}$ $\mathfrak{D}.\ \text{arc sin}\ \frac{1}{x} = \mathfrak{D}.\ 2\ \text{arc tg}\ (x - \sqrt{x^2-1})$.

22 $\mathfrak{D}.\ \text{arc sin}\ U = \mathfrak{D}.\ 2\ \text{arc tg}\ \sqrt{\frac{1+U}{1-U}}$ $\mathfrak{D}.\ \text{arc sin}\ \frac{2ax-b}{b} = \mathfrak{D}.\ 2\ \text{arc tg}\ \sqrt{\frac{ax}{b-ax}}$.

23. $\mathfrak{D}\ \text{arc sin}\ \frac{2x-b}{\sqrt{4c+b^2}} = \mathfrak{D}.\ 2\ \text{arc tg}\ \frac{x}{\sqrt{c+bx-x^2}-\sqrt{c}}$ $\mathfrak{D}.\ \text{arc sin}\ \frac{2x+1}{\sqrt{5}} = \mathfrak{D}.\ 2\ \text{arc tg}\ \frac{x}{\sqrt{1-x-x^2}-1}$.

24. $\mathfrak{D}.\ \text{arc sin}\ \frac{prx+qr}{r^2x+pq} = \mathfrak{D}.\ 2\ \text{arc tg}\ \sqrt{\frac{r+p}{r-p}\cdot\frac{rx+q}{rx-q}}$ where $r > o,\quad r^2 > p^2,\quad -1 < \frac{p}{r} < 1.$

$\mathfrak{D}.\ \text{arc sin}\ \frac{2x+15}{10x+3} = \mathfrak{D}.\ 2\ \text{arc tg}\ \sqrt{\frac{3}{2}\cdot\frac{2x+3}{2x-3}}$.

$\mathfrak{D}.\ \text{arc sin}\ \frac{15-2x}{10x-3} = \mathfrak{D}.\ 2\ \text{arc tg}\ \sqrt{\frac{2}{3}\cdot\frac{2x+3}{2x-3}}$.

25. $\mathfrak{D}.\ \text{arc sin}\ \frac{prx+qr}{r^2x+pq} = -\ \mathfrak{D}.\ 2\ \text{arc tg}\ \sqrt{\frac{p-r}{p+r}\cdot\frac{q-rx}{q+rx}}$ where $r > o, p > o,\quad p^2 > r^2,\quad -1 > \frac{p}{r} > 1.$

$\mathfrak{D}.\ \text{arc sin}\ \frac{4x+1}{x+4} = -\ \mathfrak{D}.\ 2\ \text{arc tg}\ \sqrt{\frac{3}{5}\cdot\frac{1-x}{1+x}}$.

26. $\mathfrak{D}.\ \text{arc sin}\ \frac{prx+qr}{r^2x+pq} = \mathfrak{D}.\ 2\ \text{arc tg}\ \sqrt{\frac{p-r}{p+r}\cdot\frac{q-rx}{q+rx}}$ where $r > o,\ p < o,\ p^2 > r^2,\ -1 > \frac{p}{r} > 1.$

$\mathfrak{D}.\ \text{arc sin}\ \frac{1-4x}{x-4} = \mathfrak{D}.\ 2\ \text{arc tg}\ \sqrt{\frac{5}{3}\cdot\frac{1-x}{1+x}}$

27. $\mathfrak{D}.\ \text{arc cos}\ U = \mathfrak{D}.\ \frac{1}{2}\ \text{arc cos}\ (2U^2-1)$ $\mathfrak{D}.\ \text{arc cos}\ \frac{x+1}{\sqrt{2}} = \mathfrak{D}.\ \frac{1}{2}\ \text{arc cos}\ (x^2+2x)$.

28. $\mathfrak{D}.\ \text{arc cos}\ U = -\ \mathfrak{D}.\ \text{arc cos}\ \sqrt{1-U^2}$

29. $\mathfrak{D}\ \text{arc cos}\ U = \mathfrak{D}.\ \frac{1}{2}\ \text{arc sin}\ 2U\sqrt{1-U^2}$ $\mathfrak{D}.\ \text{arc cos}\ \frac{1}{\sqrt{1+x}} = \mathfrak{D}.\ \frac{1}{2}\ \text{arc sin}\ \frac{2\sqrt{x}}{1+x}$.

30. $\mathfrak{D}.\ \text{arc cos}\ U = \mathfrak{D}.\ \text{arc sin}\ \sqrt{1-U^2}$.

31. $\mathfrak{D}.\ \text{arc cos}\ U = \mathfrak{D}\ 2\ \text{arc sin}\ \sqrt{\frac{1-U}{2}}$. $\mathfrak{D}.\ \text{arc cos}\ (1-2x^2) = \mathfrak{D}.\ 2\ \text{arc sin}\ x$.

32. $\mathfrak{D}.\ \text{arc cos}\ U = \mathfrak{D}\ \text{arc tg}\ \frac{\sqrt{1-U^2}}{U}$ $\mathfrak{D}.\ \text{arc cos}\ \sqrt{\frac{b-ax}{b}} = \mathfrak{D}.\ \text{arc tg}\ \sqrt{\frac{ax}{b-ax}}$.

33. $\mathfrak{D}\ \text{arc cos}\ U = \mathfrak{D}.\ 2\ \text{arc tg}\ \sqrt{\frac{1-U}{1+U}}$. $\mathfrak{D}.\ \text{arc cos}\ \frac{1-x^2}{1+x^2} = \mathfrak{D}.\ 2\ \text{arc tg}\ x$.

34. $\mathfrak{D}.\ \text{arc tg}\ U = \mathfrak{D}.\ \frac{1}{2}\ \text{arc tg}\ \frac{2U}{1-U^2}$. $\mathfrak{D}.\ \text{arc tg}\ \sqrt{\frac{x-a}{x+a}} = \mathfrak{D}.\ \frac{1}{2}\ \text{arc tg}\ \frac{\sqrt{x^2-a^2}}{a}$.

35. $\mathfrak{D}.\ \text{arc tg}\ U = \mathfrak{D}.\ \text{arc tg}\ \frac{U-a}{aU+1}$. $\mathfrak{D}.\ \text{arc tg}\ \frac{mx+n}{px+q} = \mathfrak{D}.\ \text{arc tg}\ \frac{(m^2+p^2)x+mn+pq}{mq-pn}$.

$\mathfrak{D}.\ \text{arc tg}\ (x-1) = \mathfrak{D}.\ \text{arc tg}\ \frac{x-2}{x}$.

36. $\mathfrak{D}.\ \text{arc tg}\ \mathrm{U} = -\ \mathfrak{D}.\ \text{arc tg}\ (-\mathrm{U}).$

37. $\mathfrak{D}.\ \text{arc tg}\ \mathrm{U} = -\ \mathfrak{D}.\ \text{arc tg}\ \frac{1}{\mathrm{U}}.$

38. $\mathfrak{D}.\ \text{arc tg}\ \frac{\mathrm{U}}{\mathrm{V}} = -\ \mathfrak{D}.\ \text{arc tg}\ \frac{\mathrm{V}}{\mathrm{U}}.$ $\qquad \mathfrak{D}.\ \text{arc tg}\ \sqrt{\frac{a+x}{a-x}} = -\ \mathfrak{D}.\ \text{arc tg}\ \sqrt{\frac{a-x}{a+x}}.$

39. $\mathfrak{D}.\ \text{arc tg}\ \frac{ax+b}{c} = \mathfrak{D}.\ \text{arc tg}\ \frac{acx}{abx+b^2+c^2}.$ $\qquad \mathfrak{D}.\ \text{arc tg}\ \frac{2x+1}{\sqrt{3}} = \mathfrak{D}.\ \text{arc tg}\ \frac{x\sqrt{3}}{x+2}.$

40. $\mathfrak{D}.\ \text{arctg}\ \frac{ax}{b\sqrt{a^2+1}-x} = \mathfrak{D}.\text{arctg}\ \frac{x\sqrt{a^2+1}-b}{ab}.$ $\qquad \mathfrak{D}.\ \text{arc tg}\ \frac{x\sqrt{3}}{2a-x} = \mathfrak{D}.\ \text{arc tg}\ \frac{2x-a}{a\sqrt{3}}.$

41. $\mathfrak{D}.\ \text{arc tg}\ \frac{2ax-b}{2\sqrt{a(c+bx-ax^2)}} = \mathfrak{D}.2\ \text{arc tg}\ \frac{x\sqrt{a}}{\sqrt{c+bx-ax^2}+\sqrt{c}}.$ $\qquad \mathfrak{D}.\text{arctg}\ \frac{x}{\sqrt{a^2-x^2}} = \mathfrak{D}.2\ \text{arc tg}\ \frac{x}{\sqrt{a^2-x^2}+a}.$

42. $\mathfrak{D}.\ \text{arc tg}\ \mathrm{U} = \mathfrak{D}.\ \text{arc sin}\ \frac{\mathrm{U}}{\sqrt{1+\mathrm{U}^2}}.$ $\qquad \mathfrak{D}.\ \text{arc tg}\ \frac{b}{a}\sqrt{\frac{x^2-a^2}{b^2-x^2}} = \mathfrak{D}.\ \text{arc sin}\ \frac{b}{x}\sqrt{\frac{x^2-a^2}{b^2-a^2}}.$

43. $\mathfrak{D}.\ \text{arc tg}\ \mathrm{U} = -\ \mathfrak{D}.\ \text{arc sin}\ \frac{1}{\sqrt{1+\mathrm{U}^2}}.$ $\qquad \mathfrak{D}.\ \text{arc tg}\ \sqrt{x^2-1} = -\ \mathfrak{D}.\ \text{arc sin}\ \frac{1}{x}.$

44. $\mathfrak{D}.\ \text{arc tg}\ \mathrm{U} = \mathfrak{D}.\ \frac{1}{2}\ \text{arc sin}\ \frac{2\mathrm{U}}{1+\mathrm{U}^2}.$ $\qquad \mathfrak{D}.\ \text{arc tg}\ \frac{2ax+b}{\sqrt{4ac-b^2}} = \mathfrak{D}.\ \frac{1}{2}\ \text{arc sin}\ \frac{(2ax+b)\sqrt{4ac-b^2}}{2a\,(ax^2+bx+c)}.$

45. $\mathfrak{D}.\ \text{arc tg}\ \mathrm{U} = \mathfrak{D}.\ \frac{1}{2}\ \text{arc sin}\ \frac{\mathrm{U}^2-1}{\mathrm{U}^2+1}.$ $\qquad \mathfrak{D}.\ \text{arc tg}\ \frac{1}{x}\sqrt{\frac{b}{a}} = \mathfrak{D}.\ \frac{1}{2}\ \text{arc sin}\ \frac{b-ax^2}{b+ax^2}.$

46. $\mathfrak{D}.\ \text{arc tg}\ \mathrm{U} = \mathfrak{D}.\ 2\ \text{arc sin}\ \sqrt{\frac{1}{2}-\frac{1}{2\sqrt{\mathrm{U}^2+1}}}.$ $\qquad \mathfrak{D}.\ \text{arc tg}\ \frac{\sqrt{1-x^2}}{x} = \mathfrak{D}.\ 2\ \text{arc sin}\ \sqrt{\frac{1-x}{2}}.$

47. $\mathfrak{D}.\ \text{arc tg}\ \mathrm{U} = \mathfrak{D}.\ \text{arc cos}\ \frac{1}{\sqrt{1+\mathrm{U}^2}}.$ $\qquad \mathfrak{D}.\ \text{arc tg}\ \sqrt{x-1} = \mathfrak{D}.\ \text{arc cos}\ \frac{1}{\sqrt{x}}.$

48. $\mathfrak{D}.\ \text{arc tg}\ \mathrm{U} = \mathfrak{D}\ \frac{1}{2}\ \text{arc cos}\ \frac{1-\mathrm{U}^2}{1+\mathrm{U}^2}.$ $\qquad \mathfrak{D}.\ \text{arc tg}\ x\sqrt{\frac{a}{b}} = \mathfrak{D}.\ \frac{1}{2}\ \text{arc cos}\ \frac{b-ax^2}{b+ax^2}.$

49. $\mathfrak{D}.\left\{\text{arc tg}\ \mathrm{U} \pm \text{arc tg}\ \mathrm{V}\right\} = \mathfrak{D}.\ \text{arc tg}\ \frac{\mathrm{U}\pm\mathrm{V}}{1\mp\mathrm{UV}}.$

Table of Contents

Algebraic Functions

Transcendental Functions

Algebraic Functions

$$\frac{mx + n}{ax + b}$$

$$\int dx = x = 2 \operatorname{arc\,tg} \sqrt{\frac{1 - \cos x}{1 + \cos x}}\ .$$

$$\int (ax + b)\, dx = \frac{a}{2} x^2 + bx = \frac{(ax + b)^2}{2a}.$$

$$\int \frac{dx}{x} = \log x = \log \left(\frac{x}{a} + \frac{x}{b}\right).$$

$$\int \frac{dx}{x + a} = \log (x + a).$$

$$\int \frac{dx}{x - a} = \log (x - a) = \log (a - x).$$

$$\int \frac{dx}{ax + b} = \frac{1}{a} \log (ax + b).$$

$$\int \frac{dx}{ax - b} = \frac{1}{a} \log (ax - b) = \frac{2}{a} \log \sqrt{ax - b} = \frac{2}{a} \log \sqrt{b - ax}.$$

$$\int \frac{dx}{b - ax} = -\frac{1}{a} \log (b - ax) = \frac{1}{a} \log \frac{a}{b - ax}.$$

$$\int \frac{x}{x + a}\, dx = x - a \log (x + a).$$

$$\int \frac{x}{ax + b}\, dx = \frac{x}{a} - \frac{b}{a^2} \log (ax + b).$$

$$\int \frac{x}{b - ax}\, dx = -\frac{1}{a} \log (b - ax) = -\frac{x}{a} - \frac{b}{a^2} \log (b - ax).$$

$$\int \frac{mx + n}{ax + b}\, dx = \frac{m}{a} x + \frac{an - bm}{a^2} \log (ax + b).$$

$$\frac{1}{ax^2 + bx + c}$$

$$\int \frac{dx}{x^2 + 1} = \operatorname{arc\,tg} x = \operatorname{arc\,tg} \frac{x - 1}{x + 1} = \operatorname{arc\,tg} \frac{x - a}{ax + 1} = \frac{1}{3} \operatorname{arc\,tg} \frac{3x - x^3}{1 - 3x^2}.$$

$$= -2 \operatorname{arc\,tg} (\sqrt{x^2 + 1} - x) = \operatorname{arc\,sin} \frac{x}{\sqrt{1 + x^2}}$$

$$\int \frac{dx}{x^2 + a^2} = \frac{1}{a} \operatorname{arc\,tg} \frac{x}{a} = \frac{1}{a} \operatorname{arc\,sin} \frac{x}{\sqrt{x^2 + a^2}} = -\frac{1}{a} \operatorname{arc\,sin} \frac{a}{\sqrt{x^2 + a^2}}$$

$$\int\frac{dx}{x^2-a^2}=\frac{1}{2a}\log\frac{x-a}{x+a}=\frac{1}{2a}\log\frac{a-x}{a+x}.$$

$$\int\frac{dx}{ax^2-b}=\frac{1}{2\sqrt{ab}}\log\frac{x\sqrt{a}-\sqrt{b}}{x\sqrt{a}+\sqrt{b}}=\frac{1}{\sqrt{ab}}\log\frac{x\sqrt{a}-\sqrt{b}}{\sqrt{ax^2-b}}=-\frac{1}{\sqrt{ab}}\log\frac{x\sqrt{a}+\sqrt{b}}{\sqrt{ax^2-b}}.$$

$$\int\frac{dx}{a^2-x^2}=\frac{1}{2a}\log\frac{a+x}{a-x}=\frac{1}{2a}\log\frac{x+a}{x-a}.$$

$$\int\frac{dx}{b-ax^2}=\frac{1}{2\sqrt{ab}}\log\frac{\sqrt{b}+x\sqrt{a}}{\sqrt{b}-x\sqrt{a}}=\frac{1}{\sqrt{ab}}\log\frac{\sqrt{b}-x\sqrt{a}}{\sqrt{b-ax^2}}.$$

$$\int\frac{dx}{(a-b)\,x^2+(a+b)}=\frac{1}{\sqrt{a^2-b^2}}\text{arc tg}\,x\sqrt{\frac{a-b}{a+b}}. \qquad a>b.$$

$$=\frac{1}{2\sqrt{b^2-a^2}}\log\frac{\sqrt{a+b}+x\sqrt{b-a}}{\sqrt{a+b}-x\sqrt{b-a}}. \qquad a<b.$$

$$\int\frac{dx}{ax^2+b}=\frac{1}{\sqrt{ab}}\text{arc tg}\,x\sqrt{\frac{a}{b}}=-\frac{1}{\sqrt{ab}}\text{art tg}\frac{\sqrt{b}}{x\sqrt{a}}=\frac{1}{2\sqrt{ab}}\text{art tg}\frac{2x\sqrt{ab}}{b-ax^2}$$

$$=\frac{1}{\sqrt{ab}}\text{arc sin}\,x\sqrt{\frac{a}{ax^2+b}}=\frac{1}{2\sqrt{ab}}\text{arc sin}\frac{2x\sqrt{ab}}{ax^2+b}=\frac{1}{2\sqrt{ab}}\text{arc cos}\frac{b-ax^2}{b+ax^2}$$

$$=\frac{1}{\sqrt{ab}}\text{arc cos}\sqrt{\frac{b}{ax^2+b}}=\frac{1}{\sqrt{ab}}\text{arc cotg}\sqrt{\frac{b}{ax^2}}=\frac{1}{\sqrt{ab}}\text{arc séc.}\sqrt{\frac{ax^2+b}{b}}$$

$$=\frac{1}{2\sqrt{ab}}\text{arc séc.}\frac{b+ax^2}{b-ax^2}=\frac{1}{2\sqrt{-ab}}\log\frac{\sqrt{b}+x\sqrt{-a}}{\sqrt{b}-x\sqrt{-a}}=\frac{1}{2\sqrt{-ab}}\log\frac{\sqrt{-b}-x\sqrt{a}}{\sqrt{-b}+x\sqrt{a}}.$$

$$\int\frac{dx}{ax^2+bx}=\frac{1}{b}\log\frac{x}{ax+b}$$

$$\int\frac{dx}{bx-ax^2}=\frac{1}{b}\log\frac{x}{ax-b}.$$

$$\int\frac{dx}{ax^2-bx}=\frac{1}{b}\log\frac{ax-b}{x}.$$

$$\int\frac{dx}{(x+1)^2}=-\frac{1}{x+1}=\frac{1}{2}\text{arc tg}\left\{\text{tg}\frac{x-1}{x+1}\right\}.$$

$$\int\frac{dx}{(1-x)^2}=\frac{1}{1-x}=\frac{1+x}{2\,(1-x)}.$$

$$\int\frac{dx}{(x+a)^2}=-\frac{1}{x+a}=\frac{x}{a\,(x+a)}=-\frac{1}{2a}\frac{a-x}{a+x}.$$

$$\int\frac{dx}{(ax+b)^2}=-\frac{1}{a\,(ax+b)}=\frac{x}{b\,(ax+b)}=\frac{mx+n}{(bm-an)\,(ax+b)}.$$

$$\int\frac{dx}{(x+a)\,(x+b)}=\frac{1}{b-a}\log\frac{x+a}{x+b}.$$

$$\int\frac{dx}{ax+b)\,(mx+n)}=\frac{1}{an-bm}\log\frac{ax+b}{mx+n}.$$

$$\int\frac{dx}{(ax+b)^2-c^2}=\frac{1}{2ac}\log\frac{ax+b-c}{ax+b+c}.$$

$$\int \frac{dx}{(ax+b)^2+c^2} = \frac{1}{ac} \text{ arc tg } \frac{ax+b}{c}.$$

$$\int \frac{dx}{(x-a)^2+(b-x)^2} = \frac{1}{b-a} \text{ arc tg } \frac{2x-a-b}{b-a} = \frac{1}{b-a} \text{ art tg } \frac{x-a}{b-x}.$$

$$\int \frac{dx}{(ax+b)^2+(mx+n)^2} = \frac{1}{an-bm} \text{ arc tg } \frac{ax+b}{mx+n}.$$

$$\int \frac{dx}{x^2+x+1} = \frac{2}{\sqrt{3}} \text{ arc tg } \frac{2x+1}{\sqrt{3}} = \frac{2}{\sqrt{3}} \text{ arc tg } \frac{x\sqrt{3}}{x+2}.$$

$$\int \frac{dx}{x^2+x-1} = \frac{1}{\sqrt{5}} \log \frac{2x+1-\sqrt{5}}{2x+1+\sqrt{5}}.$$

$$\int \frac{dx}{x^2-x+1} = \frac{2}{\sqrt{3}} \text{ arc tg } \frac{2x-1}{\sqrt{3}}.$$

$$\int \frac{dx}{x^2-x-1} = \frac{1}{\sqrt{5}} \log \frac{2x-1-\sqrt{5}}{2x-1+\sqrt{5}}.$$

$$\int \frac{dx}{1+x-x^2} = \frac{1}{\sqrt{5}} \log \frac{2x-1+\sqrt{5}}{2x-1-\sqrt{5}} = \frac{1}{\sqrt{5}} \log \frac{\sqrt{5}-1+2x}{\sqrt{5}+1-2x}.$$

$$\int \frac{dx}{x^2+5x+4} = \frac{1}{3} \log \frac{x+1}{x+4}.$$

$$\int \frac{dx}{2x^2-2x+1} = \text{arc tg } (2x+1).$$

$$\int \frac{dx}{6x-x^2-5} = \frac{1}{4} \log \frac{x-1}{5-x}.$$

$$\int \frac{dx}{x^2-2x\cos\alpha+1} = \frac{1}{\sin\alpha} \text{ arc tg } \frac{x-\cos\alpha}{\sin\alpha}.$$

$$\int \frac{dx}{5x^2+2x+2} = \frac{1}{3} \text{ arc tg } \frac{3x}{x+2} = \frac{1}{3} \text{ arc tg } \frac{5x+1}{3}.$$

$$\int \frac{dx}{4x^2+16x+17} = \frac{1}{2} \text{ arc tg } (2x+4) = \frac{1}{2} \text{ arc tg } \frac{2x}{8x+17}.$$

$$\int \frac{dx}{5x^2+26x+34} = \text{arc tg } (5x+13) = \text{arc tg } \frac{2x+5}{x+3}.$$

$$\int \frac{dx}{x^2+2abx+b^2} = \frac{1}{b\sqrt{1-a^2}} \text{ arc sin } \frac{x\sqrt{1-a^2}}{\sqrt{x^2+2abx+b^2}} = -\frac{1}{b\sqrt{1-a^2}} \text{ arc sin } \frac{b\sqrt{1-a^2}}{\sqrt{x^2+2abx+b^2}}.$$

$$\int \frac{dx}{ax^2+bx+c} = \frac{1}{\sqrt{b^2-4ac}} \log \frac{2ax+b-\sqrt{b^2-4ac}}{2ax+b+\sqrt{b^2-4ac}}. \qquad b^2 > 4ac$$

$$= \frac{2}{\sqrt{b^2-4ac}} \log \frac{2ax+b-\sqrt{b^2-4ac}}{\sqrt{ax^2+bx+c}}. \qquad \text{»}$$

$$= \frac{2}{\sqrt{4ac-b^2}} \text{ arc tg } \frac{2ax+b}{\sqrt{4ac-b^2}}. \qquad b^2 < 4ac$$

$$= \frac{1}{\sqrt{4ac-b^2}} \text{ arc sin } \frac{(2ax+b)\sqrt{4ac-b^2}}{2a(ax^2+bx+c)}. \qquad \text{»}$$

$$= \frac{2}{\sqrt{4ac - b^2}} \operatorname{arc\,sin} \frac{2ax + b}{2\sqrt{a(ax^2 + bx + c)}}. \qquad \text{»}$$

$$= -\frac{2}{2ax + b}. \qquad b^2 = 4ac.$$

$$\int \frac{dx}{c + bx - ax^2} = \frac{1}{\sqrt{4ac + b^2}} \log \frac{\sqrt{4ac + b^2} + 2ax - b}{\sqrt{4ac + b^2} - 2ax + b} = \frac{1}{\sqrt{4ac + b^2}} \log \frac{2ax - b + \sqrt{4ac + b^2}}{2ax - b - \sqrt{4ac + b^2}}.$$

$$\frac{mx + n}{ax^2 + bx + c}$$

$$\int \frac{x}{ax^2 + b}\,dx = \frac{1}{2a} \log (ax^2 + b).$$

$$\int \frac{x}{a^2x^2 - b^2}\,dx = \frac{1}{2a^2} \log (a^2x^2 - b^2) = \frac{1}{2a^2} \left\{ 2 \log (ax + b) + \log \frac{b - ax}{b + ax} \right\}.$$

$$\int \frac{x}{(1 - x)^2}\,dx = \frac{1}{1 - x} + \log (1 - x) = -\frac{1}{x - 1} + \log (x - 1).$$

$$\int \frac{x}{(ax + b)^2}\,dx = \frac{b}{a^2(ax + b)} + \frac{1}{a^2} \log (ax + b) = -\frac{x}{a(ax + b)} - \frac{1}{a^2} \log \frac{b}{ax + b}.$$

$$\int \frac{x}{(x + a)(x + b)}\,dx = \frac{1}{a - b} \log \frac{(x + a)^a}{(x + b)^b}.$$

$$\int \frac{x}{(x - a)(x - b)}\,dx = \frac{1}{a - b} \log \frac{(x - a)^a}{(x - b)^b}.$$

$$\int \frac{x}{(ax + b)(mx + n)}\,dx = -\frac{b}{an - bm} \int \frac{dx}{ax + b} + \frac{n}{an - bm} \int \frac{dx}{mx + n}.$$

$$\int \frac{x}{x^2 + x + 1}\,dx = \frac{1}{2} \log (x^2 + x + 1) - \frac{1}{\sqrt{3}} \operatorname{arc\,tg} \frac{2x + 1}{\sqrt{3}}.$$

$$\int \frac{x}{x^2 + 4x + 2}\,dx = \frac{1}{2} \log (x^2 + 4x + 2) - \frac{1}{\sqrt{2}} \log \frac{x + 2 - \sqrt{2}}{x + 2 + \sqrt{2}}.$$

$$\int \frac{x}{x^2 - 2x - 2}\,dx = \frac{1}{2} \log (x^2 - 2x - 2) + \frac{1}{2\sqrt{3}} \log \frac{x + 1 - \sqrt{3}}{x + 1 + \sqrt{3}}.$$

$$\int \frac{x}{x^2 - 2x \cos \alpha + 1}\,dx = \frac{1}{2} \log (x^2 - 2x \cos \alpha + 1) + \operatorname{cotg} \alpha \operatorname{arc\,tg} \frac{x - \cos \alpha}{\sin \alpha}.$$

$$\int \frac{x}{x^2 + 2bx + a^2}\,dx = \frac{1}{2} \log (x^2 + 2bx + a^2) - \frac{b}{\sqrt{a^2 - b^2}} \operatorname{arc\,tg} \frac{x + b}{\sqrt{a^2 - b^2}}.$$

$$\int \frac{x}{ax^2 + bx + c}\,dx = \frac{1}{2a} \log (ax^2 + bx + c) - \frac{b}{2a} \int \frac{dx}{ax^2 + bx + c}.$$

$$= \frac{1}{2a} \log (ax^2 + bx + c) - \frac{b}{2a\sqrt{b^2 - 4ac}} \log \frac{2ax + b - \sqrt{b^2 - 4ac}}{2ax + b + \sqrt{b^2 - 4ac}}. \qquad b^2 > 4ac$$

$$= \frac{1}{2a} \log (ax^2 + bx + c) - \frac{b}{a\sqrt{4ac - b^2}} \operatorname{arc\,tg} \frac{2ax + b}{\sqrt{4ac - b^2}}. \qquad b^2 < 4ac$$

$$= \frac{b}{a(2ax + b)} + \frac{1}{a} \log (2ax + b). \qquad b^2 = 4ac.$$

$$\int \frac{x}{c+bx-ax^2}\,dx = \frac{1}{2a}\log(c+bx-ax^2) + \frac{b}{2a\sqrt{b^2+4ac}}\log\frac{\sqrt{b^2+4ac}+2ax-b}{\sqrt{b^2+4ac}-2ax+b}.$$

$$\int \frac{x+1}{x^2+1}\,dx = \frac{1}{2}\log(x^2+1) + \text{arc tg}\, x.$$

$$\int \frac{\mathrm{A}x+\mathrm{A}\alpha+\mathrm{B}}{x^2+\beta^2}\,dx = \frac{\mathrm{A}}{2}\log(x^2+\beta^2) + \frac{\mathrm{A}\alpha+\mathrm{B}}{\beta}\,\text{arc tg}\,\frac{x}{\beta}.$$

$$\int \frac{mx-n}{x^2-c^2}\,dx = \frac{m}{2}\log(x^2-c^2) - \frac{n}{2c}\log\frac{x-c}{x+c}.$$

$$\int \frac{mx+n}{ax^2+b}\,dx = \frac{m}{2a}\log(ax^2+b) + \frac{n}{\sqrt{ab}}\,\text{arc tg}\, x\sqrt{\frac{a}{b}}.$$

$$\int \frac{mx+n}{ax^2-b}\,dx = \frac{m}{2a}\log(ax^2-b) + \frac{n}{2\sqrt{ab}}\log\frac{x\sqrt{a}-\sqrt{b}}{x\sqrt{a}+\sqrt{b}}.$$

$$\int \frac{2x+1}{x^2-3x}\,dx = \log(x^2-3x) - \frac{4}{3}\log\frac{x}{x-3}.$$

$$\int \frac{2a-x}{x^2-ax}\,dx = \log\frac{x-a}{x^2}.$$

$$\int \frac{ax+b}{x-x^2}\,dx = \log\frac{x^b}{(x-1)^{a+b}}.$$

$$\int \frac{mx+n}{ax^2+bx}\,dx = \frac{m}{2a}\log(ax^2+bx) + \frac{2an-bm}{2ab}\log\frac{ax}{ax+b}.$$

$$\int \frac{mx+n}{(x-a)^2}\,dx = -\frac{am+n}{x-a} + m\log(x-a).$$

$$\int \frac{mx+n}{(ax+b)^2}\,dx = \frac{bm-an}{a^2(ax+b)} + \frac{m}{a^2}\log(ax+b).$$

$$\int \frac{mx+n}{(ax+b)(px+q)}\,dx = \frac{bm-an}{bp-aq}\int\frac{dx}{ax+b} + \frac{np-mq}{bp-aq}\int\frac{dx}{px+q}.$$

$$\int \frac{x-a}{(x-a)^2+b^2}\,dx = \frac{1}{2}\log\left\{(x-a)^2+b^2\right\}.$$

$$\int \frac{mx+n}{(x-a)^2+b^2}\,dx = \frac{m}{2}\log\left\{(x-a)^2+b^2\right\} + \frac{am+n}{b}\,\text{arc tg}\,\frac{x-a}{b}.$$

$$\int \frac{mx+n}{(x+a)^2+b^2}\,dx = \frac{m}{2}\log\left\{(x+a)^2+b^2\right\} - \frac{n-am}{b}\,\text{arc cos}\,\frac{x+a}{\sqrt{(x+a)^2+b^2}}.$$

$$\int \frac{1-x\cos\alpha}{x^2-2x\cos\alpha+1}\,dx = -\frac{1}{2}\cos\alpha.\log(x^2-2x\cos\alpha+1) + \sin\alpha.\,\text{arc tg}\,\frac{x-\cos\alpha}{\sin\alpha}.$$

$$\int \frac{x+1}{x^2+x+1}\,dx = \frac{1}{2}\log(x^2+x+1) + \frac{1}{\sqrt{3}}\,\text{arc tg}\,\frac{x\sqrt{3}}{x+2}.$$

$$\int \frac{x-1}{x^2+x+1}\,dx = \frac{1}{2}\log(x^2+x+1) - \sqrt{3}\,\text{arc tg}\,\frac{2x+1}{\sqrt{3}}.$$

$$\int \frac{3x+2}{x^2-x-2}\,dx = \frac{3}{2}\log(x^2-x-2)+\frac{7}{6}\log\frac{x-2}{x+1} = \frac{1}{3}\log(x+1)+\frac{8}{3}\log(x-2).$$

$$\int \frac{x+1}{x^2+5x+6}\,dx = \log\frac{(x+3)^2}{x+2}.$$

$$\int \frac{3x+7}{2x^2-3x+5}\,dx = \frac{3}{4}\log(2x^2-3x+5)+\frac{37}{2\sqrt{31}}\text{arc tg}\,\frac{4x-3}{\sqrt{31}}.$$

$$\int \frac{mx+n}{ax^2+bx+c}\,dx = \frac{m}{2a}\log(ax^2+bx+c)+\frac{2an-bm}{2a\sqrt{b^2-4ac}}\log\frac{2ax+b-\sqrt{b^2-4ac}}{2ax+b+\sqrt{b^2-4ac}}. \qquad b^2>4ac.$$

$$= \frac{m}{2a}\log(ax^2+bx+c)+\frac{2an-bm}{a\sqrt{4ac-b^2}}\text{arc tg}\,\frac{2ax+b}{\sqrt{4ac-b^2}}. \qquad b^2<4ac.$$

$$= \frac{bm-2an}{a(2ax+b)}+\frac{m}{a}\log(2ax+b). \qquad b^2=4ac.$$

$$\int \frac{mx+n}{c+bx-ax^2}\,dx = -\frac{m}{2a}\log(c+bx-ax^2)+\frac{2an+bm}{2a\sqrt{b^2+4ac}}\log\frac{b-2ax-\sqrt{b^2+4ac}}{b-2ax+\sqrt{b^2+4ac}}.$$

$\dfrac{mx^2+n}{ax^2+bx+c}$

$$\int \frac{x^2}{a^2+x^2}\,dx = x-a\,\text{arc tg}\,\frac{x}{a} = x+a\,\text{arc sin}\,\frac{a}{\sqrt{a^2+x^2}}.$$

$$\int \frac{x^2}{x^2-a^2}\,dx = x+\frac{a}{2}\log\frac{x-a}{x+a}.$$

$$\int \frac{x^2}{ax^2+b}\,dx = \frac{x}{a}-\frac{b}{a}\int\frac{dx}{ax^2+b} = \frac{x}{a}-\frac{\sqrt{b}}{a\sqrt{a}}\text{arc tg}\,x\sqrt{\frac{a}{b}}.$$

$$\int \frac{x^2}{ax^2-b}\,dx = \frac{x}{a}+\frac{\sqrt{b}}{2a\sqrt{a}}\log\frac{x\sqrt{a}-\sqrt{b}}{x\sqrt{a}+\sqrt{b}}.$$

$$\int \frac{x^2}{b-ax^2}\,dx = -\frac{x}{a}+\frac{\sqrt{b}}{2a\sqrt{a}}\log\frac{x\sqrt{a}+\sqrt{b}}{x\sqrt{a}-\sqrt{b}}.$$

$$\int \frac{x^2}{(x+1)^2}\,dx = \frac{x(x+2)}{x+1}-2\log(x+1).$$

$$\int \frac{x^2}{(x-a)^2}\,dx = \frac{x(x-2a)}{x-a}+2a\log(x-a).$$

$$\int \frac{x^2}{(ax+b)^2}\,dx = \frac{x(ax+2b)}{a^2(ax+b)}-\frac{2b}{a^3}\log(ax+b).$$

$$= \frac{a^2x^2-2b^2}{a^3(ax+b)}-\frac{2b}{a^3}\log(ax+b).$$

$$\int \frac{x^2}{x^2+(a+b)x+ab}\,dx = x+\frac{a^2}{b-a}\int\frac{dx}{x+a}+\frac{b^2}{a-b}\int\frac{dx}{x+b}.$$

$$\int \frac{x^2}{ax^2+bx+c}\,dx = \frac{x}{a}-\frac{b}{2a^2}\log(ax^2+bx+c)+\frac{b^2-2ac}{2a^2}\int\frac{dx}{ax^2+bx+c}.$$

$$= \frac{x}{a}-\frac{b}{2a^2}\log(ax^2+bx+c)+\frac{b^2-2ac}{2a^2\sqrt{b^2-4ac}}\log\frac{2ax+b-\sqrt{b^2-4ac}}{2ax+b+\sqrt{b^2-4ac}}. \qquad b^2>4ac.$$

$$= \frac{x}{a} - \frac{b}{2a^2} \log (ax^2 + bx + c) + \frac{b^2 - 2ac}{a^2 \sqrt{4ac - b^2}} \operatorname{arc\,tg} \frac{2ax + b}{\sqrt{4ac - b^2}} \cdot \qquad b^2 < 4ac.$$

$$= \frac{2x\,(ax + b)}{a\,(2ax + b)} - \frac{b}{a^2} \log (2ax + b). \qquad b^2 = 4ac.$$

$$\int \frac{x^2 + 2}{x^2 + 1}\, dx = x + \operatorname{arc\,tg} x.$$

$$\int \frac{1 - ax^2}{1 + ax^2}\, dx = -x + \frac{2}{\sqrt{a}} \operatorname{arc\,tg} x\sqrt{a}.$$

$$\int \frac{mx^2 + n}{ax^2 + b}\, dx = \frac{m}{a} x + \frac{an - bm}{a\sqrt{ab}} \operatorname{arc\,tg} x \sqrt{\frac{a}{b}} \cdot$$

$$\int \frac{mx^2 + n}{b - ax^2}\, dx = -\frac{m}{a} x + \frac{an + bm}{2\,a\sqrt{ab}} \log \frac{\sqrt{b} + x\sqrt{a}}{\sqrt{b} - x\sqrt{a}} \cdot$$

$$\int \frac{x^2 + 1}{x^2 - x}\, dx = x + \log \frac{(x - 1)^2}{x} \cdot$$

$$\int \frac{mx^2 + n}{ax^2 + bx}\, dx = \frac{m}{a} x + \frac{n}{b} \log \frac{x}{ax + b} - \frac{mb}{a^2} \log (ax + b).$$

$$\int \frac{x^2 + 1}{(x - 1)^2}\, dx = \frac{x^2 - 2x - 1}{x - 1} + 2 \log (x - 1).$$

$$\int \frac{mx^2 + n}{(ax + b)^2}\, dx = \frac{max^2 + 2bmx - an}{a^2\,(ax + b)} - \frac{2bm}{a^3} \log (ax + b).$$

$$\int \frac{x^2 - a^2}{x^4 + 6a^2x^2 + a^4}\, dx = \frac{1}{2a} \operatorname{arc\,tg} \frac{x^2 + a^2}{2ax} \cdot$$

$$\int \frac{mx^2 + n}{ax^2 + bx + c}\, dx = \frac{m}{a} x - \frac{bm}{2a^2} \log (ax^2 + bx + c) + \frac{2a\,(an - cm) + b^2m}{2a^2\sqrt{b^2 - 4ac}} \log \frac{2ax + b - \sqrt{b^2 - 4ac}}{2ax + b + \sqrt{b^2 - 4ac}} \cdot \quad b^2 > 4ac$$

$$= \frac{m}{a} x - \frac{bm}{2a^2} \log (ax^2 + bx + c) + \frac{2a\,(an - cm) + b^2m}{a^2\sqrt{4ac - b^2}} \operatorname{arc\,tg} \frac{2ax + b}{\sqrt{4ac - b^2}} \cdot \qquad b^2 < 4ac$$

$$= \frac{2\,(amx^2 + bmx - an)}{a\,(2ax + b)} - \frac{bm}{a^2} \log (2ax + b). \qquad b^2 = 4ac.$$

$$\int \frac{mx^2 + n}{c + bx - ax^2}\, dx = -\frac{m}{a} x - \frac{bm}{2a^2} \log (c + bx - ax^2) + \frac{2a\,(an + cm) + b^2m}{2a^2\sqrt{4ac + b^2}} \log \frac{\sqrt{4ac + b^2} + 2ax - b}{\sqrt{4ac + b^2} - 2ax + b} \cdot$$

$\dfrac{mx^2 + nx + p}{ax^2 + bx + c}$

$$\int \frac{mx^2 + nx}{ax^2 + b}\, dx = \frac{m}{a} x + \frac{n}{2a} \log (ax^2 + b) - \frac{m}{a}\sqrt{\frac{b}{a}} \operatorname{arc\,tg} x \sqrt{\frac{a}{b}} \cdot$$

$$\int \frac{mx^2 + nx}{b - ax^2}\, dx = -\frac{m}{a} x - \frac{n}{2a} \log (b - ax^2) + \frac{m}{2a}\sqrt{\frac{b}{a}} \log \frac{x\sqrt{a} + \sqrt{b}}{x\sqrt{a} - \sqrt{b}} \cdot$$

$$\int \frac{x\,(x + 4)}{(x + 2)^2}\, dx = \frac{(x + 4)^2}{x + 2} = \frac{x^2}{x + 2} \cdot$$

$$\int \frac{mx^2 + nx}{(ax + b)^2}\, dx = \frac{mx\,(ax + 2b) + bn}{a^2\,(ax + b)} + \frac{an - 2bm}{a^3} \log (ax + b).$$

$$\int \frac{mx^2+nx}{ax^2+bx+c}\,dx = \frac{m}{a}x + \frac{an-bm}{2a^2}\log(ax^2+bx+c) + \frac{b^2m-2acm-abn}{2a^2\sqrt{b^2-4ac}}\log\frac{2ax+b-\sqrt{b^2-4ac}}{2ax+b+\sqrt{b^2-4ac}}. \qquad b^2 > 4ac.$$

$$= \frac{m}{a}x + \frac{an-bm}{2a^2}\log(ax^2+bx+c) + \frac{b^2m-2acm-abn}{a^2\sqrt{4ac-b^2}}\operatorname{arc\,tg}\frac{2ax+b}{\sqrt{4ac-b^2}}. \qquad b^2 < 4ac.$$

$$= \frac{m}{a}x + \frac{b(an-bm)+2acm}{a^2(2ax+b)} + \frac{an-bm}{a^2}\log(2ax+b). \qquad b^2 = 4ac.$$

$$\int \frac{x^2-6x-16}{(x-3)^2}\,dx = \frac{(x+2)^2}{x-3} = \frac{x^2+16}{x-3}.$$

$$\int \frac{2x^2+6x+15}{(2x+3)^2}\,dx = \frac{x^2+5x}{2x+3} = \frac{2x^2-15}{2(2x+3)}.$$

$$\int \frac{mx^2+nx+p}{(ax+b)^2}\,dx = \frac{m}{a^2}x - \frac{b(bm-an)+a^2p}{a^3(ax+b)} + \frac{an-2bm}{a^3}\log(ax+b).$$

$$\int \frac{mx^2+nx+p}{ax^2+bx+c}\,dx = \frac{m}{a}x + \frac{an-bm}{2a^2}\log(ax^2+bx+c)$$

$$+ \frac{2a(ap-cm)-b(an-bm)}{2a^2\sqrt{b^2-4ac}}\log\frac{2ax+b-\sqrt{b^2-4ac}}{2ax+b+\sqrt{b^2-4ac}}. \qquad b^2 > 4ac.$$

$$= \frac{m}{a}x + \frac{an-bm}{2a^2}\log(ax^2+bx+c)$$

$$+ \frac{2a(ap-cm)-b(an-bm)}{a^2\sqrt{4ac-b^2}}\operatorname{arc\,tg}\frac{2ax+b}{\sqrt{4ac-b^2}}. \qquad b^2 < 4ac.$$

$$= \frac{m}{a}x + \frac{b(an-bm)-2a(ap-cm)}{a^2(2ax+b)} + \frac{an-bm}{a^2}\log(2ax+b). \qquad b^2 = 4ac.$$

$\frac{X^2}{ax+b}$

$$\int \frac{x^2}{ax+b}\,dx = \frac{x^2}{2a} - \frac{bx}{a^2} + \frac{b^2}{a^3}\log(ax+b).$$

$$\int \frac{mx^2+nx}{ax+b}\,dx = \frac{mx^2}{2a} + \frac{an-bm}{a^2}x - \frac{b(an-bm)}{a^3}\log(ax+b).$$

$$\int \frac{(x-a)(x-b)}{x-c}\,dx = \frac{x^2}{2} + (c-a-b)x + (c-a)(c-b)\log(x-c).$$

$$\int \frac{mx^2+nx+p}{ax+b}\,dx = \frac{mx^2}{2a} + \frac{an-bm}{a^2}x + \frac{a^2p-abn+b^2m}{a^3}\log(ax+b).$$

$X^3, \frac{1}{X^3}$

$$\int x^5\,dx = \frac{x^4}{4} = \frac{1}{4}\left(x^2-ax+\frac{a^2}{2}\right)\left(x^2+ax+\frac{a^2}{2}\right).$$

$$\int (x-a)(x-b)(x-c)\,dx = \frac{x^4}{4} - (a+b+c)\frac{x^3}{3} + (bc+ac+ab)\frac{x^2}{2} - abcx.$$

$$\int \frac{dx}{x^3+1} = \frac{1}{6}\log\frac{(x+1)^2}{x^2-x+1} + \frac{1}{\sqrt{3}}\operatorname{arc\,tg}\frac{2x-1}{\sqrt{3}}.$$

$$= \frac{1}{3}\log(x+1) - \frac{1}{3}\cos\frac{\pi}{3}\log(1-2x\cos\frac{\pi}{3}+x^2) + \frac{2}{3}\sin\frac{\pi}{3}\operatorname{arc\,tg}\frac{x\sin\frac{\pi}{3}}{1-x\cos\frac{\pi}{3}}$$

$$\int\frac{dx}{x^3+a^3} = \frac{1}{6a^2}\log\frac{(x+a)^2}{x^2-ax+a^2} + \frac{1}{a^2\sqrt{3}}\operatorname{arc\,tg}\frac{2x-a}{a\sqrt{3}}.$$

$$= \frac{1}{6a^2}\log\frac{(x+a)^2}{x^2-ax+a^2} + \frac{1}{a^2\sqrt{3}}\operatorname{arc\,tg}\frac{x\sqrt{3}}{2a-x}.$$

$$\int\frac{dx}{ax^3+b} = \frac{k}{6b}\log\frac{(x+k)^2}{x^2-kx+k^2} + \frac{k}{b\sqrt{3}}\operatorname{arc\,tg}\frac{x\sqrt{3}}{2k-x}. \qquad k = \left(\frac{b}{a}\right)^{\frac{1}{3}}.$$

$$\int\frac{dx}{x^3-1} = \frac{1}{3}\log(x-1) - \frac{1}{6}\log(x^2+x+1) - \frac{1}{\sqrt{3}}\operatorname{arc\,tg}\frac{2x+1}{\sqrt{3}}.$$

$$\int\frac{dx}{ax^3-b} = \frac{1}{6ak^2}\log\frac{(x-k)^2}{x^2+kx+k^2} + \frac{1}{ak^2\sqrt{3}}\operatorname{arc\,tg}\frac{k\sqrt{3}}{2x+k} \qquad k = \left(\frac{b}{a}\right)^{\frac{1}{3}}$$

$$\int\frac{dx}{1-x^3} = -\frac{1}{6}\log\frac{(1-x)^2}{x^2+x+1} + \frac{1}{\sqrt{3}}\operatorname{arc\,tg}\frac{x\sqrt{3}}{2+x}.$$

$$= -\frac{1}{3}\log(1-x) - \frac{1}{3}\cos\frac{2}{3}\pi\log(1-2x\cos\frac{2}{3}\pi+x^2) + \frac{2}{3}\sin\frac{2}{3}\pi\operatorname{arc\,tg}\frac{x\sin\frac{2}{3}\pi}{1-x\cos\frac{2}{3}\pi}.$$

$$\int\frac{dx}{b-ax^3} = -\frac{1}{3bk}\log\frac{1-kx}{\sqrt{1+kx+k^2x^2}} + \frac{1}{bk\sqrt{3}}\operatorname{arc\,tg}\frac{kx\sqrt{3}}{2+kx}. \qquad k = \left(\frac{b}{a}\right)^{\frac{1}{3}}.$$

$$\int\frac{dx}{x^3+x} = \log\frac{x}{\sqrt{x^2+1}}.$$

$$\int\frac{dx}{ax^3+bx} = \frac{1}{2b}\log\frac{x^2}{ax^2+b}.$$

$$\int\frac{dx}{ax^3+bx^2} = -\frac{1}{bx} + \frac{a}{b^2}\log\frac{ax+b}{x}.$$

$$\int\frac{dx}{x^3-7x+6} = -\frac{1}{4}\log(x-1) + \frac{1}{5}\log(x-2) + \frac{1}{20}\log(x+3) = \frac{1}{20}\log\frac{(x-2)^4(x+3)}{(x-1)^5}.$$

$$\int\frac{dx}{x^3-37x+84} = \frac{1}{10}\log\frac{x+7}{x-3} + \frac{1}{11}\log\frac{x-4}{x+7}.$$

$$\int\frac{dx}{x^3-13x+12} = \frac{1}{70}\log\frac{(x-3)^5(x+4)^2}{(x-1)^7}.$$

$$\int\frac{dx}{x(ax+b)^2} = \frac{1}{b(ax+b)} - \frac{1}{b^2}\log\frac{ax+b}{x}.$$

$$\int\frac{dx}{x(x+1)(x+2)} = \frac{1}{2}\log x(x+2) - \log(x+1).$$

$$\int\frac{dx}{x(x+a)(x+b)} = \frac{1}{ab}\log x + \frac{1}{a(a-b)}\log(x+a) + \frac{1}{b(b-a)}\log(x+b).$$

$$\int\frac{dx}{x(x^2+x+1)} = \frac{1}{2}\log\frac{x^2}{x^2+x+1} - \frac{1}{\sqrt{3}}\operatorname{arc\,tg}\frac{2x+1}{\sqrt{3}}.$$

$$\int\frac{dx}{x(2x^2+7x+6)} = \frac{1}{6}\log x - \frac{2}{3}\log(2x+3) + \frac{1}{2}\log(x+2).$$

$$\int \frac{dx}{x(ax^2+bx+c)} = \frac{1}{2c}\log\frac{x^2}{ax^2+bx+c} - \frac{b}{2c}\int\frac{dx}{ax^2+bx+c}.$$

$$= \frac{1}{2c}\log\frac{x^2}{ax^2+bx+c} - \frac{b}{2c\sqrt{b^2-4ac}}\log\frac{2ax+b-\sqrt{b^2-4ac}}{2ax+b+\sqrt{b^2-4ac}}. \qquad b^2 > 4ac.$$

$$= \frac{1}{2c}\log\frac{x^2}{ax^2+bx+c} - \frac{b}{c\sqrt{4ac-b^2}}\text{arc tg}\frac{2ax+b}{\sqrt{4ac-b^2}}. \qquad b^2 < 4ac.$$

$$\int\frac{dx}{(ax+b)^3} = -\frac{1}{2a(ax+b)^2}.$$

$$\int\frac{dx}{(a+x)^2(b+x)} = \frac{1}{(a-b)(x+a)} + \frac{1}{(a-b)^2}\log\frac{x+b}{x+a}.$$

$$\int\frac{dx}{(x-1)^2(x-2)} = \frac{1}{x-1} + \log\frac{x-2}{x-1}.$$

$$\int\frac{dx}{(x^2+a)(x+b)} = \frac{1}{b^2+a}\log\frac{x+b}{\sqrt{x^2+a}} + \frac{b}{b^2+a}\int\frac{dx}{x^2+a}.$$

$$\int\frac{dx}{(x^2+1)(x+1)} = \frac{1}{2}\log\frac{x+1}{\sqrt{x^2+1}} + \frac{1}{2}\text{arc tg } x.$$

$$\int\frac{dx}{(x^2+1)(x-a)} = \frac{1}{a^2+1}\log\frac{x-a}{\sqrt{x^2+1}} - \frac{a}{a^2+1}\text{arc tg } x.$$

$$\int\frac{dx}{(x^2-1)(x-a)} = \frac{1}{2(a^2-1)}\log\frac{(x-a)^2(x+1)^a}{(x^2-1)(x-1)^a}.$$

$$\int\frac{dx}{(x^2+ax+b)(x+c)} = \frac{1}{c^2-ac+b}\,\frac{1}{2}\log\frac{(x+c)^2}{x^2+ax+b} + \frac{2c-a}{2(c^2-ac+b)}\int\frac{dx}{x^2+ax+b}.$$

$$\int\frac{dx}{(x-a)(x-b)(2x-a-b)} = \frac{1}{(a-b)^2}\log\frac{(x-a)(x-b)}{(2x-a-b)^2}.$$

$$\int\frac{dx}{(x-a)(x-b)(3x-2a-b)} = \frac{1}{2(a-b)^2}\log\frac{(x-a)^2(x-b)}{(3x-2a-b)^3}.$$

$$\int\frac{dx}{(x+a)(x+b)(x+c)} = \frac{1}{(b-a)(c-a)}\log(x+a) + \frac{1}{(a-b)(c-b)}\log(x+b) + \frac{1}{(b-c)(a-c)}\log(x+c).$$

$$\int\frac{dx}{x^3-6x^2+11x-6} = \frac{1}{2}\log(x-1) - \log(x-2) + \frac{1}{2}\log(x-3).$$

$$\int\frac{dx}{x^3-x^2-x+1} = -\frac{1}{2(x-1)} + \frac{1}{4}\log\frac{x+1}{x-1}.$$

$$\int\frac{dx}{x^3+x^2-x+15} = \frac{1}{20}\log(x+3) - \frac{1}{40}\log(x^2-2x+5) + \frac{1}{10}\text{arc tg}\frac{x-1}{2}.$$

$$\int\frac{dx}{ax^3+bx^2+cx+d} = \frac{\log(x-x_1)}{3ax_1^2+2bx_1+c} + \frac{\log(x-x_2)}{3ax_2^2+2bx_2+c} + \frac{\log(x-x_3)}{3ax_3^2+2bx_3+c}.$$

x_1, x_2, x_3 being the roots of $X^3 = 0$.

$$= \frac{1}{3ax_1^2+2bx_1+c}\left[\log(x-x_1) - \frac{1}{2}\log\left\{(x-\alpha)^2+\beta^2\right\} + \frac{\alpha-x_1}{\beta}\text{arc tg}\frac{x-\alpha}{\beta}\right].$$

$x_1, \quad x_2 = \alpha + \beta\sqrt{-1}, \quad x_3 = \alpha - \beta\sqrt{-1}$ being the roots of $X^3 = 0$.

$$\frac{mx+n}{X^3}$$

$$\int \frac{x}{x^3+1}\,dx = \frac{1}{6}\log\frac{x^2-x+1}{(x+1)^2} + \frac{1}{\sqrt{3}}\text{arc tg}\frac{2x-1}{\sqrt{3}}.$$

$$\int \frac{x}{x^3-1}\,dx = \frac{1}{3}\log(x-1) - \frac{1}{3}\log\sqrt{x^2+x+1} + \frac{1}{\sqrt{3}}\text{arc tg}\frac{2x+1}{\sqrt{3}}.$$

$$\int \frac{x}{a^3+x^3}\,dx = -\frac{1}{6a}\log\frac{(x+a)^2}{x^2-ax+a^2} + \frac{1}{a\sqrt{3}}\text{arc tg}\frac{2x-a}{a\sqrt{3}}.$$

$$= \frac{1}{6a}\log\frac{x^2-ax+a^2}{(x+a)^2} + \frac{1}{a\sqrt{3}}\text{arc tg}\frac{x\sqrt{3}}{2a-x}.$$

$$\int \frac{x}{x^3-8}\,dx = \frac{1}{6}\left\{\log(x-2) - \frac{1}{2}\log(x^2+2x+4) + \sqrt{3}\,\text{arc tg}\frac{x+1}{\sqrt{3}}\right\}.$$

$$\int \frac{x}{ax^3+b}\,dx = -\frac{1}{6ak}\log\frac{(x+k)^2}{x^2-kx+k^2} + \frac{1}{ak\sqrt{3}}\text{arc tg}\frac{x\sqrt{3}}{2k-x}. \qquad k = \left(\frac{b}{a}\right)^{\frac{1}{3}}.$$

$$\int \frac{x}{ax^3-b}\,dx = \frac{1}{3a\,k}\left\{\log(x-k) - \frac{1}{2}\log(x^2+kx+k^2) + \sqrt{3}\,\text{arc tg}\frac{2x+k}{k\sqrt{3}}\right\}. \qquad k = \left(\frac{b}{a}\right)^{\frac{1}{3}}.$$

$$\int \frac{x}{(ax+b)^3}\,dx = \frac{x^2}{2b\,(ax+b)^2} = -\frac{2ax+b}{2a^2\,(ax+b)^2}.$$

$$\int \frac{x}{(x+a)^2\,(x+b)}\,dx = -\frac{a}{(a-b)\,(x+a)} - \frac{b}{(a-b)^2}\log\frac{x+b}{x+a}.$$

$$\int \frac{x}{(x^2+a)\,(x+b)}\,dx = \frac{b}{a+b^2}\log\frac{\sqrt{x^2+a}}{x+b} + \frac{\sqrt{a}}{a+b^2}\text{arc tg}\frac{x}{\sqrt{a}}.$$

$$\int \frac{x}{(mx+n)^2\,(ax+b)}\,dx = \frac{n}{amn-bm^2}\left\{-\frac{1}{mx+n} - \frac{bm}{n\,(bm-an)}\log\frac{mx+n}{ax+b}\right\}.$$

$$\int \frac{x}{(x-20)\,(x-8)\,(x-5)}\,dx = \frac{1}{9}\log\frac{(x-20)\,(x-5)}{(x-8)^2}.$$

$$\int \frac{x}{(x+a)\,(x+b)\,(x+c)}\,dx = -\frac{a\log(x+a)}{(b-a)\,(c-a)} - \frac{b\log(x+b)}{(a-b)\,(c-b)} - \frac{c\log(x+c)}{(a-c)\,(b-c)}.$$

$$\int \frac{x}{x^3-6x^2+11x-6}\,dx = \frac{1}{2}\log\frac{(x-1)\,(x-3)^3}{(x-2)^4}.$$

$$\int \frac{x}{x^3-6x^2+13x-10}\,dx = 2\log(x-2) - \log(x^2-4x+5) + \text{arc tg}\,(x-2).$$

$$\int \frac{x}{ax^3+bx^2+cx+d}\,dx = \frac{x_1\log(x-x_1)}{3ax_1^2+2bx_1+c} + \frac{x_2\log(x-x_2)}{3ax_2^2+2bx_2+c} + \frac{x_3\log(x-x_3)}{3ax_3^2+2bx_3+c}.$$

x_1, x_2, x_3 being the roots of $X^3 = 0$.

$$= \frac{1}{3ax_1^2+2bx_1+c}\left[x_1\log(x-x_1) - \frac{1}{2}x_1\log\left\{(x-\alpha)^2+\beta^2\right\} + \frac{\alpha^2-\alpha x_1+\beta^2}{\beta}\text{arc tg}\frac{x-\alpha}{\beta}\right].$$

$x_1, \quad x_2 = \alpha + \beta\sqrt{-1}, \quad x_3 = \alpha - \beta\sqrt{-1}$ being the roots of $X^3 = 0$.

$$\int \frac{x+a}{x^3+a^3}\,dx = \frac{2}{a\sqrt{3}} \text{ arc tg} \frac{2x-a}{a\sqrt{3}}.$$

$$\int \frac{ax+b}{x^3-1}\,dx = \frac{a+b}{3}\log(x-1) - \frac{a+b}{6}\log(x^2+x+1) + \frac{a-b}{\sqrt{3}} \text{ arc tg} \frac{2x+1}{\sqrt{3}}.$$

$$\int \frac{mx+n}{ax^3+b}\,dx = m\int \frac{x}{ax^3+b}\,dx + n\int \frac{dx}{ax^3+b}.$$

$$\int \frac{2x+3}{x^3+x^2-2x}\,dx = \log \frac{(x-1)^{\frac{5}{3}}}{x^{\frac{3}{2}}(x+2)^{\frac{1}{6}}}$$

$$\int \frac{x-3}{(x+1)^3}\,dx = \frac{1-x}{(1+x)^2}.$$

$$\int \frac{3x-2}{(x-2)^2(x+2)}\,dx = -\frac{1}{x-2} + \frac{1}{2}\log\frac{x-2}{x+2}.$$

$$\int \frac{mx+n}{x+1)^2(x+3)}\,dx = \frac{m-n}{2(x+1)} - \frac{3m-n}{4}\log\frac{x+3}{x+1}.$$

$$\int \frac{x+3}{4x^3+9x^2+18x+17}\,dx = \int \frac{8(x+3)}{(3x+1)^3+5(x+3)^3}\,dx = 4\int \frac{dz}{z^3+5}. \qquad \frac{3x+1}{x+3} = z.$$

$\dfrac{mx^2+n}{X^3}$

$$\int \frac{x^2}{ax^3+b}\,dx = \frac{1}{3a}\log(ax^3+b).$$

$$\int \frac{x^2}{x^3-7x+6}\,dx = \frac{1}{3}\int \frac{3x^2-7+7}{x^3-7x+6}\,dx = \frac{1}{3}\log(x^3-7x+6) + \frac{7}{60}\log\frac{(x-2)^4(x+3)}{(x-1)^5}.$$

$$\int \frac{x^2}{(x-a)^3}\,dx = -\frac{a^2}{2(x-a)^2} - \frac{2a}{x-a} + \log(x-a).$$

$$\int \frac{x^2}{(ax+b)^3}\,dx = \frac{4abx+3b^2}{2a^3(ax+b)^2} + \frac{1}{a^3}\log(ax+b).$$

$$\int \frac{x^2}{(1-ax)^3}\,dx = \frac{1}{2a^3(1-ax)^2} - \frac{2}{a^3(1-ax)} - \frac{1}{a^3}\log(1-ax).$$

$$\int \frac{x^2}{(x-a)^2(x+a)}\,dx = -\frac{a}{2(x-a)} + \frac{3}{4}\log(x-a) + \frac{1}{4}\log(x+a).$$

$$\int \frac{x^2}{x^3+5x^2+8x+4}\,dx = \frac{4}{x+2} + \log(x+1).$$

$$\int \frac{x^2}{x^3+x^2+x+1}\,dx = \frac{1}{2}\log(x+1)\sqrt{x^2+1} - \frac{1}{2}\text{ arc tg } x.$$

$$\int \frac{x^2}{ax^3+bx^2+cx+d}\,dx = \frac{1}{3a}\log\Phi - \frac{2b}{3a}\int\frac{x}{\Phi}\,dx - \frac{c}{3a}\int\frac{dx}{\Phi}.$$

$$\int \frac{1-2x^2}{1-x^3}\,dx = \frac{1}{3}\log(x-1) + \frac{5}{6}\log(1+x+x^2) - \frac{1}{\sqrt{3}}\text{ arc cotg}\frac{2x+1}{\sqrt{3}}.$$

$$\int \frac{2x^2+1}{x^3+x}\,dx = \log x\sqrt{x^2+1}.$$

$$\int \frac{x^2+1}{x^3-x}\,dx = \log \frac{x^2-1}{x}.$$

$$\int \frac{x^2+1}{x-x^3}\,dx = \log \frac{x}{1-x^2}.$$

$$\int \frac{a^3+bx^2}{a^2x-x^3}\,dx = a\log x - \frac{a+b}{2}\log(a^2-x^2).$$

$$\int \frac{x^2+1}{(x-2)^3}\,dx = -\frac{4}{x-2} - \frac{5}{2(x-2)^2} + \log(x-2).$$

$$\int \frac{x^2-1}{(x+2)^3}\,dx = \frac{8x+13}{2(x+2)^2} + \log(x+2).$$

$$\int \frac{x^2+a^2}{(x-a)^3}\,dx = -\frac{a(2x-a)}{(x-a)^2} + \log(x-a).$$

$$\int \frac{x^2+1}{x^3-3x^2+4x-2}\,dx = \log \frac{(x-1)^2}{\sqrt{x^2-2x+2}} + 2\,\mathrm{arc\,tg}\,(x-1).$$

$$\int \frac{3x^2+4}{x^3+x^2-8x-12}\,dx = \frac{16}{5(x+2)} + \frac{44}{25}\log(x+2) + \frac{31}{25}\log(x-3).$$

$$\frac{mx^2+nx+p}{X^3}$$

$$\int \frac{x^2+2x+3}{x^3+1}\,dx = \frac{2}{3}\log(x+1) + \frac{1}{6}\log(x^2-x+1) + \frac{5}{\sqrt{3}}\,\mathrm{arc\,tg}\,\frac{2x-1}{\sqrt{3}}.$$

$$\int \frac{2x^2-x+3}{x^3+1}\,dx = 2\log(x+1) + \frac{2}{\sqrt{3}}\,\mathrm{arc\,tg}\,\frac{2x-1}{\sqrt{3}}.$$

$$\int \frac{4x^2-6x+1}{2x^3-x^2}\,dx = \frac{1}{x} + 4\log x - 2\log(2x-1).$$

$$\int \frac{x^2-5x+3}{(x-2)^3}\,dx = \frac{1}{x-2} + \frac{3}{2(x-2)^2} + \log(x-2).$$

$$\int \frac{x^2-6x+1}{(x-3)^3}\,dx = \frac{4}{(x-3)^2} + \log(x-3).$$

$$\int \frac{mx^2+nx+p}{(x-a)^3}\,dx = \frac{2am+n}{a-x} - \frac{p+a^2m+an}{2(x-a)^2} + m\log(x-a).$$

$$\int \frac{x^2-5x+3}{(x-2)^2(x+1)}\,dx = \frac{1}{x-2} + \log(x+1).$$

$$\int \frac{x^2+x-1}{(x+1)(x-1)^2}\,dx = -\frac{1}{2(x-1)} + \frac{1}{4}\log\frac{(x-1)^5}{x+1}.$$

$$\int \frac{x^2-x+1}{(x^2+1)(x+1)}\,dx = \frac{3}{2}\log(x+1) - \frac{1}{4}\log(x^2+1) - \frac{1}{2}\,\mathrm{arc\,tg}\,x.$$

$$\int \frac{x^2-x+4}{(x^2-1)(x+2)}\,dx = -3\log(x+1) + \frac{2}{3}\log(x-1) + \frac{10}{3}\log(x+2).$$

$$\int \frac{x^2 - x - 1}{(x^2 + x + 1)(x - 1)}\,dx = \frac{1}{3}\log\frac{(x^2 + x + 1)^2}{1 - x}.$$

$$\int \frac{x^2 - 5x + 3}{(x^2 - 2x + 5)(x + 1)}\,dx = \frac{9}{8}\log(x + 1) - \frac{1}{16}\log(x^2 - 2x + 5) - \frac{11}{8}\operatorname{arc\,tg}\frac{x - 1}{2}.$$

$$\int \frac{x^2 - 5x + 3}{(x + 1)(x - 1)(x - 2)}\,dx = \frac{2}{3}\log(x + 1) + \frac{1}{2}\log(x - 1) - \log(x - 2).$$

$$\int \frac{x^2 - x + 1}{x^3 + x^2 + x + 1}\,dx = \frac{1}{4}\log\frac{(x + 1)^6}{x^2 + 1} - \frac{1}{2}\operatorname{arc\,tg} x.$$

$$\int \frac{2x^2 - 3x - 1}{x^3 - 2x^2 - x + 2}\,dx = \log(x - 1)(x + 1)^{\frac{2}{3}}(x - 2)^{\frac{1}{3}}.$$

$$\frac{Y^3}{X^3}$$

$$\int \frac{x^3}{ax^3 + b}\,dx = \frac{x}{a} - \frac{b}{a}\int\frac{dx}{ax^3 + b}.$$

$$\int \frac{x^3}{(x - a)^3}\,dx = \Phi - \frac{3a^2}{\Phi} - \frac{a^3}{2\Phi^2} + 3a\log\Phi.$$

$$\int \frac{x^3}{x^3 - 37x + 84}\,dx = x - \frac{27}{10}\log(x - 3) + \frac{64}{11}\log(x - 4) - \frac{343}{110}\log(x + 7)$$

$$\int \frac{x^3}{(ax + b)^3}\,dx = \frac{\Phi}{a^4} - \frac{3b^2}{a^4\Phi} + \frac{b^3}{2a^4\Phi^2} - \frac{3b}{a^4}\log\Phi.$$

$$\int \frac{x^3}{(x - 1)(x - 2)(x - 3)}\,dx = x + \frac{1}{2}\log\frac{(x - 1)(x - 3)^{27}}{(x - 2)^{16}}.$$

$$\int \frac{x^3 + 2x^2 - 1}{x^2(x - 1)}\,dx = \frac{x^2 - 1}{x} + \log x + 2\log(x - 1).$$

$$\int \frac{5x^3 + 3x - 1}{x^3 + 3x + 1}\,dx = -\frac{x}{(x^3 + 3x + 1)^2}.$$

$$\int \frac{(ax + b)^3}{x^3}\,dx = a^3x - \frac{3ab^2}{x} - \frac{b^3}{2x^2} + 3a^2b\log x.$$

$$\frac{Y^3}{X^{3-a}}$$

$$\int \frac{x^3}{ax + b}\,dx = \frac{x^3}{3a} - \frac{bx^2}{2a^2} + \frac{b^2}{a^3}x - \frac{b^3}{a^4}\log(ax + b).$$

$$\int \frac{x^3}{ax^2 + b}\,dx = \frac{x^2}{2a} - \frac{b}{2a^2}\log(ax^2 + b).$$

$$\int \frac{x^3}{(ax + b)^2}\,dx = \frac{\Phi^2}{2a^4} - \frac{3b\Phi}{a^4} + \frac{b^3}{a^4\Phi} + \frac{3b^2}{a^4}\log\Phi.$$

$$\int \frac{x^3}{3x^2 + 6x + 4}\,dx = \frac{x^2}{6} - \frac{2}{3}x + \frac{4}{9}\log(3x^2 + 6x + 4).$$

$$\int \frac{x^3}{ax^2 + bx + c}\,dx = \frac{x^2}{2a} - \frac{bx}{a^2} + \frac{b^2 - ac}{2a^3}\log\Phi + \frac{b(3ac - b^2)}{2a^3}\int\frac{dx}{\Phi}$$

$$\int \frac{mx^3 + n}{ax + b} dx = \frac{mx^3}{3a} - \frac{bm}{2a^2} x^2 + \frac{b^2m}{a^3} x + \frac{a^3n - b^3m}{a^4} \log (ax + b).$$

$$\int \frac{x^3 + 1}{x^2 - 3x + 2} dx = \frac{x^2}{2} + 3x + 9 \log (x - 2) - 2 \log (x - 1).$$

$$\int \frac{x^3 + 2}{x^2 + 4x + 5} dx = \frac{x^2}{2} - 4x + \frac{11}{2} \log (x^2 + 4x + 5).$$

$$\int \frac{mx^3 + nx}{ax + b} dx = \frac{mx^3}{3a} - \frac{bm}{2a^2} x^2 + \frac{a^2n - b^2m}{a^3} - \frac{b(a^2n + b^2m)}{a^4} \log (ax + b).$$

$$\int \frac{x^3 + 3x}{x^2 + 3x + 4} dx = \frac{x^2}{2} - 3x + 4 \log (x^2 + 3x + 4).$$

$$\int \frac{mx^3 + nx + p}{ax + b} dx = \frac{mx^3}{3a} - \frac{bm}{2a^2} x^2 + \frac{a^2n + b^2m}{a^3} x + \frac{a^2(ap - bn) - b^3m}{a^4} \log (ax + b).$$

$$\int \frac{x^3 - 12x - 14}{(x - 2)^2} dx = \frac{(x + 1)(x + 2)(x + 3)}{2(x - 2)}.$$

$$\int \frac{mx^3 + nx^2}{ax + b} dx = \frac{mx^3}{3a} + \frac{an - bm}{a^2} \left\{ \frac{x^2}{2} - \frac{b}{a} x + \frac{b^2}{a^2} \log (ax + b) \right\}.$$

$$\int \frac{x^3 + 3x^2}{x^2 + 2x + 2} dx = \frac{x^2}{2} + x - 2 \log (x^2 + 2x + 2) + 2 \operatorname{arc\,tg} (x + 1).$$

$$\int \frac{x(x - a)^2}{x + a} dx = \frac{x^3}{3} - \frac{3a}{2} x^2 + 4a^2x - 4a^3 \log (x + a).$$

$$\int \frac{mx^3 + nx^2 + px + q}{ax + b} dx = \frac{mx^3}{3a} + \frac{an - bm}{2a^2} x^2 + \frac{a^2p - b(an - bm)}{a^3} x$$

$$+ \frac{a^2(aq - bp) + b^2(an - bm)}{a^4} \log (ax + b).$$

$$\int \frac{mx^3 + nx^2 + px + q}{(ax + b)^2} dx = (amx - 5bm + 2an) \frac{ax + b}{2a^4} + \frac{b^3m - ab^2n + a^2bp - a^3q}{a^4(ax + b)}$$

$$+ \frac{3b^2m + a^2p - 2abn}{a^4} \log (ax + b).$$

$$\int \frac{mx^3 + nx^2 + px + q}{ax^2 + bx + c} dx = \frac{mx^2}{2a} + \frac{an - bm}{a^2} x + \frac{a(ap - cm) - b(an - bm)}{2a^3} \log (ax^2 + bx + c)$$

$$+ \frac{2a^3q - ab(ap - cm) + (an - bm)(b^2 - 2ac)}{2a^3\sqrt{b^2 - 4ac}} \log \frac{2ax + b - \sqrt{b^2 - 4ac}}{2ax + b + \sqrt{b^2 - 4ac}}. \quad b^2 > 4ac.$$

$$= \frac{mx^2}{2a} + \frac{an - bm}{a^2} x + \frac{a(ap - cm) - b(an - bm)}{2a^3} \log (ax^2 + bx + c)$$

$$+ \frac{2a^3q - ab(ap - cm) + (an - bm)(b^2 - 2ac)}{a^3\sqrt{4ac - b^2}} \operatorname{arc\,tg} \frac{2ax + b}{\sqrt{4ac - b^2}}. \qquad b^2 < 4ac.$$

$$= (2amx - 5bm + 4an) \frac{2ax + b}{8a^3} + \frac{b^3m - 2ab^2n + 4a^2bp - 8a^3q}{4a^3(2ax + b)}$$

$$+ \frac{3b^2m + 4a^2p - 4abn}{4a^3} \log (2ax + b). \qquad b^2 = 4ac.$$

$$\frac{1}{\mathbf{X}^4\,(2)}$$

$$\int \frac{dx}{x^4 + 4} = \frac{1}{16} \log \frac{x^2 + 2x + 2}{x^2 - 2x + 2} + \frac{1}{8} \operatorname{arc\,tg} \frac{2x}{2 - x^2}.$$

$$\int \frac{dx}{x^4 + a^4} = \frac{1}{4a^3\sqrt{2}} \log \frac{x^2 + ax\sqrt{2} + a^2}{x^2 - ax\sqrt{2} + a^2} + \frac{1}{2a^3\sqrt{2}} \operatorname{arc\,tg} \frac{ax\sqrt{2}}{a^2 - x^2}.$$

$$\int \frac{dx}{x^4 - a^4} = \frac{1}{4a^3} \log \frac{x - a}{x + a} - \frac{1}{2a^3} \operatorname{arc\,tg} \frac{x}{a}.$$

$$\int \frac{dx}{ax^4 + b} = \frac{k}{4b\sqrt{2}} \log \frac{x^2 + kx\sqrt{2} + k^2}{x^2 - kx\sqrt{2} + k^2} + \frac{k}{2b\sqrt{2}} \operatorname{arc\,tg} \frac{kx\sqrt{2}}{k^2 - x^2}. \qquad k = \left(\frac{b}{a}\right)^{\frac{1}{4}}.$$

$$\int \frac{dx}{ax^4 - b} = \frac{k}{4b} \log \frac{k - x}{k + x} - \frac{k}{2b} \operatorname{arc\,tg} \frac{x}{k}. \qquad k = \left(\frac{b}{a}\right)^{\frac{1}{4}}$$

$$\int \frac{dx}{1 - x^4} = \frac{1}{4} \log \frac{1 + x}{1 - x} + \frac{1}{2} \operatorname{arc\,tg} x.$$

$$\int \frac{dx}{a^4 - x^4} = \frac{1}{4a^3} \log \frac{x + a}{x - a} + \frac{1}{2a^3} \operatorname{arc\,tg} \frac{x}{a}.$$

$$\int \frac{dx}{b^4 - a^4x^4} = \frac{1}{4ab^3} \log \frac{b + ax}{b - ax} + \frac{1}{2ab^3} \operatorname{arc\,tg} \frac{ax}{b}.$$

$$\int \frac{dx}{ax^4 + bx} = \frac{1}{3b} \log \frac{x^3}{ax^3 + b}.$$

$$\int \frac{dx}{x^4 + a^2x^2} = -\frac{1}{a^2x} - \frac{1}{a^3} \operatorname{arc\,tg} \frac{x}{a}.$$

$$\int \frac{dx}{x^2(ax^2 + b)} = -\frac{1}{bx} - \frac{a}{b} \int \frac{dx}{ax^2 + b}.$$

$$\int \frac{dx}{x^4 - bx^2} = \frac{1}{bx} + \frac{1}{2b\sqrt{b}} \log \frac{x - \sqrt{b}}{x + \sqrt{b}}.$$

$$\int \frac{dx}{x^4 - x^3} = \frac{1}{x} + \frac{1}{2x^2} + \log \frac{x - 1}{x}.$$

$$\int \frac{dx}{ax^4 + bx^3} = \frac{a}{b^2x} - \frac{1}{2bx^2} - \frac{a^2}{b^3} \log \frac{ax + b}{x}.$$

$$\frac{1}{ax^4 + bx^2 + c}$$

$$\int \frac{dx}{(x^2 + 1)^2} = \frac{x}{2(x^2 + 1)} + \frac{1}{2} \operatorname{arc\,tg} x.$$

$$\int \frac{dx}{(x^2 + a^2)^2} = \frac{x}{2a^2(x^2 + a^2)} + \frac{1}{2a^3} \operatorname{arc\,tg} \frac{x}{a}.$$

$$\int \frac{dx}{(ax^2 + b)^2} = \frac{x}{2b(ax^2 + b)} + \frac{1}{2b} \int \frac{dx}{ax^2 + b}.$$

$$\int \frac{dx}{(x^2 - a)^2} = -\frac{x}{2a(x^2 - a)} + \frac{1}{2a\sqrt{a}} \log \frac{x + \sqrt{a}}{\sqrt{x^2 - a}}.$$

$$\int \frac{dx}{(ax^2-b)^2} = -\frac{x}{2b(ax^2-b)} + \frac{1}{4b\sqrt{ab}} \log \frac{x\sqrt{a}+\sqrt{b}}{x\sqrt{a}-\sqrt{b}}.$$

$$\int \frac{dx}{(1-x^2)^2} = \frac{x}{2(1-x^2)} + \frac{1}{4} \log \frac{1+x}{1-x}.$$

$$\int \frac{dx}{(b-ax^2)^2} = \frac{x}{2b(b-ax^2)} + \frac{1}{4b\sqrt{ab}} \log \frac{\sqrt{b}+x\sqrt{a}}{\sqrt{b}-x\sqrt{a}}.$$

$$\int \frac{dx}{(x^2-1)(x^2-4)} = \frac{1}{12} \log \frac{(x+2)(x+1)^2}{(x+2)(x-1)^2}.$$

$$\int \frac{dx}{(1+a^2x^2)(1+b^2x^2)} = \frac{1}{a^2-b^2}(a \text{ arc tg } ax - b \text{ arc tg } bx).$$

$$\int \frac{dx}{(x^2+a)(x^2+b)} = \frac{1}{b-a}\int \frac{dx}{x^2+a} - \frac{1}{b-a}\int \frac{dx}{x^2+b}.$$

$$\int \frac{dx}{x^4-7x^2+12} = \frac{1}{4} \log \frac{x-2}{x+2} + \frac{1}{2\sqrt{3}} \log \frac{x+\sqrt{3}}{x-\sqrt{3}}.$$

$$\int \frac{dx}{x^4+7x^2+12} = \frac{1}{\sqrt{3}} \text{ arc tg } \frac{x}{\sqrt{3}} - \frac{1}{2} \text{ arc tg } \frac{x}{2}.$$

$$\int \frac{dx}{x^4+5x^2+4} = \frac{1}{3} \text{ arc tg } x - \frac{1}{6} \text{ arc tg } \frac{x}{2}.$$

$$\int \frac{dx}{4x^4+5x^2+1} = \frac{2}{3} \text{ arc tg } 2x - \frac{1}{3} \text{ arc tg } x.$$

$$\int \frac{dx}{x^4-2a^2x^2\cos 2\alpha+a^4} = -\frac{1}{8a^3\cos\alpha} \log \frac{x^2-2ax\cos\alpha+a^2}{x^2+2ax\cos\alpha+a^2} + \frac{1}{4a^3\sin\alpha} \text{ arc tg } \frac{2ax\sin\alpha}{a^2-x^2}.$$

$$\int \frac{dx}{b^4m^2x^4+b^2(m^2+1)x^2+1} = \frac{1}{b(m^2-1)}\left\{ m \text{ arc tg } bmx - \text{arc tg } bx \right\}$$

$$\int \frac{dx}{ax^4+bx^2+c} = \frac{2a}{\sqrt{b^2-4ac}}\left\{ \int \frac{dx}{2ax^2+\mathrm{A}} - \int \frac{dx}{2ax^2+\mathrm{B}} \right\}.$$

in which $\mathrm{A} = b-\sqrt{b^2-4ac}$, $\mathrm{B} = b+\sqrt{b^2-4ac}$, $b^2 > 4ac$.

$$= \frac{1}{4ak^3}\left\{ \frac{1}{2\cos\frac{\alpha}{2}} \log \frac{x^2+2kx\cos\frac{\alpha}{2}+k^2}{x^2-2kx\cos\frac{\alpha}{2}+k^2} + \frac{1}{\sin\frac{\alpha}{2}} \text{ arc tg } \frac{2kx\sin\frac{\alpha}{2}}{k^2-x^2} \right\}.$$

in which $k = \left(\frac{c}{a}\right)^{\frac{1}{4}}$, $\cos\alpha = -\frac{b}{2\sqrt{ac}}$, $b^2 < 4ac$.

$$= \frac{x}{2c+bx^2} + \frac{1}{\sqrt{2bc}} \text{ arc tg } x\sqrt{\frac{b}{2c}}. \qquad b^2 = 4ac.$$

$\frac{1}{\mathrm{X}^4}$

$$\int \frac{dx}{x^2(x-a)^2} = \frac{a-2x}{a^2x(x-a)} + \frac{2}{a^3} \log \frac{x}{x-a}.$$

$$\int \frac{dx}{x^2(ax+b)^2} = -\frac{b+2ax}{b^2x(ax+b)} + \frac{2a}{b^3} \log \frac{ax+b}{x}.$$

$$\int \frac{dx}{x^2(ax^2+bx+c)} = -\frac{1}{cx} - \frac{b}{2c^2}\log\frac{x^2}{\Phi} + \frac{b^2-2ac}{2c^2}\int\frac{dx}{\Phi}.$$

$$\int \frac{dx}{x^2(x-1)(x-2)} = -\frac{1}{2x} + \frac{1}{4}\log\frac{x^3(x-2)}{(x-1)^4}.$$

$$\int \frac{dx}{x(x-a)^3} = \frac{2x-3a}{2a^2(x-d)^2} + \frac{1}{a^3}\log\frac{x-a}{x}.$$

$$\int \frac{dx}{x(ax+b)^3} = \frac{1}{b^2\Phi} + \frac{1}{2b\Phi^2} + \frac{1}{b^3}\log\frac{x}{\Phi}.$$

$$\int \frac{dx}{x(x^3+x^2+x+1)} = \frac{1}{4}\log\frac{x^4}{(1+x)^2(1+x^2)} + \frac{1}{2}\operatorname{arc\,tg} x.$$

$$\int \frac{dx}{x(ax^3+bx^2+cx+d)} = \frac{1}{3d}\log\frac{x^3}{\Phi} - \frac{b}{3d}\int\frac{x}{\Phi}dx - \frac{2c}{3d}\int\frac{dx}{\Phi}.$$

$$\int \frac{dx}{(ax+b)^4} = -\frac{1}{3a(ax+b)^3}.$$

$$\int \frac{dx}{(x^2+x+1)^2} = \frac{2x+1}{3\Phi} + \frac{4}{3\sqrt{3}}\operatorname{arc\,tg}\frac{2x+1}{\sqrt{3}}.$$

$$\int \frac{dx}{(x^2-x+1)^2} = \frac{2x-1}{3\Phi} + \frac{4}{3\sqrt{3}}\operatorname{arc\,tg}\frac{2x-1}{\sqrt{3}}.$$

$$\int \frac{dx}{(x^2-2x\cos\alpha+1)^2} = \frac{x-\cos\alpha}{2\Phi\sin^2\alpha} + \frac{1}{2\sin^3\alpha}\operatorname{arc\,tg}\frac{x-\cos\alpha}{\sin\alpha}.$$

$$\int \frac{dx}{(ax^2+bx+c)^2} = \frac{2ax+b}{(4ac-b^2)\Phi} + \frac{2a}{4ac-b^2}\int\frac{dx}{\Phi}.$$

$$\int \frac{dx}{(x+b)^2(x^2+a)} = -\frac{1}{(b^2+a)(x+b)} + \frac{1}{(b^2+a)^2}\left\{b\log\frac{(x+b)^2}{x^2+a} + \frac{b^2-a}{\sqrt{a}}\operatorname{arc\,tg}\frac{x}{\sqrt{a}}\right\}.$$

$$\int \frac{dx}{(x+a)^2(x+b)^2} = -\frac{1}{(a-b)^2}\left(\frac{1}{x+a} + \frac{1}{x+b}\right) - \frac{2}{(a-b)^3}\log\frac{x+b}{x+a}.$$

$$\int \frac{dx}{x(x^2+a)(x+b)} = \frac{1}{ab}\int(x+b)(x^2+a)\,dx - \frac{1}{b(b^2+a)}\int x(x^2+a)\,dx - \frac{1}{a(b^2+a)}\int x(x+b)(bx+a)\,dx.$$

$$\int \frac{dx}{(x-a)^2(x-b)^2} = -\frac{2x-a-b}{(a-b)^2(x-a)(x-b)} + \frac{2}{(a-b)^3}\log\frac{x-b}{x-a}$$

$$\int \frac{dx}{(x-a)^2(x-b)(x-c)} = \frac{1}{(a-b)(a-c)(x-a)} + \frac{2a-b-c}{(a-b)^2(a-c)^2}\log(x-a) + \frac{\log(x-b)}{(a-b)^2(b-c)} - \frac{\log(x-c)}{(a-c)^2(b-c)}.$$

$\frac{mx+n}{X^4}$

$$\int \frac{x}{ax^4+b}dx = \frac{1}{2\sqrt{ab}}\operatorname{arc\,tg} x^2\sqrt{\frac{a}{b}}.$$

$$\int \frac{x}{ax^4-b}dx = \frac{1}{4\sqrt{ab}}\log\frac{x^2\sqrt{a}-\sqrt{b}}{x^2\sqrt{a}+\sqrt{b}}.$$

$$\int \frac{x}{b-ax^4}dx = \frac{1}{4\sqrt{ab}}\log\frac{\sqrt{b}+x^2\sqrt{a}}{\sqrt{b}-x^2\sqrt{a}}.$$

$$\int \frac{x}{x^4 - 4x + 3}\,dx = -\frac{1}{6(x-1)} + \frac{1}{18}\log(x-1) - \frac{1}{36}\log(x^2+2x+3) - \frac{5}{18\sqrt{2}}\operatorname{arc\,tg}\frac{x+1}{\sqrt{2}}.$$

$$\int \frac{x}{(ax^2+b)^2}\,dx = -\frac{1}{2a(ax^2+b)} = \frac{x^2}{2b(ax^2+b)} = -\frac{1}{4ab}\cdot\frac{b-ax^2}{b+ax^2} = \frac{1}{2(bn-am)}\cdot\frac{nx^2+m}{ax^2+b}.$$

$$\int \frac{x}{(b-ax^2)^2}\,dx = \frac{1}{2a(b-ax^2)} = \frac{1}{2(bn+am)}\cdot\frac{m+nx^2}{b-ax^2}.$$

$$\int \frac{x}{(x^2+a)(x^2+b)}\,dx = \frac{1}{2(b-a)}\log\frac{x^2+a}{x^2+b}.$$

$$\int \frac{x}{x^4+(a+b)x^2+ab}\,dx = \frac{1}{b-a}\int\frac{x}{x^2+a}\,dx - \frac{1}{b-a}\int\frac{x}{x^2+b}\,dx.$$

$$\int \frac{x}{ax^4+bx^2+c}\,dx = \frac{1}{2\sqrt{b^2-4ac}}\log\frac{2ax^2+b-\sqrt{b^2-4ac}}{2ax^2+b+\sqrt{b^2-4ac}}. \qquad b^2 > 4ac.$$

$$= \frac{1}{\sqrt{4ac-b^2}}\operatorname{arc\,tg}\frac{2ax^2+b}{\sqrt{4ac-b^2}}. \qquad b^2 < 4ac.$$

$$= -\frac{1}{2ax^2+b}. \qquad b^2 = 4ac.$$

$$\int \frac{x}{x^4-5x^3+20x-6}\,dx = \frac{1}{9}\log(x^2-5x+4) + \frac{1}{36}\log(x+2) - \frac{1}{4}\log(x-2).$$

$$\int \frac{x}{(ax+b)^4}\,dx = -\frac{1}{2a^2(ax+b)^2} + \frac{b}{3a^2(ax+b)^3} = \frac{1}{2b^2}\left(\frac{x}{\Phi}\right)^2 - \frac{a}{3b^2}\left(\frac{x}{\Phi}\right)^3.$$

$$\int \frac{x}{(x+a)^2(x+b)^2}\,dx = \frac{1}{(a-b)^2}\left\{\frac{a}{x+a} + \frac{b}{x+b}\right\} + \frac{a+b}{(a-b)^3}\log\frac{x+b}{x+a}.$$

$$\int \frac{x}{(b+x)^2(a+x^2)}\,dx = \frac{b}{(a+b^2)(b+x)} + \frac{1}{(a+b^2)^2}\left\{\frac{a-b^2}{2}\log\frac{(b+x)^2}{a+x} + 2ab\int\frac{dx}{a+x^2}\right\}.$$

$$\int \frac{x}{(x^2+x+1)^2}\,dx = -\frac{x+2}{3(x^2+x+1)} - \frac{2}{3\sqrt{3}}\operatorname{arc\,tg}\frac{2x+1}{\sqrt{3}}.$$

$$\int \frac{x}{(x^2-x\sqrt{3}+1)^2}\,dx = \frac{x\sqrt{3}-2}{x^2-x\sqrt{3}+1} + 2\sqrt{3}\operatorname{arc\,tg}(2x-\sqrt{3}).$$

$$\int \frac{x}{(x^2-2x\cos\alpha+1)^2}\,dx = -\frac{x-\cos\alpha+\sin^2\alpha}{2\sin^2\alpha(x^2-2x\cos\alpha+1)} + \frac{\cos\alpha}{2\sin^3\alpha}\operatorname{arc\,tg}\frac{x-\cos\alpha}{\sin\alpha}.$$

$$\int \frac{x}{(ax^2+bx+c)^2}\,dx = -\frac{bx+2c}{(4ac-b^2)(ax^2+bx+c)} - \frac{b}{4ac-b^2}\int\frac{dx}{ax^2+bx+c}.$$

$$= -\frac{1}{2a(ax^2+bx+c)} - \frac{b}{2a}\int\frac{dx}{(ax^2+bx+c)^2}.$$

$$\int \frac{mx+n}{x^4-1}\,dx = \frac{m}{4}\log\frac{x^2-1}{x^2+1} + \frac{n}{4}\log\frac{x-1}{x+1} + \frac{n}{2}\operatorname{arc\,tg} x.$$

$$\int \frac{mx+n}{(x^2+1)^2}\,dx = \frac{nx-m}{2(x^2+1)} + \frac{n}{2}\operatorname{arc\,tg} x.$$

$$\int \frac{x+2}{(x^2+x+1)^2}\,dx = \frac{x}{x^2+x+1} + \frac{2}{\sqrt{3}}\operatorname{arc\,tg}\frac{2x+1}{\sqrt{3}}.$$

$$\int \frac{mx+n}{(ax^2+bx+c)^2}\,dx = \frac{(2an-bm)x+bn-2cm}{(4ac-b^2)\,\Phi} + \frac{2an-bm}{4ac-b^2}\int\frac{dx}{\Phi}.$$

$\frac{Y^2}{X^4}$

$$\int \frac{x^2}{x^4+a^4}\,dx = \frac{1}{4a\sqrt{2}}\log\frac{x^2-ax\sqrt{2}+a^2}{x^2+ax\sqrt{2}+a^2} + \frac{1}{2a\sqrt{2}}\operatorname{arc\,tg}\frac{ax\sqrt{2}}{a^2-x^2}.$$

$$\int \frac{x^2}{x^4-a^4}\,dx = \frac{1}{4a}\log\frac{x-a}{x+a} + \frac{1}{2a}\operatorname{arc\,tg}\frac{x}{a}.$$

$$\int \frac{x^2}{ax^4+b}\,dx = \frac{1}{4ak\sqrt{2}}\log\frac{x^2-kx\sqrt{2}+k^2}{x^2+kx\sqrt{2}+k^2} + \frac{1}{2ak\sqrt{2}}\operatorname{arc\,tg}\frac{kx\sqrt{2}}{k^2-x^2}. \qquad k=\left(\frac{b}{a}\right)^{\frac{1}{4}}.$$

$$\int \frac{x^2}{ax^4-b}\,dx = \frac{1}{4ak}\log\frac{x-k}{x+k} + \frac{1}{2ak}\operatorname{arc\,tg}\frac{x}{k}. \qquad k=\left(\frac{b}{a}\right)^{\frac{1}{4}}.$$

$$\int \frac{x^2}{(x^2+1)^2}\,dx = -\frac{x}{2(x^2+1)} + \frac{1}{2}\operatorname{arc\,tg} x.$$

$$\int \frac{x^2}{(x^2-a)^2}\,dx = -\frac{x}{2(x^2-a)} + \frac{1}{4\sqrt{a}}\log\frac{x-\sqrt{a}}{x+\sqrt{a}}.$$

$$\int \frac{x^2}{(ax^2+b)^2}\,dx = -\frac{x}{2a(ax^2+b)} + \frac{1}{2a}\int\frac{dx}{ax^2+b}.$$

$$= -\frac{x}{2a(ax^2+b)} + \frac{1}{2a\sqrt{ab}}\operatorname{arc\,tg} x\sqrt{\frac{a}{b}}.$$

$$\int \frac{x^2}{(ax^2-b)^2}\,dx = -\frac{x}{2a(ax^2-b)} + \frac{1}{4a\sqrt{ab}}\log\frac{x\sqrt{a}-\sqrt{b}}{x\sqrt{a}+\sqrt{b}}.$$

$$\int \frac{x^2}{(b-ax^2)^2}\,dx = \frac{x}{2a(b-ax^2)} - \frac{1}{4a\sqrt{ab}}\log\frac{\sqrt{b}+x\sqrt{a}}{\sqrt{b}-x\sqrt{a}}.$$

$$\int \frac{x^2}{(1+x^2)(1+4x^2)}\,dx = \frac{1}{6}\operatorname{arc\,tg}\frac{2x^3}{1+3x^2}.$$

$$\int \frac{x^2}{(1+a^2x^2)(1+b^2x^2)}\,dx = \frac{1}{a(b^2-a^2)}\left\{\operatorname{arc\,tg} ax - \frac{a}{b}\operatorname{arc\,tg} bx\right\}.$$

$$\int \frac{x^2}{(x^2+a)(x^2+b)}\,dx = \frac{a}{a-b}\int\frac{dx}{x^2+a} - \frac{b}{a-b}\int\frac{dx}{x^2+b}.$$

$$\int \frac{x^2}{x^4+x^2-2}\,dx = \frac{1}{6}\log\frac{x-1}{x+1} + \frac{\sqrt{2}}{3}\operatorname{arc\,tg}\frac{x}{\sqrt{2}}.$$

$$\int \frac{x^2}{3x^4+5x^2+2}\,dx = \operatorname{arc\,tg} x - \sqrt{\frac{2}{3}}\operatorname{arc\,tg} x\sqrt{\frac{3}{2}}.$$

$$\int \frac{x^2}{x^4-2a^2x^2\cos 2\alpha+a^4}\,dx = \frac{1}{8a\cos\alpha}\log\frac{x^2-2ax\cos\alpha+a^2}{x^2+2ax\cos\alpha+a^2} + \frac{1}{4a\sin\alpha}\operatorname{arc\,tg}\frac{2ax\sin\alpha}{a^2-x^2}.$$

$$\int \frac{x^2}{ax^4+bx^2+c}\,dx = \frac{A-b}{A}\int\frac{dx}{2ax^2+b-A} + \frac{A+b}{A}\int\frac{dx}{2ax^2+b+A},$$

in which $A=\sqrt{b^2-4ac}$, $b^2>4ac$.

$$= \frac{1}{8ak\cos\frac{\varepsilon}{2}} \log \frac{x^2 - 2kx\cos\frac{\varepsilon}{2} + k^2}{x^2 + 2kx\cos\frac{\varepsilon}{2} + k^2} + \frac{1}{4ak\sin\frac{\varepsilon}{2}} \text{arc tg} \frac{2kx\sin\frac{\varepsilon}{2}}{k^2 - x^2},$$

in which $k = \left(\frac{c}{a}\right)^{\frac{1}{4}}$, $\cos\varepsilon = -\frac{b}{2\sqrt{ac}}$, $b^2 < 4ac.$

$$= -\frac{2cx}{b(2c + bx^2)} + \frac{2c}{b\sqrt{2bc}} \text{arc tg}\, x\sqrt{\frac{b}{2c}}. \qquad b^2 = 4ac.$$

$$\int \frac{x^2}{(ax+b)^4}\,dx = -\frac{1}{a^3\Phi} + \frac{b}{a^3\Phi^2} - \frac{b^2}{3a^3\Phi^3}.$$

$$\int \frac{x^2}{(x+a)^2(x^2+a^2)}\,dx = -\frac{1}{2(x+a)} + \frac{1}{4a}\log\frac{x^2+a^2}{(x+a)^2}.$$

$$\int \frac{x^2}{(x^2+x+1)^2}\,dx = \frac{1-x}{3(x^2+x+1)} + \frac{4}{3\sqrt{3}}\text{arc tg}\frac{2x+1}{\sqrt{3}}.$$

$$\int \frac{x^2}{(ax^2+bx+c)^2}\,dx = \frac{(b^2-2ac)x + bc}{a(4ac-b^2)\Phi} + \frac{2c}{4ac-b^2}\int\frac{dx}{\Phi} = -\frac{x}{a\Phi} + \frac{c}{a}\int\frac{dx}{\Phi^2}.$$

$$\int \frac{x^2+1}{x^4+1}\,dx = \frac{1}{\sqrt{2}}\text{arc tg}(x\sqrt{2}-1) + \frac{1}{\sqrt{2}}\text{arc tg}(x\sqrt{2}+1) = \frac{1}{\sqrt{2}}\text{arc tg}\frac{x\sqrt{2}}{1-x^2}.$$

$$\int \frac{a^2-x^2}{(a^2+x^2)^2}\,dx = \frac{x}{a^2+x^2}.$$

$$\int \frac{x^2+2}{(x^2+1)^2}\,dx = \frac{x}{2(x^2+1)} + \frac{3}{2}\text{arc tg}\,x.$$

$$\int \frac{mx^2+n}{(ax^2+b)^2}\,dx = \frac{(an-bm)x}{2ab(ax^2+b)} + \frac{an+bm}{2ab\sqrt{ab}}\text{arc tg}\, x\sqrt{\frac{a}{b}}$$

$$\int \frac{mx^2+n}{(ax^2-b)^2}\,dx = -\frac{(an+bm)x}{2ab(ax^2-b)} + \frac{bm-an}{4ab\sqrt{ab}}\log\frac{x\sqrt{a}-\sqrt{b}}{x\sqrt{a}+\sqrt{b}}.$$

$$\int \frac{mx^2+n}{(b-ax^2)^2}\,dx = \frac{(an+bm)x}{2ab(b-ax^2)} + \frac{an-bm}{4ab\sqrt{ab}}\log\frac{\sqrt{b}+x\sqrt{a}}{\sqrt{b}-x\sqrt{a}}.$$

$$\int \frac{x^2+1}{x^4+x^2+1}\,dx = \frac{1}{\sqrt{3}}\text{arc tg}\frac{x\sqrt{3}}{1-x^2} = \frac{1}{\sqrt{3}}\text{arc tg}\frac{2x+1}{\sqrt{3}} + \frac{1}{\sqrt{3}}\text{arc tg}\frac{2x-1}{\sqrt{3}}.$$

$$\int \frac{b-cx^2}{(b+cx^2)^2+a^2x^2}\,dx = \frac{1}{a}\text{arc tg}\frac{ax}{b+cx^2}.$$

$$\int \frac{x^2-a^2}{x^4+6a^2x^2+a^4}\,dx = \frac{1}{2a}\text{arc tg}\frac{x^2+a^2}{2ax}.$$

$$\int \frac{x^2+1}{(x^2+x+1)^2}\,dx = \frac{x+2}{3(x^2+x+1)} + \frac{8}{3\sqrt{3}}\text{arc tg}\frac{2x+1}{\sqrt{3}}.$$

$$\int \frac{1-x^2}{(x^2+x+1)^2}\,dx = \frac{x}{x^2+x+1}.$$

$$\int \frac{1-x^2}{(x^2-x+1)^2}\,dx = \frac{x}{x^2-x+1} = \frac{x^2+x+1}{2(x^2-x+1)}.$$

$$\int \frac{x^2-2}{(x^2-2x+1)^2}\,dx = \frac{1+3x-3x^2}{3(x-1)^3}.$$

$$\int \frac{mx^2+n}{(ax^2+bx+c)^2}\,dx = \frac{b\,(an+cm)+(mb^2-2acm+2a^2n)x}{a\,(4ac-b^2)\,\Phi} + \frac{2\,(an+cm)}{4ac-b^2}\int\frac{dx}{\Phi}.$$

$$\int \frac{x^2-x+2}{x^4-5x^2+4}\,dx = \frac{1}{3}\log\frac{(x+1)^2\,(x-2)}{(x-1)\,(x+2)^2}.$$

$$\int \frac{x^2-x+1}{(x^2+x+1)^2}\,dx = \frac{2\,(x+2)}{3\,(x^2+x+1)} + \frac{10}{3\sqrt{3}}\operatorname{arc\,tg}\frac{2x+1}{\sqrt{3}}.$$

$$\frac{Y^3}{X^4}$$

$$\int \frac{x^3}{ax^4+b}\,dx = \frac{1}{4a}\log\,(ax^4+b).$$

$$\int \frac{x^3}{(ax^2+b)^2}\,dx = \frac{b}{2a^2\,(ax^2+b)} + \frac{1}{2a^2}\log\,(ax^2+b).$$

$$\int \frac{x^3}{(x^2-a)\,(x^2-b)}\,dx = \frac{1}{2\,(a-b)}\log\frac{(x^2-a)^a}{(x^2-b)^b}.$$

$$\int \frac{x^3}{(x^2-2)\,(3-x^2)}\,dx = \log\frac{x^2-2}{(6-2x^2)^{\frac{3}{2}}}.$$

$$\int \frac{x^3}{x^4+3x^2+2}\,dx = \log\frac{x^2+2}{\sqrt{x^2+1}}.$$

$$\int \frac{x^3}{(ax+b)^4}\,dx = \frac{1}{a^4}\left\{\frac{3b}{\Phi} - \frac{3b^2}{2\Phi^2} + \frac{b^3}{3\Phi^3} + \log\Phi\right\}.$$

$$\int \frac{x^3}{(x^2+x+1)^2}\,dx = \frac{2x+1}{3\Phi} + \frac{1}{2}\log\Phi - \frac{5}{3\sqrt{3}}\operatorname{arc\,tg}\frac{2x+1}{\sqrt{3}}.$$

$$\int \frac{x^3}{(ax^2+bx+c)^2}\,dx = \frac{(3ac-b^2)\,bx+(2ac-b^2)\,c}{a^2\,(4ac-b^2)\,\Phi} + \frac{1}{2a^2}\log\Phi - \frac{b\,(6ac-b^2)}{2a^2\,(4ac-b^2)}\int\frac{dx}{\Phi}.$$

$$\int \frac{x^3}{(x-1)^2\,(x^2-2x+2)}\,dx = -\frac{1}{x-1} + \log\frac{(x-1)^3}{x^2-2x+2} + 2\operatorname{arc\,tg}\,(x-1).$$

$$\int \frac{x^3+1}{x\,(ax^3+b)}\,dx = \frac{1}{3b}\log x^3 + \frac{b-a}{3ab}\log\,(ax^3+b).$$

$$\int \frac{x^3+1}{x^4-3x^3+3x^2-x}\,dx = -\frac{x}{(x-1)^2} + \log\frac{(x-1)^2}{x}.$$

$$\int \frac{x^3+1}{(x-1)^4}\,dx = -\frac{2}{3\,(x-1)^3} - \frac{3}{2\,(x-1)^2} - \frac{3}{x-1} + \log\,(x-1).$$

$$\int \frac{x^3-2}{x^4+x^3-3x^2-x+2}\,dx = \frac{1}{6\,(x-1)} + \frac{23}{36}\log\,(x-1) - \frac{3}{4}\log\,(x+1) + \frac{10}{9}\log\,(x+2).$$

$$\int \frac{2x^3-x}{(1-x^2)^2}\,dx = \frac{1}{2\,(1-x^2)} + \log\,(1-x^2).$$

$$\int \frac{2x^3-7x}{x^4-7x^2+8}\,dx = \frac{1}{2}\log\,(x^4-7x^2+8).$$

$$\int \frac{bx-2ax^3}{ax^4-bx^2+c}\,dx = \frac{1}{2}\log\frac{1}{ax^4-bx^2+c}.$$

$$\int \frac{x(2x^2 - x + 5)}{(x^2+1)^2} dx = \frac{x-3}{2(x^2+1)} + \log(x^2+1) - \frac{1}{2} \text{arc tg } x.$$

$$\int \frac{x^3 - 4x^2 + 1}{(x-2)^4} dx = -\frac{29 - 30x + 6x^2}{3(x-2)^3} + \log(x-2).$$

$$\int \frac{x^3 + 4x^2 + 6x}{x^4 + 2x^3 + 3x^2 + 4x + 2} dx = \frac{1}{x+1} + \frac{1}{3} \log \frac{(x^2+2)^2}{x+1} + \frac{4\sqrt{2}}{3} \text{arc tg } \frac{x}{\sqrt{2}}.$$

$$\int \frac{2x^3 + 7x^2 + 6x + 2}{x^4 + 3x^3 + 2x^2} dx = -\frac{1}{x} + \log(x^2 + x)\sqrt{\frac{x}{x+2}}.$$

$$\int \frac{(x-2)^3}{(x+1)^4} dx = \frac{9}{x+1} - \frac{27}{2(x+1)^2} + \frac{9}{(x+1)^3} + \log(x+1).$$

$$\int \frac{(x+1)^3}{(x-1)^4} dx = -\frac{6x^2 - 6x + \frac{8}{3}}{(x-1)^3} + \log(x-1).$$

$\frac{Y^4}{X^4}$

$$\int \frac{x^4}{x^4 - a^2} dx = \frac{1}{2} \int \frac{x^2}{x^2 + a} dx + \frac{1}{2} \int \frac{x^2}{x^2 - a} dx$$

$$\int \frac{x^4}{ax^4 + b} dx = \frac{x}{a} - \frac{b}{a} \int \frac{dx}{ax^4 + b}.$$

$$\int \frac{x^4}{(x^2-1)^2} dx = x + \frac{1}{2} \log \frac{x-1}{x+1} + \int \frac{x^2}{(x^2-1)^2} dx.$$

$$\int \frac{x^4}{(1 - ax^2)^2} dx = \frac{x(3 - 2ax^2)}{2a^2(1 - ax^2)} - \frac{3}{4a^2\sqrt{a}} \log \frac{1 + x\sqrt{a}}{1 - x\sqrt{a}}$$

$$\int \frac{x^4}{(x^2 + a^2)^2} dx = -\frac{x^3}{2(x^2 + a^2)} + \frac{3}{2} x - \frac{3a}{2} \text{arc tg } \frac{x}{a}.$$

$$\int \frac{x^4}{(ax^2 + b)^2} dx = \frac{x}{a^2} + \frac{bx}{2a^2\Phi} - \frac{3b}{2a^2} \int \frac{dx}{\Phi}$$

$$\int \frac{x^4}{x^4 - 2x^2 + 1} dx = x - \frac{x}{2(x^2 - 1)} + \frac{3}{4} \log \frac{x-1}{x+1}.$$

$$\int \frac{x^4}{ax^4 + bx^2 + c} dx = \frac{x}{a} - \frac{c}{a} \int \frac{dx}{ax^4 + bx^2 + c} - \frac{b}{a} \int \frac{x^2}{ax^4 + bx^2 + c} dx.$$

$$\int \frac{x^4}{(ax+b)^4} dx = \frac{1}{a^5} \left\{ \Phi - \frac{6b^2}{\Phi} + \frac{2b^3}{\Phi^2} - \frac{b^4}{3\Phi^3} - 4b \log \Phi \right\}$$

$$\int \frac{x^4}{(ax^2 + bx + c)^2} dx = \frac{x}{a^2} + 2\left(\frac{c}{a^2} - \frac{b^2}{a^3}\right) \frac{x}{\Phi} + \frac{b^3}{2a^4\Phi} - \frac{b}{a^3} \log \Phi - \left\{ \frac{b^4}{2a^4} - \frac{3b^2c}{a^3} + \frac{3c^2}{a^2} \right\} \int \frac{dx}{\Phi^2}.$$

$\frac{Y^4}{X^{4-a}}$

$$\int \frac{x^4}{ax+b} dx = \frac{x^4}{4a} - \frac{bx^3}{3a^2} + \frac{b^2x^2}{2a^3} - \frac{b^3x}{a^4} + \frac{b^4}{a^5} \log(ax+b).$$

$$\int \frac{x^4}{x^2+1} dx = \frac{x^3}{3} - x + \text{arc tg } x.$$

$$\int \frac{x^4}{ax^2+b}\,dx = \frac{x^3}{3a} - \frac{bx}{a^2} + \frac{b^2}{a^2}\int \frac{dx}{ax^2+b}.$$

$$\int \frac{x^4}{ax^2-b}\,dx = \frac{x^3}{3a} + \frac{bx}{a^2} + \frac{b\sqrt{b}}{2a^2\sqrt{a}} \log \frac{x\sqrt{a}-\sqrt{b}}{x\sqrt{a}+\sqrt{b}}.$$

$$\int \frac{x^4}{b-ax^2}\,dx = -\frac{x^3}{3a} - \frac{bx}{a^2} + \frac{b\sqrt{b}}{2a^2\sqrt{a}} \log \frac{\sqrt{b}+x\sqrt{a}}{\sqrt{b}-x\sqrt{a}}.$$

$$\int \frac{x^4}{(ax+b)^2}\,dx = \frac{1}{a^5}\left\{ \frac{\Phi^3}{3} - 2b\,\Phi^2 + 6b^2\,\Phi - \frac{b^4}{\Phi} - 4b^3 \log \Phi \right\}.$$

$$\int \frac{x^4}{(x-1)(x-2)}\,dx = \frac{x^3}{3} + \frac{3}{2}x^2 + 7x + 16\log(x-2) - \log(x-1).$$

$$\int \frac{x^4}{ax^2+bx+c}\,dx = \frac{x^3}{3a} - \frac{bx^2}{2a^2} + \frac{b^2-ac}{a^3}x + \frac{(2ac-b^2)\,b}{2a^4}\log\Phi + \frac{2a^2c^2-4ab^2c+b^4}{2a^4}\int\frac{dx}{\Phi}.$$

$$\int \frac{(ax^2+b)^2}{x}\,dx = \frac{1}{4}(ax^2+b)^2 + \frac{b}{2}(ax^2+b) + \frac{b^2}{2}\log x^2.$$

$$\frac{1}{X^5}$$

$$\int \frac{dx}{1+x^5} = -\frac{1}{5}\cos\frac{\pi}{5}\log\left(1-2x\cos\frac{\pi}{5}+x^2\right) + \frac{2}{5}\sin\frac{\pi}{5}\operatorname{arc\,tg}\frac{x\sin\frac{\pi}{5}}{1-x\cos\frac{\pi}{5}}$$

$$-\frac{1}{5}\cos\frac{3\pi}{5}\log\left(1-2x\cos\frac{3\pi}{5}+x^2\right) + \frac{2}{5}\sin\frac{3\pi}{5}\operatorname{arc\,tg}\frac{x\sin\frac{3\pi}{5}}{1-x\cos\frac{3\pi}{5}}$$

$$+\frac{1}{5}\log(1+x).$$

$$\int \frac{dx}{1-x^5} = -\frac{1}{5}\cos\frac{2\pi}{5}\log\left(1-2x\cos\frac{2\pi}{5}+x^2\right) + \frac{2}{5}\sin\frac{2\pi}{5}\operatorname{arc\,tg}\frac{x\sin\frac{2\pi}{5}}{1-x\cos\frac{2\pi}{5}}$$

$$-\frac{1}{5}\cos\frac{4\pi}{5}\log\left(1-2x\cos\frac{4\pi}{5}+x^2\right) + \frac{2}{5}\sin\frac{4\pi}{5}\operatorname{arc\,tg}\frac{x\sin\frac{4\pi}{5}}{1-x\cos\frac{4\pi}{5}}$$

$$-\frac{1}{5}\log(1-x).$$

$\int \frac{dx}{a+bx^5}$ See p. 26.

$$\int \frac{dx}{x(ax^4+b)} = \frac{1}{2b}\log\frac{x^2}{\sqrt{ax^4+b}}.$$

$$\int \frac{dx}{x^2(ax^3+b)} = -\frac{1}{bx} + \frac{1}{6bk}\log\frac{(x+k)^2}{x^2-kx+k^2} - \frac{1}{bk\sqrt{3}}\operatorname{arc\,tg}\frac{x\sqrt{3}}{2k-x}. \qquad k = \left(\frac{b}{a}\right)^{\frac{1}{3}}$$

$$\int \frac{dx}{x^3(ax^2+b)} = -\frac{1}{2bx^2} + \frac{a}{2b^2}\log\frac{ax^2+b}{x^2}.$$

$$\int \frac{dx}{x^4(ax+b)} = -\frac{a^2}{b^3x} + \frac{a}{2b^2x^2} - \frac{1}{3bx^3} + \frac{a^3}{b^4}\log\frac{ax+b}{x}.$$

$$\int \frac{dx}{x(x^2-a^2)^2} = \frac{1}{2a^2(x^2-a^2)} + \frac{1}{2a^4}\log\frac{x^2-a^2}{x^2}.$$

$$\int\frac{dx}{x(ax^2+b)^2}=\frac{1}{2b\,(ax^2+b)}+\frac{1}{2b^2}\log\frac{x^2}{ax^2+b}=-\frac{ax^2}{2b^2\Phi}+\frac{1}{b^2}\log\frac{x}{\sqrt{\Phi}}\cdot$$

$$\int\frac{dx}{x^3(x-1)^2}=-\frac{1}{2x^2}-\frac{2}{x}-\frac{1}{x-1}+3\log\frac{x}{x-1}\cdot$$

$$\int\frac{dx}{x^3(ax+b)^2}=\frac{a^2}{b^3(ax+b)}-\frac{1}{2b^2x^2}+\frac{2a}{b^3x}+\frac{3a^2}{b^4}\log\frac{x}{ax+b}\cdot$$

$$\int\frac{dx}{x^3(ax^2+bx+c)}=-\frac{1}{2cx^2}+\frac{b}{c^2x}+\left(\frac{b^2}{2c^3}-\frac{a}{2c^2}\right)\log\frac{x^2}{\Phi}+\left(\frac{3\,ab}{2c^2}-\frac{b^3}{2c^3}\right)\int\frac{dx}{\Phi}\cdot$$

$$\int\frac{dx}{x(x^2+a)(x^2+b)}=\frac{1}{2b\,(a-b)}\log\frac{x^2}{x^2+b}-\frac{1}{2a\,(a-b)}\log\frac{x^2}{x^2+a}\cdot$$

$$\int\frac{dx}{x^5+2x^3+3x^2}=-\frac{1}{3x}+\frac{1}{90}\log\frac{(x^2-x+3)(x+1)^{18}}{x^{20}}-\frac{13}{45\sqrt{11}}\operatorname{arc\,tg}\frac{2x-1}{\sqrt{11}}\cdot$$

$$\int\frac{dx}{x^2(ax+b)^3}=-\frac{a}{2b^2\Phi^2}-\frac{2a}{b^3\Phi}-\frac{1}{b^3x}+\frac{3a}{b^4}\log\frac{\Phi}{x}\cdot$$

$$\int\frac{dx}{(x^3-1)(x^2+1)}=\frac{1}{6}\log\frac{(x-1)(x^2+1)^{\frac{3}{2}}}{(x^2+x+1)^2}-\frac{1}{2}\operatorname{arc\,tg}x\cdot$$

$$\int\frac{dx}{x^5-10x^2+15x-6}=-\int\frac{z^3}{10z^2+5z+1}\,dz.\qquad z=\frac{1}{x-1}\cdot$$

$$\int\frac{dx}{x(ax+b)^4}=\frac{1}{b^3\Phi}+\frac{1}{2b^2\Phi^2}+\frac{1}{3b\Phi^3}+\frac{1}{b^4}\log\frac{x}{\Phi}\cdot$$

$$\int\frac{dx}{x(ax^2+bx+c)^2}=\frac{2ac-abx-b^2}{c\,(4ac-b^2)\,\Phi}+\frac{1}{2c^2}\log\frac{x^2}{\Phi}-\frac{b\,(6ac-b^2)}{2c^2(4ac-b^2)}\int\frac{dx}{\Phi}\cdot$$

$$\int\frac{dx}{(x-1)(x^2+1)^2}=-\frac{x-1}{4\,(x^2+1)}+\frac{1}{4}\log\frac{x-1}{\sqrt{x^2+1}}-\frac{1}{2}\operatorname{arc\,tg}x\cdot$$

$$\int\frac{dx}{(x+1)(x^2+1)^2}=\frac{x+1}{4\,(x^2+1)}+\frac{1}{8}\log\frac{(x+1)^2}{x^2+1}+\frac{1}{2}\operatorname{arc\,tg}x.$$

$$\int\frac{dx}{(x-a)^2(x-b)^3}=\frac{(x-a)^2-5\,(x-a)(x-b)-2\,(x-b)^2}{2\,(a-b)^3(x-a)(x-b)^2}-\frac{3}{(a-b)^4}\log\frac{x-a}{x-b}\cdot$$

$$\int\frac{dx}{x^5+x^4+2x^3+2x^2+x+1}=\frac{x+1}{4\,(x^2+1)}+\frac{1}{4}\log\frac{x+1}{\sqrt{x^2+1}}+\frac{1}{2}\operatorname{arc\,tg}x.$$

$\frac{Y^{1\ldots5}}{X^5}$

$\int\frac{x}{ax^5+b}\,dx.$ See p. 26.

$\int\frac{x^2}{ax^5+b}\,dx.$ See p. 26.

$$\int\frac{x^2}{(ax+b)^5}\,dx=\frac{1}{a^3}\left\{-\frac{1}{2\Phi^2}+\frac{2b}{3\Phi^3}-\frac{b^2}{4\Phi^4}\right\}$$

$$\int\frac{x(x+1)}{(1-x)(1+x^2)^2}\,dx=\frac{1}{2}\int\frac{dx}{1-x}-\int\frac{dx}{(1+x^2)^2}+\frac{1}{2}\int\frac{x+1}{1+x^2}\,dx.$$

$$\int \frac{3x^2 + x - 2}{(x-1)^3(x^2+1)}\,dx = -\frac{1}{2(x-1)^2} - \frac{5}{2(x-1)} + \frac{3}{4}\log\frac{x^2+1}{(x-1)^2} - \operatorname{arc\,tg} x.$$

$$\left\{\begin{aligned}
\int \frac{dx}{ax^5+b} &= \frac{2k}{5a}\left\{\frac{1}{2}\log(x+k) - P_0\cos\frac{\pi}{5} + P_1\cos\frac{2\pi}{5} + Q_0\sin\frac{\pi}{5} + Q_1\sin\frac{2\pi}{5}\right\},\\
\int \frac{x}{ax^5+b}\,dx &= \frac{2k^2}{5a}\left\{-\frac{1}{2}\log(x+k) - P_0\cos\frac{2\pi}{5} + P_1\cos\frac{\pi}{5} + Q_0\sin\frac{2\pi}{5} - Q_1\sin\frac{\pi}{5}\right\},\\
\int \frac{x^2}{ax^5+b}\,dx &= \frac{2k^3}{5a}\left\{\frac{1}{2}\log(x+k) + P_0\cos\frac{2\pi}{5} - P_1\cos\frac{\pi}{5} + Q_0\sin\frac{2\pi}{5} - Q_1\sin\frac{\pi}{5}\right\},\\
\int \frac{x^3}{ax^5+b}\,dx &= \frac{2k^4}{5a}\left\{-\frac{1}{2}\log(x+k) + P_0\cos\frac{\pi}{5} - P_1\cos\frac{2\pi}{5} + Q_0\sin\frac{\pi}{5} + Q_1\sin\frac{2\pi}{5}\right\},
\end{aligned}\right.$$

in which $k = \left(\frac{b}{a}\right)^{\frac{1}{5}}$,

$$P_0 = \frac{1}{2}\log\left(x^2 - 2kx\cos\frac{\pi}{5} + k^2\right), \quad Q_0 = \operatorname{arc\,tg}\frac{x\sin\frac{\pi}{5}}{k - x\cos\frac{\pi}{5}},$$

$$P_1 = \frac{1}{2}\log\left(x^2 + 2kx\cos\frac{2\pi}{5} + k^2\right), \quad Q_1 = \operatorname{arc\,tg}\frac{x\sin\frac{2\pi}{5}}{k + x\cos\frac{2\pi}{5}}.$$

$$\int \frac{x^3}{(ax+b)^5}\,dx = \frac{1}{a^4}\left\{-\frac{1}{\Phi} + \frac{3b}{2\Phi^2} - \frac{b^2}{\Phi^3} + \frac{b^3}{4\Phi^4}\right\}.$$

$$\int \frac{2x^3+3}{x^5-9x}\,dx = -\frac{1}{3}\log x + \frac{1}{12}\log(x^4-9) + \frac{1}{2\sqrt{3}}\log\frac{x-\sqrt{3}}{x+\sqrt{3}} + \frac{1}{\sqrt{3}}\operatorname{arc\,tg}\frac{x}{\sqrt{3}}.$$

$$\int \frac{x^3-1}{x^5-2x^4+8x^2-12x+8}\,dx = -\frac{7x+2}{20(x^2-2x+2)} + \frac{9}{100}\log\frac{\sqrt{x^2-2x+2}}{x+2} + \frac{19}{50}\operatorname{arc\,tg}(x-1).$$

$$\int \frac{x^3+x^2+2}{x^5-2x^3+x}\,dx = -\frac{x+3}{2(x^2-1)} + 2\log x - \frac{3}{4}\log(x-1) - \frac{5}{4}\log(x+1).$$

$$\int \frac{x^3+3x+2}{x^2(x-1)^3}\,dx = \frac{2}{x} + \frac{6}{x-1} - \frac{3}{(x-1)^2} + 9\log\frac{x-1}{x}.$$

$$\int \frac{x^4}{x^5+a^5}\,dx = \frac{1}{5}\log(x^5+a^5).$$

$$\int \frac{x^4}{(ax+b)^5}\,dx = \frac{1}{a^5}\left\{\frac{4b}{\Phi} - \frac{3b^2}{\Phi^2} + \frac{4b^3}{3\Phi^3} - \frac{b^4}{4\Phi^4} + \log\Phi\right\}.$$

$$\int \frac{x^4+1}{x(x^2+1)^2}\,dx = \frac{1}{1+x^2} + \log x.$$

$$\int \frac{x^4+1}{x^3(x^2+x+1)}\,dx = \frac{1}{x} - \frac{1}{2x^2} + \frac{1}{2}\log(x^2+x+1) + \frac{1}{\sqrt{3}}\operatorname{arc\,tg}\frac{2x+1}{\sqrt{3}}$$

$$\int \frac{x^4-5x^3-30x^2-36x}{(x+1)^3(x^2-4)}\,dx = \frac{1-x}{(1+x)^2} + \log(x+1) + 2\log\frac{x+2}{x-2}.$$

$$\int \frac{5-3x+6x^2+5x^3-x^4}{x^5-x^4-2x^3+2x^2+x-1}\,dx = -\frac{2x^2+7x-3}{2(x^3-x^2-x+1)} + \log\frac{x-1}{(x+1)^2}.$$

$$\int \frac{x^5}{ax^5+b}\,dx = \frac{x}{a} - \frac{b}{a}\int\frac{dx}{ax^5+b}.$$

$$\int \frac{x^5}{(ax+b)^5}\,dx = \frac{1}{a^6}\left\{ \Phi - \frac{10b^2}{\Phi} + \frac{5b^3}{\Phi^2} - \frac{5b^4}{3\Phi^3} + \frac{b^5}{4\Phi^4} - 5b \log \Phi \right\}.$$

$$\frac{Y^5}{X^{5-a}}$$

$$\int \frac{x^5}{ax+b}\,dx = \frac{1}{a^6}\left\{ \frac{\Phi^5}{5} - \frac{5b\Phi^4}{4} + \frac{10b^2\Phi^3}{3} - 5b^3\Phi^2 + 5b^4\Phi - b^5 \log \Phi \right\}.$$

$$\int \frac{x^5}{1-x^2}\,dx = -\frac{x^4+2x^2}{4} - \frac{1}{2}\log(1-x^2).$$

$$\int \frac{x^5}{ax^2+b}\,dx = \frac{x^4}{4a} - \frac{bx^2}{2a^2} + \frac{b^2}{2a^3}\log(ax^2+b).$$

$$\int \frac{x^5}{x^3-1}\,dx = \frac{x^3}{3} + \frac{1}{3}\log(x^3-1).$$

$$\int \frac{x^5}{x^4+1}\,dx = \frac{x^2}{2} - \frac{1}{2}\operatorname{arc\,tg} x^2.$$

$$\int \frac{x^5}{(1-x^2)^2}\,dx = \frac{x^2}{2} + \frac{1}{2(1-x^2)} + \log(1-x^2).$$

$$\int \frac{x^5}{ax^2+bx+c}\,dx = \frac{x^4}{4a} - \frac{bx^3}{3a^2} + \left(\frac{b^2}{a^3} - \frac{c}{a^2}\right)\frac{x^2}{2} - \left(\frac{b^3}{a^4} - \frac{2bc}{a^3}\right)x + \left(\frac{b^4}{2a^5} - \frac{3b^2c}{2a^4} + \frac{c^2}{2a^3}\right)\log \Phi$$
$$-\left(\frac{b^5}{2a^5} - \frac{5b^3c}{2a^4} + \frac{5bc^2}{2a^3}\right)\int \frac{dx}{\Phi}.$$

$$\int \frac{x^5}{(ax+b)^3}\,dx = \frac{1}{a^6}\left\{ \frac{\Phi^3}{3} - \frac{5b\Phi^2}{2} + 10\,b^2\Phi - \frac{5b^4}{\Phi} + \frac{b^5}{2\Phi^2} - 10b^3 \log \Phi \right\}.$$

$$\int \frac{x^5}{(ax+b)^4}\,dx = \frac{1}{a^6}\left\{ \frac{\Phi^2}{2} - 5b\Phi + \frac{10b^3}{\Phi} - \frac{5b^4}{2\Phi^2} + \frac{b^5}{3\Phi^3} + 10b^2 \log \Phi \right\}$$

$$\frac{1}{X^6}$$

$$\int \frac{dx}{1-x^6} = \frac{1}{12}\log\frac{(1+x+x^2)(1+x)^2}{(1-x+x^2)(1-x)^2} + \frac{1}{2\sqrt{3}}\operatorname{arc\,tg}\frac{x\sqrt{3}}{1-x^2}.$$

$$\int \frac{dx}{x^6-a^6} = \frac{1}{3a^5}\left\{ \log\frac{x-a}{x+a}\sqrt{\frac{x^2-ax+a^2}{x^2+ax+a^2}} - \sqrt{3}\operatorname{arc\,tg}\frac{2x-a}{a\sqrt{3}} - \sqrt{3}\operatorname{arc\,tg}\frac{2x+a}{a\sqrt{3}} \right\}.$$

$$\int \frac{dx}{a+bx^6} = \frac{k}{a}\left\{ \frac{1}{4\sqrt{3}}\log\frac{x^2+kx\sqrt{3}+k^2}{x^2-kx\sqrt{3}+k^2} + \frac{1}{6}\operatorname{arc\,tg}\frac{kx}{k^2-x^2} + \frac{1}{3}\operatorname{arc\,tg}\frac{x}{k} \right\}. \qquad k = \left(\frac{a}{b}\right)^{\frac{1}{6}}.$$

$$\int \frac{dx}{a-bx^6} = \frac{k}{2a}\left\{ \frac{1}{6}\log\frac{x^2+kx+k^2}{x^2-kx+k^2} + \frac{1}{3}\log\frac{x+k}{x-k} + \frac{1}{\sqrt{3}}\operatorname{arc\,tg}\frac{kx\sqrt{3}}{k^2-x^2} \right\}. \qquad k = \left(\frac{a}{b}\right)^{\frac{1}{6}}.$$

$$\int \frac{dx}{x^2(x^4-1)} = \frac{1}{x} + \frac{1}{4}\log\frac{x-1}{x+1} + \frac{1}{2}\operatorname{arc\,tg} x.$$

$$\int \frac{dx}{x^2(a+bx^4)} = -\frac{1}{ax} - \frac{b}{a}\int \frac{x^2}{a+bx^4}\,dx.$$

$$\int \frac{dx}{x^3(a+bx^3)} = -\frac{1}{2ax^2} - \frac{b}{a}\int \frac{dx}{a+bx^3}.$$

$$\int \frac{dx}{x^4(1+x^2)} = \frac{3x^2-1}{3x^3} + \text{arc tg } x.$$

$$\int \frac{dx}{x^4(a+bx^2)} = -\frac{1}{3ax^3} + \frac{b}{a^2x} + \frac{b^2}{a^2}\int \frac{dx}{a+bx^2}.$$

$$\int \frac{dx}{x^5(a+bx)} = -\frac{1}{4ax^4} + \frac{b}{3a^2x^3} - \frac{b^2}{2a^3x^2} + \frac{b^3}{a^4x} - \frac{b^4}{a^5}\log\frac{a+bx}{x}.$$

$$\int \frac{dx}{(x^3+1)^2} = \frac{x}{3(x^2+1)} + \frac{1}{9}\log\frac{(x+1)^2}{x^2-x+1} + \frac{2}{3\sqrt{3}}\text{arc tg }\frac{2x-1}{\sqrt{3}}.$$

$$\int \frac{dx}{(a+bx^3)^2} = \frac{x}{3a(a+bx^3)} + \frac{2}{3a}\int \frac{dx}{a+bx^3}.$$

$$\int \frac{dx}{x^2(a^2+x^2)^2} = -\frac{1}{a^4x} - \frac{x}{2a^4(a^2+x^2)} - \frac{3}{2a^5}\text{arc tg }\frac{x}{a}.$$

$$\int \frac{dx}{x^2(x^2-a^2)^2} = -\frac{1}{a^4x} + \frac{1}{2a^5}\log\frac{x+a}{x-a} + \frac{1}{a^2}\int \frac{dx}{(x^2-a^2)^2}.$$

$$\int \frac{dx}{x^2(x^2-b)^2} = \frac{3x^2-2b}{2b^2x(b-x^2)} + \frac{3}{4b^2\sqrt{b}}\log\frac{x+\sqrt{b}}{x-\sqrt{b}}.$$

$$\int \frac{dx}{x^2(a+bx^2)^2} = -\frac{bx}{2a^2(a+bx^2)} - \frac{1}{a^2x} - \frac{3b}{2a^2}\int \frac{dx}{a+bx^2}.$$

$$\int \frac{dx}{x^4(a+bx)^2} = -\frac{b^3}{a^4\Phi} - \frac{1}{3a^2x^3} + \frac{b}{a^3x^2} - \frac{3b^2}{a^4x} + \frac{4b^3}{a^5}\log\frac{\Phi}{x}.$$

$$\int \frac{dx}{x^4(a+bx+cx^2)} = \frac{3bx-2a}{6a^2x^3} + \frac{ac-b^2}{a^3x} + \frac{2abc-b^3}{2a^4}\log\frac{x^2}{\Phi} + \frac{2a^2c^2-4ab^2c+b^4}{2a^4}\int \frac{dx}{\Phi}.$$

$$\int \frac{dx}{x^2(a+bx^2+cx^4)} = -\frac{1}{ax} - \frac{b}{a}\int \frac{dx}{\Phi} - \frac{c}{a}\int \frac{x^2}{\Phi}\,dx.$$

$\int \frac{dx}{a+bx^3+cx^6}$ See p. 31.

$$\int \frac{dx}{(1+x^2)^3} = \frac{x}{4(1+x^2)^2} + \frac{3x}{8(1+x^2)} + \frac{3}{8}\text{arc tg } x.$$

$$\int \frac{dx}{(a^2+x^2)^3} = \frac{x(5a^2+3x^2)}{8a^4\Phi^2} + \frac{3}{8a^5}\text{arc tg }\frac{x}{a}.$$

$$\int \frac{dx}{(a^2-x^2)^3} = \frac{x(5a^2-3x^2)}{8a^4\Phi^2} + \frac{3}{16a^5}\log\frac{a+x}{a-x}.$$

$$\int \frac{dx}{(a+bx^2)^3} = \frac{x}{4a\Phi^2} + \frac{3x}{8a^2\Phi} + \frac{3}{8a^2}\int \frac{dx}{\Phi}.$$

$$\int \frac{dx}{x^3(a+bx)^3} = \frac{b^2}{2a^3\Phi^2} + \frac{3b^2}{a^4\Phi} - \frac{1}{2a^3x^2} + \frac{3b}{a^4x} - \frac{6b^2}{a^5}\log\frac{\Phi}{x}.$$

$$\int \frac{dx}{(x^2+1)^2(x^2-1)} = -\frac{x}{4(x^2+1)} + \frac{1}{8}\log\frac{x-1}{x+1} - \frac{1}{2}\text{arc tg } x.$$

$$\int \frac{dx}{x^6-x^5-x^4+x^3} = -\frac{1}{2x^2} - \frac{1}{x} + \frac{1}{2(1-x)} + 2\log x - \frac{7}{4}\log(1-x) - \frac{1}{4}\log(1+x).$$

$$\int \frac{dx}{x^2(a+bx+cx^2)^2} = -\frac{1}{a^2x} + \left[\frac{b^3}{a^2} - \frac{3bc}{a} + \left(\frac{b^2}{a^2} - \frac{2c}{a}\right) cx\right]\frac{1}{(4ac-b^2)\Phi} + \frac{b}{a^2}\log\frac{\Phi}{x^2}$$
$$-\left(\frac{b^4}{a^3} - \frac{6b^2c}{a^2} + \frac{6c^2}{a}\right)\frac{1}{4ac-b^2}\int\frac{dx}{\Phi}.$$

$$\int \frac{dx}{x^2(a+bx)^4} = -\frac{b}{3a^2\Phi^3} - \frac{b}{a^3\Phi^2} - \frac{3b}{a^4\Phi} - \frac{1}{a^4x} + \frac{4b}{a^5}\log\frac{\Phi}{x}.$$

$$\int \frac{dx}{(a+bx+cx^2)^3} = \frac{2cx+b}{4ac-b^2}\left\{\frac{1}{2\Phi^2} + \frac{3c}{(4ac-b^2)\Phi}\right\} + \frac{6c^2}{(4ac-b^2)^2}\int\frac{dx}{\Phi}.$$

$$\int \frac{dx}{x(a+bx)^5} = \frac{1}{4a\Phi^4} + \frac{1}{3a^2\Phi^3} + \frac{1}{2a^3\Phi^2} + \frac{1}{a^4\Phi} - \frac{1}{a^5}\log\frac{\Phi}{x}.$$

$$\int \frac{dx}{(x-1)^3(x-2)^2x} = -\frac{2x-1}{2(x-1)^2} - \frac{1}{2(x-2)} + 2\log(x-1) - \frac{1}{4}\log x - \frac{7}{4}\log(x-2).$$

$$\int \frac{dx}{(x^2-2x\cos\alpha+1)^3} = \frac{x-\cos\alpha}{4\Phi^2\sin^2\alpha} + \frac{3(x-\cos\alpha)}{8\Phi\sin^4\alpha} + \frac{3}{8\sin^5\alpha}\operatorname{arc\,tg}\frac{x-\cos\alpha}{\sin\alpha}.$$

$\dfrac{Y^{1,2}}{X^6}$

$$\int \frac{x}{a+bx^6}\,dx = \frac{1}{2}\int\frac{dz}{a+bz^3}. \qquad z = x^2.$$

$$\int \frac{x}{(a+bx^3)^2}\,dx = \frac{x^2}{3a\Phi} + \frac{1}{3a}\int\frac{x}{\Phi}\,dx.$$

$\displaystyle\int \frac{x}{a+bx^3+cx^6}\,dx.$ See p. 31.

$$\int \frac{x}{(a+bx^2)^3}\,dx = -\frac{1}{4b(a+bx^2)^2}.$$

$$\int \frac{x}{(x^2+x+1)^3}\,dx = -\frac{x^3+3x^2+3x+2}{6(x^2+x+1)^2} - \frac{2}{3\sqrt{3}}\operatorname{arc\,tg}\frac{2x+1}{\sqrt{3}}.$$

$$\int \frac{x}{(a+bx+cx^2)^3}\,dx = -\frac{1}{4c\Phi^2} - \frac{b}{2c}\int\frac{dx}{\Phi^3}$$
$$= -\frac{bx+2a}{2(4ac-b^2)\Phi^2} - \frac{3b(2cx+b)}{2(4ac-b^2)^2\Phi} - \frac{3bc}{(4ac-b^2)^2}\int\frac{dx}{\Phi}.$$

$$\int \frac{x+1}{(x^2+4x+6)^3}\,dx = -\frac{x+4}{8\Phi^2} - \frac{3(x+2)}{32\Phi} - \frac{3}{32\sqrt{2}}\operatorname{arc\,tg}\frac{x+\sqrt{2}}{\sqrt{2}}.$$

$$\int \frac{x+1}{(x^2+5x+4)^3}\,dx = \frac{x+1}{6\Phi^2} - \frac{2x+5}{18\Phi} - \frac{1}{27}\log\frac{x+1}{x+4}.$$

$$\int \frac{x^2}{a^6+x^6}\,dx = \frac{1}{3a^3}\operatorname{arc\,tg}\left(\frac{x}{a}\right)^3.$$

$$\int \frac{x^2}{a^2-x^6}\,dx = \frac{1}{6a}\log\frac{x^3+a}{x^3-a}$$

$$\int \frac{x^2}{a+bx^6}\,dx = \frac{1}{3}\int\frac{dz}{a+bz^2} \qquad z = x^2.$$

$$\int \frac{x^2}{(a+bx^3)^2}\,dx = -\frac{1}{3b\,(a+bx^3)} = \frac{x^3}{3a\,(a+bx^3)}$$

$$\int \frac{x^2}{x^6-10x^3+9}\,dx = \frac{1}{24}\log\frac{x^3-9}{x^3-1}.$$

$$\int \frac{x^2}{ax^6+bx^3+c}\,dx = \frac{1}{3\sqrt{b^2-4ac}}\log\frac{2ax^3+b-\sqrt{b^2-4ac}}{2ax^3+b+\sqrt{b^2-4ac}}. \qquad b^2>4ac.$$

$$= \frac{2}{3\sqrt{4ac-b^2}}\,\text{arc tg}\,\frac{2ax^3+b}{\sqrt{4ac-b^2}}. \qquad b^2<4ac.$$

$$\int \frac{x^2}{(a^2+x^2)^3}\,dx = \frac{x}{8a^2\Phi} - \frac{x}{4\Phi^2} + \frac{1}{8a^3}\,\text{arc tg}\,\frac{x}{a}.$$

$$\int \frac{x^2}{(x^2-a^2)^3}\,dx = \frac{x}{8a^2\Phi} - \frac{x^3}{4a^2\Phi^2} - \frac{1}{16a^3}\log\frac{x-a}{x+a}.$$

$$\int \frac{x^2}{(a+bx^2)^3}\,dx = \frac{x}{8ab\Phi} - \frac{x}{4b\Phi^2} + \frac{1}{8ab}\int\frac{dx}{\Phi}.$$

$$\int \frac{x^2}{(x^2+4)(x^2+2)(x^2+1)}\,dx = -\frac{1}{3}\,\text{arc tg}\,\frac{3x}{2-x^2} + \frac{1}{\sqrt{2}}\,\text{arc tg}\,\frac{x}{\sqrt{2}}.$$

$$\int \frac{x^2}{(x^2+9)(x^2+4)(x^2+1)}\,dx = -\frac{1}{24}\,\text{arc tg}\,x + \frac{2}{15}\,\text{arc tg}\,\frac{x}{2} - \frac{3}{40}\,\text{arc tg}\,\frac{x}{3}.$$

$$\int \frac{x^2}{(x^2+a^2)(x^2+a)(x^2+1)}\,dx = \frac{1}{(a-1)^2}\left\{-\frac{1}{a+1}\,\text{arc tg}\,\frac{(a+1)\,x}{a-x^2} + \frac{1}{\sqrt{a}}\,\text{arc tg}\,\frac{x}{\sqrt{a}}\right\}.$$

$$\int \frac{x^2}{(x^2+a^2)(x^2+b^2)(x^2+1)}\,dx = -\frac{1}{(a^2-1)(b^2-1)}\,\text{arc tg}\,x + \frac{a}{(b^2-a^2)(a^2-1)}\,\text{arc tg}\,\frac{x}{a} - \frac{b}{(b^2-a^2)(b^2-1)}\,\text{arc tg}\,\frac{x}{b}.$$

$$\int \frac{x^2}{(a+bx)^6}\,dx = \frac{1}{b^3}\left\{-\frac{1}{3\Phi^3} + \frac{a}{2\Phi^4} - \frac{a^2}{5\Phi^5}\right\}.$$

$$\int \frac{x^2}{(a+bx+cx^2)^3}\,dx = \frac{(b^2-2ac)\,x+ab}{2c\,(4ac-b^2)\,\Phi^2} + \frac{(2ac+b^2)(2cx+b)}{2c\,(4ac-b^2)^2\,\Phi} + \frac{2ac+b^2}{(4ac-b^2)^2}\int\frac{dx}{\Phi}.$$

$$\int \frac{x^2-x+1}{(x^2+x+1)^3}\,dx = \frac{4x^3+6x^2+7x+4}{3\,(x^2+x+1)^2} + \frac{8\sqrt{3}}{9}\,\text{arc tg}\,\frac{2x+1}{\sqrt{3}}.$$

$$\int \frac{(x^2+1)^2}{(x-1)^6}\,dx = \frac{-15x^4+30x^3-40x^2+20x-7}{15\,(x-1)^5}.$$

$\dfrac{Y^{3,4,5,6}}{X^6}$

$$\int \frac{x^3}{a+bx^6}\,dx = \frac{1}{2}\int\frac{z}{a+bz^3}\,dz. \qquad z=x^2$$

$$\int \frac{x^3}{(a+bx^3)^2}\,dx = -\frac{x}{3b\Phi} + \frac{1}{3b}\int\frac{dx}{\Phi}.$$

$\displaystyle\int \frac{x^3}{a+bx^3+cx^6}\,dx.$ See p. 31.

$$\int \frac{x^3}{(a+bx^2)^3}\,dx = \frac{x^4}{4a\Phi^2} = -\frac{a+2bx^2}{4b^2\Phi^2}.$$

$$\int \frac{x^3}{x^6 - 2x^5 + x^4 + x^2 - 2x + 1}\, dx = -\frac{1}{2(x-1)} + \frac{1}{2}\log(x-1) + \frac{\sqrt{2}-1}{8}\log(x^2 + x\sqrt{2} + 1)$$

$$-\frac{\sqrt{2}+1}{8}\log(x^2 - x\sqrt{2} + 1) - \frac{\sqrt{2}-1}{4}\operatorname{arc\,tg}(x\sqrt{2}+1) - \frac{\sqrt{2}+1}{4}\operatorname{arc\,tg}(x\sqrt{2}-1).$$

$$\int \frac{x^3}{(a+bx)^6}\, dx = \frac{1}{b^4}\left\{ -\frac{1}{2\Phi^2} + \frac{a}{\Phi^3} - \frac{3a^2}{4\Phi^4} + \frac{a^3}{5\Phi^5} \right\}.$$

$$\int \frac{x^3}{(a+bx+cx^2)^3}\, dx = \frac{(3ac-b^2)bx + (2ac-b^2)a}{c^2(4ac-b^2)\Phi} + \frac{1}{2c^2}\log\Phi - \frac{b(6ac-b^2)}{2c^2(4ac-b^2)}\int \frac{dx}{\Phi}$$

$$\int \frac{x^4}{a+bx^6}\, dx = \frac{k^5}{a}\left\{ -\frac{1}{4\sqrt{3}}\log\frac{x^2 + kx\sqrt{3} + k^2}{x^2 - kx\sqrt{3} + k^2} + \frac{1}{6}\operatorname{arc\,tg}\frac{kx}{k^2 - x^2} + \frac{1}{3}\operatorname{arc\,tg}\frac{x}{k} \right\},$$

in which $k = \left(\frac{a}{b}\right)^{\frac{1}{6}}$

$$\int \frac{x^4}{(27+8x^3)^2}\, dx = \frac{1}{24}\left[-\frac{x^2}{27+8x^3} - \frac{1}{36}\log\frac{\left(x+\frac{3}{2}\right)^2}{x^2 - \frac{3}{2}x + \frac{9}{4}} + \frac{\sqrt{3}}{18}\operatorname{arc\,tg}\frac{4x-3}{3\sqrt{3}} \right].$$

$$\int \frac{x^4}{(a+bx^3)^2}\, dx = -\frac{x^2}{3b\Phi} + \frac{2}{3b}\int \frac{x}{\Phi}\, dx.$$

$$\int \frac{x^4}{(x^2-a)^3}\, dx = -\frac{5x}{8\Phi} - \frac{ax}{4\Phi^2} + \frac{3}{8}\int \frac{dx}{\Phi}.$$

$$\int \frac{x^4}{(a+bx^2)^3}\, dx = \frac{ax}{4b^2\Phi^2} - \frac{5x}{8b^2\Phi} + \frac{3}{8b^2}\int \frac{dx}{\Phi}.$$

$$\int \frac{x^4}{(a+bx+cx^2)^3}\, dx = -\left\{ \frac{ax}{c^2} + \frac{bx^2}{2c^2} + \frac{x^3}{c} \right\}\frac{1}{\Phi^2} + \frac{a^2}{c^2}\int \frac{dx}{\Phi^3}.$$

$$\int \frac{x^4}{(a+bx)^6}\, dx = \frac{1}{b^5}\left\{ -\frac{1}{\Phi} + \frac{2a}{\Phi^2} - \frac{2a^2}{\Phi^3} + \frac{a^3}{\Phi^4} - \frac{a^4}{5\Phi^5} \right\}.$$

$$\int \frac{x^4+1}{x^6+1}\, dx = \frac{1}{3}\operatorname{arc\,tg}\frac{3x(1-x^2)}{x^4 - 4x^2 + 1}.$$

$$\int \frac{x^4+1}{(x^3-1)^2}\, dx = -\frac{2}{9(x-1)} - \frac{x+2}{9(x^2+x+1)} + \frac{4\sqrt{3}}{9}\operatorname{arc\,tg}\frac{2x+1}{\sqrt{3}}.$$

$$\int \frac{ax^4 + bx^2 + c}{(1+x^2)^3}\, dx = a\int \frac{dx}{1+x^2} + (b-2a)\int \frac{dx}{(1+x^2)^2} + (a+c-b)\int \frac{dx}{(1+x^2)^3}.$$

$$\int \frac{(x^2+1)^2}{(x-1)^6}\, dx = -\frac{15x^4 - 30x^3 + 40x^2 - 20x + 7}{15(x-1)^5}.$$

$$\left\{\begin{aligned}
&\int \frac{dx}{a+bx^3+cx^6} = \frac{2c}{k}\int \frac{dx}{2cx^3+A} - \frac{2c}{k}\int \frac{dx}{2cx^3+B},\\
&\int \frac{x}{a+bx^3+cx^6}\, dx = \frac{2c}{k}\int \frac{x}{2cx^3+A}\, dx - \frac{2c}{k}\int \frac{x}{2cx^3+B}\, dx,\\
&\int \frac{x^3}{a+bx^3+cx^6}\, dx = \frac{k-b}{k}\int \frac{dx}{2cx^3+A} + \frac{k+b}{k}\int \frac{dx}{2cx^3+B},\\
&\int \frac{x^4}{a+bx^3+cx^6}\, dx = \frac{k-b}{k}\int \frac{x}{2cx^3+A}\, dx + \frac{k+b}{k}\int \frac{x}{2cx^3+B}\, dx,
\end{aligned}\right.$$

in which $k = \sqrt{b^2 - 4ac}$, $A = b - k$, $B = b + k$, $b^2 > 4ac$.

$$\int \frac{dx}{a + bx^3 + cx^6} = \frac{1}{3cq^5 \sin \varepsilon}\left(-P_0 \sin 2\varepsilon_0 - P_1 \sin 2\varepsilon_1 - P_2 \sin 2\varepsilon_2 + Q_0 \cos 2\varepsilon_0 + Q_1 \cos 2\varepsilon_1 + Q_2 \cos 2\varepsilon_2\right).$$

$$\int \frac{x}{a + bx^3 + cx^6}\, dx = \frac{1}{3cq^4 \sin \varepsilon}\left(-P_0 \sin \varepsilon_0 - P_1 \sin \varepsilon_1 - P_2 \sin \varepsilon_2 + Q_0 \cos \varepsilon_0 + Q_1 \cos \varepsilon_1 + Q_2 \cos \varepsilon_2\right),$$

$$\int \frac{x^3}{a + bx^3 + cx^6}\, dx = \frac{1}{3cq^2 \sin \varepsilon}\left(+P_0 \sin \varepsilon_0 + P_1 \sin \varepsilon_1 + P_2 \sin \varepsilon_2 + Q_0 \cos \varepsilon_0 + Q_1 \cos \varepsilon_1 + Q_2 \cos \varepsilon_2\right),$$

$$\int \frac{x^4}{a + bx^3 + cx^6}\, dx = \frac{1}{3cq \sin \varepsilon}\left(+P_0 \sin 2\varepsilon_0 + P_1 \sin 2\varepsilon_1 + P_2 \sin 2\varepsilon_2 + Q_0 \cos 2\varepsilon_0 + Q_1 \cos 2\varepsilon_1 + Q_2 \cos 2\varepsilon_2\right),$$

in which $b^2 < 4ac$, $q = \left(\frac{a}{c}\right)^{\frac{1}{6}}$, $\cos \varepsilon = -\frac{b}{2\sqrt{ac}}$,

$$\varepsilon_0 = \frac{1}{3}\varepsilon, \qquad \varepsilon_1 = \frac{1}{3}(2\pi + \varepsilon), \qquad \varepsilon_2 = \frac{1}{3}(4\pi + \varepsilon),$$

$$P_0 = \frac{1}{2}\log(x^2 - 2qx\cos\varepsilon_0 + q^2), \qquad P_1 = \frac{1}{2}\log(x^2 - 2qx\cos\varepsilon_1 + q^2),$$

$$P_2 = \frac{1}{2}\log(x^2 - 2qx\cos\varepsilon_2 + q^2),$$

$$Q_0 = \text{arc tg}\,\frac{x\sin\varepsilon_0}{q - x\cos\varepsilon_0}, \qquad Q_1 = \text{arc tg}\,\frac{x\sin\varepsilon_1}{q - x\cos\varepsilon_1}, \qquad Q_2 = \text{arc tg}\,\frac{x\sin\varepsilon_2}{q - x\cos\varepsilon_2}.$$

$$\int \frac{x^5}{(x^3 + a)^2}\, dx = \frac{a}{3\Phi} + \frac{1}{3}\log\Phi.$$

$$\int \frac{x^5}{(x^2 - a)^3}\, dx = -\int \frac{(1 + az^2)^2}{z}\, dz. \qquad x^2 - a = \frac{1}{z^2}.$$

$$\int \frac{x^5}{(1 - x^2)^3}\, dx = \frac{1}{4(1 - x^2)^2} - \frac{1}{1 - x^2} - \frac{1}{2}\log(1 - x^2).$$

$$\int \frac{x^5}{(a + bx)^6}\, dx = \frac{1}{b^6}\left\{\frac{5a}{\Phi} - \frac{5a^2}{\Phi^2} + \frac{10a^3}{3\Phi^3} - \frac{5a^4}{4\Phi^4} + \frac{a^5}{5\Phi^5} + \log\Phi\right\}.$$

$$\int \frac{x^5 + 1}{x^6 + x^4}\, dx = \frac{1}{x} - \frac{1}{3x^3} + \frac{1}{2}\log(x^2 + 1) + \text{arc tg}\, x.$$

$$\int \frac{x^5 + 4}{(x^2 - 2x + 2)^3}\, dx = -\frac{4x^2 - 8x + 7}{(x^2 - 2x + 2)^2} + \frac{1}{2}\log(x^2 - 2x + 2) + 5\,\text{arc tg}\,(x - 1).$$

$$\int \frac{x^6}{(a + bx^3)^2}\, dx = \frac{x}{b^2} + \frac{ax}{3b^2\Phi} - \frac{4a}{3b^2}\int \frac{dx}{\Phi}.$$

$$\int \frac{x^6}{(a + bx^2)^3}\, dx = \frac{x}{b^3}\left\{1 + \frac{9a}{8\Phi} - \frac{a^2}{4\Phi^2}\right\} - \frac{15a}{8b^3}\int \frac{dx}{\Phi}.$$

$$\int \frac{x^6}{(a + bx)^6}\, dx = \frac{1}{b^7}\left\{\Phi - \frac{15a^2}{\Phi} + \frac{10a^3}{\Phi^2} - \frac{5a^4}{\Phi^3} + \frac{3a^5}{2\Phi^4} - \frac{a^6}{5\Phi^5} - 6a\log\Phi\right\}.$$

$\frac{Y^6}{X^{6-a}}$

$$\int \frac{x^6}{a+bx}\,dx = \frac{1}{b^7}\left\{\frac{\Phi^6}{6} - \frac{6a\Phi^5}{5} + \frac{15a^2\Phi^4}{4} - \frac{20a^3\Phi^3}{3} + \frac{15a^4\Phi^2}{2} - 6a^5\Phi + a^6 \log \Phi\right\}.$$

$$\int \frac{x^6}{a+bx^2}\,dx = \frac{x^5}{5b} - \frac{ax^3}{3b^2} + \frac{a^2x}{b^3} - \frac{a^3}{b^3}\int\frac{dx}{\Phi}.$$

$$\int \frac{x^6}{a+bx^3}\,dx = \frac{x^4}{4b} - \frac{ax}{b^2} + \frac{a^2}{b^2}\int\frac{dx}{\Phi}.$$

$$\int \frac{x^6}{a+bx^4}\,dx = \frac{x^3}{3b} - \frac{a}{b}\int\frac{x^2}{\Phi}\,dx.$$

$$\int \frac{x^6}{a+bx^5}\,dx = \frac{x^2}{2b} - \frac{a}{b}\int\frac{x}{\Phi}\,dx.$$

$$\int \frac{x^6}{(a+bx)^2}\,dx = \frac{1}{b^7}\left\{\frac{\Phi^5}{5} - \frac{3a\Phi^4}{2} + 5a^2\Phi^3 - 10a^3\Phi^2 + 15a^4\Phi - \frac{a^6}{\Phi} - 6a^5\log\Phi\right\}.$$

$$\int \frac{x^6}{(a+bx^2)^2}\,dx = \frac{x}{b^3}\left\{\frac{bx^2}{3} - 2a - \frac{a^2}{2\Phi}\right\} + \frac{5a^2}{2b^3}\int\frac{dx}{\Phi}.$$

$$\int \frac{x^6}{a+bx^2+cx^4}\,dx = -\frac{bx}{c^2} + \frac{x^3}{3c} + \frac{ab}{c^2}\int\frac{dx}{\Phi} + \frac{b^2-ac}{c^2}\int\frac{x^2}{\Phi}\,dx.$$

$$\int \frac{x^6}{(a+bx)^3}\,dx = \frac{1}{b^7}\left\{\frac{\Phi^4}{4} - 2a\Phi^3 + \frac{15a^2\Phi^2}{2} - 20a^3\Phi + \frac{6a^5}{\Phi} - \frac{a^6}{2\Phi^2} + 15a^4\log\Phi\right\}.$$

$$\int \frac{x^6}{(a+bx)^4}\,dx = \frac{1}{b^7}\left\{\frac{\Phi^3}{3} - 3a\Phi^2 + 15a^2\Phi - \frac{15a^4}{\Phi} + \frac{3a^5}{\Phi^2} - \frac{a^6}{3\Phi^3} - 20a^3\log\Phi\right\}.$$

$$\int \frac{x^6}{(a+bx)^5}\,dx = \frac{1}{b^7}\left\{\frac{\Phi^2}{2} - 6a\Phi + \frac{20a^3}{\Phi} - \frac{15a^4}{2\Phi^2} + \frac{2a^5}{\Phi^3} - \frac{a^6}{4\Phi^4} + 15a^2\log\Phi\right\}.$$

X^7

$$\int \frac{dx}{x(3+5x^6)} = \frac{1}{18}\log\frac{x^6}{3+5x^6}.$$

$$\int \frac{dx}{x^2(a+bx^5)} = -\frac{1}{ax} - \frac{b}{a}\int\frac{x^3}{\Phi}\,dx.$$

$$\int \frac{dx}{x^3(a+bx^4)} = -\frac{1}{2ax^2} - \frac{b}{a}\int\frac{x}{\Phi}\,dx.$$

$$\int \frac{dx}{x^4(a+bx^3)} = -\frac{1}{3ax^3} + \frac{b}{3a^2}\log\frac{\Phi}{x^3}.$$

$$\int \frac{dx}{x^5(x^2-a)} = \frac{1}{2a^2x^2} + \frac{1}{4ax^4} + \frac{1}{2a^3}\log\frac{x^2-a}{x^2}.$$

$$\int \frac{dx}{x^6(a+bx)} = -\frac{1}{5ax^5} + \frac{b}{4a^2x^4} - \frac{b^2}{3a^3x^3} + \frac{b^3}{2a^4x^2} - \frac{b^4}{a^5x} + \frac{b^5}{a^6}\log\frac{\Phi}{x}.$$

$$\int \frac{dx}{x(1+x^3)^2} = \frac{1}{3(1+x^3)} + \frac{1}{3}\log\frac{x^3}{1+x^3}.$$

$$\int \frac{dx}{x(a+bx^3)^2} = \frac{1}{3a(a+bx^3)} - \frac{1}{3a^2}\log\frac{a+bx^3}{x^3}.$$

$$\int \frac{dx}{x^5(a+bx)^2} = \frac{b^4}{a^5(a+bx)} - \frac{1}{4a^2x^4} + \frac{2b}{3a^3x^3} - \frac{3b^2}{2a^4x^2} + \frac{4b^3}{a^5x} - \frac{5b^4}{a^6}\log\frac{a+bx}{x}.$$

$$\int \frac{dx}{x^4(a+bx)^3} = -\frac{b^3}{2a^4\Phi^2} - \frac{4b^3}{a^5\Phi} - \frac{1}{3a^3x^3} + \frac{3b}{2a^4x^2} - \frac{6b^2}{a^5x} + \frac{10b^3}{a^6}\log\frac{\Phi}{x}.$$

$$\int \frac{dx}{x^3(a+bx)^4} = \frac{b^2}{3a^3\Phi^3} + \frac{3b^2}{2a^4\Phi^2} + \frac{6b^2}{a^5\Phi} - \frac{1}{2a^4x^2} + \frac{4b}{a^5x} - \frac{10b^2}{a^6}\log\frac{\Phi}{x}.$$

$$\int \frac{dx}{x^3(1+x+x^2)^2} = \frac{16x^3+11x^2+9x-3}{6x^2\Phi} + \frac{1}{2}\log\frac{x^2}{\Phi} + \frac{13}{3\sqrt{3}}\operatorname{arc\,tg}\frac{2x+1}{\sqrt{3}}.$$

$$\int \frac{dx}{x^2(a+bx)^5} = -\frac{b}{4a^2\Phi^4} - \frac{2b}{3a^3\Phi^3} - \frac{3b}{2a^4\Phi^2} - \frac{4b}{a^5\Phi} - \frac{1}{a^5x} + \frac{5b}{a^6}\log\frac{\Phi}{x}.$$

$$\int \frac{dx}{x(a+bx)^6} = \frac{1}{5a\Phi^5} + \frac{1}{4a^2\Phi^4} + \frac{1}{3a^3\Phi^3} + \frac{1}{2a^4\Phi^2} + \frac{1}{a^5\Phi} - \frac{1}{a^6}\log\frac{\Phi}{x}.$$

$$\int \frac{dx}{x(a+bx+cx^2)^3} = \frac{1}{2a^2\Phi}\left(1+\frac{a}{2\Phi}\right) - \frac{b(2cx+b)}{2a^2(4ac-b^2)\Phi}\left\{1+3q+\frac{a}{2\Phi}\right\} + \frac{1}{2a^3}\log\frac{x^2}{\Phi}$$
$$-\frac{b}{2a^3}\left\{1+2q+6q^2\right\}\int\frac{dx}{\Phi}, \qquad \text{in which} \qquad q = \frac{ac}{4ac-b^2}.$$

$$\int \frac{dx}{x(a+bx+cx^2+dx^3)^2} = \frac{1}{3a\Phi} + \frac{1}{a}\int\frac{dx}{x\Phi} - \frac{c}{3a}\int\frac{x}{\Phi^2}dx - \frac{2b}{3a}\int\frac{dx}{\Phi^2}.$$

$$\int \frac{x}{(a+bx)^7}dx = \frac{1}{b^2}\left\{-\frac{1}{5\Phi^5} + \frac{a}{6\Phi^6}\right\}.$$

$$\int \frac{x^2}{(a+bx)^7}dx = \frac{1}{b^3}\left\{-\frac{1}{4\Phi^4} + \frac{2a}{5\Phi^5} - \frac{a^2}{6\Phi^6}\right\}.$$

$$\int \frac{x^2+1}{(x-1)^4(x^3+1)}dx = -\frac{1}{3(x-1)^3} + \frac{1}{4(x-1)^2} + \frac{5}{8}\log(x-1) + \frac{1}{24}\log(x+1) - \frac{1}{3}\log(x^2-x+1).$$

$$\int \frac{x^2+x+1}{x^2(x^2+1)^2(x-1)}dx = \frac{3x^2-x+4}{4x(x^2+1)} - \frac{1}{8}\log\frac{(x^2+1)^5(x-1)^6}{x^{16}}.$$

$$\int \frac{x^3}{(a+bx)^7}dx = \frac{1}{b^4}\left\{-\frac{1}{3\Phi^3} + \frac{3a}{4\Phi^4} - \frac{3a^2}{5\Phi^5} + \frac{a^3}{6\Phi^6}\right\}.$$

$$\int \frac{x^4}{(a+bx)^7}dx = \frac{1}{b^5}\left\{-\frac{1}{2\Phi^2} + \frac{4a}{3\Phi^3} - \frac{3a^2}{2\Phi^4} + \frac{4a^3}{5\Phi^5} - \frac{a^4}{6\Phi^6}\right\}.$$

$$\int \frac{x^5}{(a+bx)^7}dx = \frac{1}{b^6}\left\{-\frac{1}{\Phi} + \frac{5a}{2\Phi^2} - \frac{10a^2}{3\Phi^3} + \frac{5a^3}{2\Phi^4} - \frac{a^4}{\Phi^5} + \frac{a^5}{6\Phi^6}\right\}.$$

$$\int \frac{x^6}{(a+bx)^7}dx = \frac{1}{b^7}\left\{\frac{6a}{\Phi} - \frac{15a^2}{2\Phi^2} + \frac{20a^3}{3\Phi^3} - \frac{15a^4}{4\Phi^4} + \frac{6a^5}{5\Phi^5} - \frac{a^6}{6\Phi^6} + \log\Phi\right\}.$$

$$\int \frac{x^7}{a+bx^3}dx = \frac{x^5}{5b} - \frac{ax^2}{2b^2} + \frac{a^2}{b^2}\int\frac{x}{a+bx^3}dx.$$

$$\int \frac{x^7}{a+bx^5}dx = \frac{x^3}{3b} - \frac{a}{b}\int\frac{x^2}{a+bx^5}dx.$$

$$\int \frac{x^7}{(a+bx^3)^2}\,dx = \frac{x^2}{2b^2} + \frac{ax^2}{3b^2\Phi} - \frac{5a}{3b^2}\int \frac{x}{\Phi}\,dx.$$

$$\int \frac{x^3(x^4-x^2-1)}{x+1}\,dx = \frac{4}{7}x^7 - \frac{2}{3}x^6 - \frac{4}{3}x^3 + 2x^2 - 4x + 4\log\frac{1}{x+1}.$$

$\frac{1}{X^8}$

$$\int \frac{dx}{x^3(a+bx^5)} = -\frac{1}{2ax^2} - \frac{b}{a}\int \frac{x^2}{a+bx^5}\,dx.$$

$$\int \frac{dx}{x^4(a+bx^4)} = -\frac{1}{3ax^3} - \frac{b}{a}\int \frac{dx}{a+bx^4}.$$

$$\int \frac{dx}{x^5(a+bx^3)} = -\frac{1}{4ax^4} + \frac{b}{a^2x} + \frac{b^2}{a^2}\int \frac{x}{a+bx^3}\,dx.$$

$$\int \frac{dx}{x^6(a+bx^2)} = -\frac{1}{5ax^5} + \frac{b}{3a^2x^3} - \frac{b^2}{a^3x} - \frac{b^3}{a^3}\int \frac{dx}{a+bx^2}.$$

$$\int \frac{dx}{x^7(a+bx)} = -\frac{1}{6ax^6} + \frac{b}{5a^2x^5} - \frac{b^2}{4a^3x^4} + \frac{b^3}{3a^4x^3} - \frac{b^4}{2a^5x^2} + \frac{b^5}{a^6x} - \frac{b^6}{a^7}\log\frac{a+bx}{x}.$$

$$\int \frac{dx}{(x^4-1)^2} = -\frac{x}{4(x^4-1)} + \frac{3}{16}\log\frac{x+1}{x-1} + \frac{3}{8}\,\text{arc tg}\,x.$$

$$\int \frac{dx}{(1-x^4)^2} = \frac{x}{4(1-x^4)} + \frac{3}{16}\log\frac{1+x}{1-x} + \frac{3}{8}\,\text{arc tg}\,x.$$

$$\int \frac{dx}{(a+bx^4)^2} = \frac{x}{4a\Phi} + \frac{3}{4a}\int \frac{dx}{\Phi}.$$

$$\int \frac{dx}{x^2(a+bx^3)^2} = -\frac{1}{a^2x} - \frac{bx^2}{3a^2\Phi} - \frac{4b}{3a^2}\int \frac{x}{\Phi}\,dx.$$

$$\int \frac{dx}{x^4(a+bx^2)^2} = -\frac{1}{3a^2x^3} + \frac{2b}{a^3x} + \frac{b^2x}{2a^3\Phi} + \frac{5b^2}{2a^3}\int \frac{dx}{\Phi}.$$

$$\int \frac{dx}{x^6(a+bx)^2} = -\frac{b^5}{a^6\Phi} - \frac{1}{5a^2x^5} + \frac{b}{2a^3x^4} - \frac{b^2}{a^4x^3} + \frac{2b^3}{a^5x^2} - \frac{5b^4}{a^6x} + \frac{6b^5}{a^7}\log\frac{\Phi}{x}.$$

$$\int \frac{dx}{x^4(a+bx^2+cx^4)} = -\frac{1}{3ax^3} + \frac{b}{a^2x} + \frac{b^2-ac}{a^2}\int \frac{dx}{\Phi} + \frac{bc}{a^2}\int \frac{x^2}{\Phi}\,dx.$$

$$\int \frac{dx}{x^2(x-1)^3} = \frac{15x^4-25x^2+8}{8x(x^2-1)^2} - \frac{15}{16}\log\frac{x+1}{x-1}.$$

$$\int \frac{dx}{x^2(a+bx^2)^3} = -\frac{1}{a^3x} - \frac{bx}{a^2}\left\{\frac{1}{4\Phi^2} + \frac{7}{8a\Phi}\right\} - \frac{15b}{8a^3}\int \frac{dx}{\Phi}.$$

$$\int \frac{dx}{x^5(a+bx)^3} = \frac{b^4}{2a^5\Phi^2} + \frac{5b^4}{a^6\Phi} - \frac{1}{4a^3x^4} + \frac{b}{a^4x^3} - \frac{3b^2}{a^5x^2} + \frac{10b^3}{a^6x} - \frac{15b^4}{a^7}\log\frac{\Phi}{x}.$$

$$\int \frac{dx}{x^8+x^7-x^4-3} = \frac{2-2x-5x^2}{4x^2(1+x)} + \frac{1}{8}\log\frac{x^2-1}{x^2+1} + \log\frac{x+1}{x} - \frac{1}{4}\,\text{arc tg}\,x.$$

$$\int \frac{dx}{(a^2+x^2)^4} = \frac{x}{6a^2\Phi^3} + \frac{5x}{24a^4\Phi^2} + \frac{5x}{16a^6\Phi} + \frac{5}{16a^7}\,\text{arc tg}\,\frac{x}{a}.$$

$$\int\frac{dx}{(a+bx^2)^4}=\frac{x}{6a\Phi^3}+\frac{5x}{24a^2\Phi^2}+\frac{5x}{16a^3\Phi}+\frac{5}{16a^3}\int\frac{dx}{\Phi}.$$

$$\int\frac{dx}{x^4(a+bx)^4}=-\frac{b^3}{3a^4\Phi^3}-\frac{2b^3}{a^5\Phi^2}-\frac{10b^3}{a^6\Phi}-\frac{1}{3a^4x^3}+\frac{2b}{a^5x^2}-\frac{10b^2}{a^6x}+\frac{20b^3}{a^7}\log\frac{\Phi}{x}.$$

$$\int\frac{dx}{(a+bx^2+cx^4)^2}=-\frac{bcx^3+(b^2-2ac)x}{2a(4ac-b^2)\Phi}-\frac{bc}{2a(4ac-b^2)}\int\frac{x^2}{\Phi}dx+\frac{6ac-b^2}{2a(4ac-b^2)}\int\frac{dx}{\Phi}.$$

$$\int\frac{dx}{x^3(a+bx)^5}=\frac{b^2}{4a^3\Phi^4}+\frac{b^2}{a^4\Phi^3}+\frac{3b^2}{a^5\Phi^2}+\frac{10b^2}{a^6\Phi}-\frac{1}{2a^5x^2}+\frac{5b}{a^6x}-\frac{15b^2}{a^7}\log\frac{\Phi}{x}.$$

$$\int\frac{dx}{x^2(a+bx)^6}=-\frac{b}{5a^2\Phi^5}-\frac{b}{2a^3\Phi^4}-\frac{b}{a^4\Phi^3}-\frac{2b}{a^5\Phi^2}-\frac{5b}{a^6\Phi}-\frac{1}{a^6x}+\frac{6b}{a^7}\log\frac{\Phi}{x}.$$

$$\int\frac{dx}{x(a+bx)^7}=\frac{1}{6a\Phi^6}+\frac{1}{5a^2\Phi^5}+\frac{1}{4a^3\Phi^4}+\frac{1}{3a^4\Phi^3}+\frac{1}{2a^5\Phi^2}+\frac{1}{a^6\Phi}-\frac{1}{a^7}\log\frac{\Phi}{x}.$$

$$\int\frac{dx}{(a+bx+cx^2)^4}=\frac{2cx+b}{4ac-b^2}\left\{\frac{1}{3\Phi^3}+\frac{5c}{3(4ac-b^2)\Phi^2}+\frac{10c^2}{(4ac-b^2)^2\Phi}\right\}+\frac{20c^3}{(4ac-b^2)^3}\int\frac{dx}{\Phi}.$$

$\dfrac{Y}{X^8}$

$$\int\frac{x}{(ax^4+b)^2}dx=\frac{x^2}{4b\Phi}+\frac{1}{2b}\int\frac{x}{\Phi}dx.$$

$$\int\frac{x}{(a+bx+cx^2)^4}dx=-\frac{bx+2a}{3(4ac-b^2)\Phi^3}-\frac{5b(2cx+b)}{3(4ac-b^2)^2}\left\{\frac{1}{2\Phi^2}+\frac{3c}{(4ac-b^2)\Phi}\right\}-\frac{10bc^2}{(4ac-b^2)^3}\int\frac{dx}{\Phi}.$$

$$\int\frac{x+1}{(x^2+1)(x^2+x+1)^3}dx=-\frac{4x^3-3x^2+2x-3}{6(x^2+x+1)^2}-\frac{1}{2}\log\frac{x^2+x+1}{x^2+1}+\frac{1}{2\sqrt{3}}\operatorname{arc\,tg}\frac{2x+1}{\sqrt{3}}-\operatorname{arc\,tg}x.$$

$$\int\frac{2x+1}{x(x-2)^3(x^2+1)^2}dx=-\frac{1}{8}\int\frac{dx}{x}+\frac{1}{10}\int\frac{dx}{(x-2)^3}-\frac{17}{100}\int\frac{dx}{(x-2)^2}+\frac{1000}{173}\int\frac{dx}{x-2}+\int\frac{cx+D}{(x^2+1)^2}dx+\int\frac{c_1x+D_1}{x^2+1}dx$$

where $c=-\frac{4}{25}$, $D=-\frac{3}{25}$, $c_1=-\frac{1}{125}$, $D_1=-\frac{2}{25}$.

$$\int\frac{x^2}{(x^4+1)^2}dx=\frac{x^3}{4\Phi}+\frac{1}{16\sqrt{2}}\log\frac{x^2-x\sqrt{2}+1}{x^2+x\sqrt{2}+1}+\frac{1}{8\sqrt{2}}\operatorname{arc\,tg}\frac{x\sqrt{2}}{1-x^2}.$$

$$\int\frac{x^2}{(a+bx^4)^2}dx=\frac{x^3}{4a\Phi}+\frac{1}{4a}\int\frac{x^2}{\Phi}dx.$$

$$\int\frac{x^2}{(1+x^2)^4}dx=\frac{x}{16\Phi}+\frac{x}{24\Phi^2}-\frac{x}{6\Phi^3}+\frac{1}{16}\operatorname{arc\,tg}x=\frac{3x^5+8x^3-3x}{48\Phi^3}+\frac{1}{16}\operatorname{arc\,tg}x.$$

$$\int\frac{x^2}{(a+bx^2)^4}dx=\frac{x}{b}\left\{-\frac{1}{6\Phi^3}+\frac{1}{24a\Phi^2}+\frac{1}{16a^2\Phi}\right\}+\frac{1}{16a^2b}\int\frac{dx}{\Phi}.$$

$$\int\frac{x^2}{(a+bx+cx^2)^4}dx=\frac{(b^2-2ac)x+ab}{3c(4ac-b^2)\Phi^3}+\frac{2(ac+b^2)(2cx+b)}{3c(4ac-b^2)^2}\left\{\frac{1}{2\Phi^2}+\frac{3c}{(4ac-b^2)\Phi}\right\}+\frac{4c(ac+b^2)}{(4ac-b^2)^3}\int\frac{dx}{\Phi}.$$

$$\int\frac{x^2}{(a+bx)^8}dx=\frac{1}{b^3}\left\{-\frac{1}{5\Phi^5}+\frac{a}{3\Phi^6}-\frac{a^2}{7\Phi^7}\right\}.$$

$$\int\frac{x^3}{(x^4-x^2+1)^2}dx=\frac{x^2-2}{6\Phi}+\frac{1}{3\sqrt{3}}\operatorname{arc\,tg}\frac{2x^2-1}{\sqrt{3}}.$$

$$\int \frac{x^3}{(a+bx^2)^4}\,dx = -\frac{1}{a^2}\int \frac{z^2-b}{z^7}\,dz. \qquad a+bx^2 = z^2x^2.$$

$$\int \frac{x^3}{(a+bx)^8}\,dx = \frac{1}{b^4}\left\{-\frac{1}{4\Phi^4}+\frac{3a}{5\Phi^5}-\frac{a^2}{2\Phi^6}+\frac{c^3}{7\Phi^7}\right\}.$$

$$\int \frac{x^3}{(a+bx+cx^2)^4}\,dx = \left\{-\frac{x^2}{c}-\frac{b^3x}{3c^2(4ac-b^2)}-\frac{a(4ac+b^2)}{3c^2(4ac-b^2)}\right\}\frac{1}{4\Phi^3}-\frac{b(6ac+b^2)}{6c^2(4ac-b^2)}\int\frac{dx}{\Phi^3}$$

$$\int \frac{x^4}{(ax^4-1)^2}\,dx = \frac{1}{a}\int\frac{dx}{ax^4-1}+\frac{1}{a}\int\frac{dx}{(ax^4-1)^2}.$$

$$\int \frac{x^4}{(a+bx^4)^2}\,dx = -\frac{x}{4b\Phi}+\frac{1}{4b}\int\frac{dx}{\Phi}.$$

$$\int \frac{x^4}{(a+bx^2)^4}\,dx = \frac{x}{b^2}\left\{\frac{a}{6\Phi^3}-\frac{7}{24\Phi^2}+\frac{1}{16a\Phi}\right\}+\frac{1}{16ab^2}\int\frac{dx}{\Phi}.$$

$$\int \frac{x^4}{(a+bx)^8}\,dx = \frac{1}{b^5}\left\{-\frac{1}{3\Phi^3}+\frac{a}{\Phi^4}-\frac{6a^2}{5\Phi^5}+\frac{2a^3}{3\Phi^6}-\frac{a^4}{7\Phi^7}\right\}.$$

$$\int \frac{x^4}{(a+bx+cx^2)^4}\,dx = -\left\{\frac{x^3}{3c}+\frac{ax}{5c^2}-\frac{ab}{15c^3}\right\}\frac{1}{\Phi^3}+\frac{a^2c+ab^2}{5c^3}\int\frac{dx}{\Phi^4}.$$

$$\int \frac{x^4+x^2+1}{(x^2+1)^4}\,dx = -\frac{x^3-3x}{24\Phi^3}+\frac{7x}{16\Phi}+\frac{7}{16}\operatorname{arc\,tg} x.$$

$$\int \frac{2x^4-5x^3+1}{(x^2+1)^3(x^2-3x+2)}\,dx = \frac{58x+11}{100(x^2+1)}-\frac{6x+7}{20(x^2+1)^2}+\frac{1}{4}\log(x-1)-\frac{7}{125}\log(x-2)-\frac{97}{1000}\log(x^2+1)+\frac{29}{50}\operatorname{arc\,tg} x.$$

$$\int \frac{x^5}{(a+bx)^8}\,dx = \frac{1}{b^6}\left\{-\frac{1}{2\Phi^2}+\frac{5a}{3\Phi^3}-\frac{5a^2}{2\Phi^4}+\frac{2a^3}{\Phi^5}-\frac{5a^4}{6\Phi^6}+\frac{a^5}{7\Phi^7}\right\}.$$

$$\int \frac{x^5}{(1+x^4)(b^2+a^2x^4)}\,dx = \frac{1}{2(a^2-b^2)}\left\{\operatorname{arc\,tg} x^2-\frac{b}{a}\operatorname{arc\,tg}\frac{a}{b}x^2\right\}.$$

$$\int \frac{x^5}{(a+bx+cx^2)^4}\,dx = -\left\{\frac{x^4}{2c}+\frac{bx^3}{6c^2}+\frac{ax^2}{2c^2}+\frac{a^2}{6c^3}\right\}\frac{1}{\Phi^3}-\frac{a^2b}{2c^3}\int\frac{dx}{\Phi^4}.$$

$$\int \frac{x^6}{(a+bx^4)^2}\,dx = -\frac{x^3}{4b\Phi}+\frac{3}{4b}\int\frac{x^2}{\Phi}\,dx.$$

$$\int \frac{x^6}{(a+bx^2)^4}\,dx = \frac{x}{b^3}\left\{-\frac{a^2}{6\Phi^3}+\frac{13a}{24\Phi^2}-\frac{11}{16\Phi}\right\}+\frac{5}{16b^3}\int\frac{dx}{\Phi}.$$

$$\int \frac{x^6}{(a+bx^2+cx^4)^2}\,dx = \frac{(b^2-2ac)x^3+abx}{2c(4ac-b^2)\Phi}-\frac{ab}{2c(4ac-b^2)}\int\frac{dx}{\Phi}+\frac{6ac-b^2}{2c(4ac-b^2)}\int\frac{x^2}{\Phi}\,dx.$$

$$\int \frac{x^6}{(a+bx)^8}\,dx = \frac{1}{b^7}\left\{-\frac{1}{\Phi}+\frac{3a}{\Phi^2}-\frac{5a^2}{\Phi^3}+\frac{5a^3}{\Phi^4}-\frac{3a^4}{\Phi^5}+\frac{a^5}{\Phi^6}-\frac{a^6}{7\Phi^7}\right\}.$$

$$\int \frac{x^6-x^4-x^2}{x^8-2x^6-x^4-2x^2+1}\,dx = \frac{1}{4\sqrt{5}}\log\frac{x^2-x\sqrt{5}+1}{x^2+x\sqrt{5}+1}+\frac{1}{2\sqrt{3}}\operatorname{arc\,tg}\frac{2x+1}{\sqrt{3}}+\frac{1}{2\sqrt{3}}\operatorname{arc\,tg}\frac{2x-1}{\sqrt{3}}.$$

$$\int \frac{x^8}{(a+bx^4)^2}\,dx = \frac{x}{b^2}+\frac{ax}{4b^2\Phi}-\frac{5a}{4b^2}\int\frac{dx}{\Phi}.$$

$$\int \frac{x^8}{(a+bx^2)^4}\,dx = \frac{x}{b^4}+\frac{x}{b^4}\left\{\frac{a^3}{6\Phi^3}-\frac{19a^2}{24\Phi^2}+\frac{29a}{16\Phi}\right\}-\frac{35a}{16b^4}\int\frac{dx}{\Phi}.$$

$\frac{Y^8}{X^{8-a}}$

$$\int \frac{x^8}{a+bx^2}\,dx = \frac{x^7}{7b} - \frac{ax^5}{5b^2} + \frac{a^2x^5}{3b^3} - \frac{a^3x}{b^4} + \frac{a^4}{b^4}\int\frac{dx}{\Phi}.$$

$$\int \frac{x^8}{a+bx^5}\,dx = \frac{x^4}{4b} - \frac{a}{b}\int\frac{x^3}{\Phi}\,dx.$$

$$\int \frac{x^8}{(a+bx^2)^2}\,dx = \frac{x^5}{5b^2} - \frac{2ax^3}{3b^3} + \frac{3a^2x}{b^4} + \frac{a^3x}{2b^4\Phi} - \frac{7a^3}{2b^4}\int\frac{dx}{\Phi}.$$

X^n

$$\int \frac{dx}{x^6(a+bx^3)} = -\frac{1}{5ax^5} + \frac{b}{2a^2x^2} + \frac{b^2}{a^2}\int\frac{dx}{\Phi}.$$

$$\int \frac{dx}{x^3(a+bx^3)^2} = -\frac{1}{2a^2x^2} - \frac{bx}{3a^2\Phi} - \frac{5b}{3a^2}\int\frac{dx}{\Phi}.$$

$$\int \frac{dx}{(a+bx^3)^3} = \frac{x}{a}\left\{\frac{1}{6\Phi^2} + \frac{5}{18a\Phi}\right\} + \frac{5}{9a^2}\int\frac{dx}{\Phi}.$$

$$\int \frac{dx}{x^3(x^2-5)^5} = \frac{1}{5^2}\left\{\frac{1}{5^3x^2} + \frac{4}{5^3\Phi} - \frac{3}{2.5^2\Phi^2} + \frac{2}{3.5\Phi^3} - \frac{1}{4\Phi^4}\right\} + \frac{1}{5^5}\log\frac{\Phi}{x^2}.$$

$$\int \frac{x}{(a+bx^3)^3}\,dx = \frac{x^2}{a}\left\{\frac{1}{6\Phi^2} + \frac{2}{9a\Phi}\right\} + \frac{2}{9a^2}\int\frac{x}{\Phi}\,dx.$$

$$\int \frac{x^3}{(a+bx^3)^3}\,dx = -\frac{x}{b}\left\{\frac{1}{6\Phi^2} - \frac{1}{18a\Phi}\right\} + \frac{1}{9ab}\int\frac{dx}{\Phi}.$$

$$\int \frac{x^4}{(a+bx^3)^3}\,dx = -\frac{x^2}{b}\left\{\frac{1}{6\Phi^2} - \frac{1}{9a\Phi}\right\} + \frac{1}{9ab}\int\frac{x}{\Phi}\,dx.$$

$$\int \frac{x^2(1-2x^2)}{(x^3+1)^3}\,dx = \int\frac{3-2z}{3z^3}\,dz. \qquad x^3+1=z.$$

$$\int \frac{5x^3+3x-1}{(x^3+3x+1)^3}\,dx = -\frac{x}{(x^3+3x+1)^2}.$$

$$\int \frac{x^5}{(5-7x^3)^3}\,dx = \frac{x^6}{30(5-7x^3)^2}.$$

$$\int \frac{dx}{x^4(5+3x^3)^2} = \frac{1}{15}\left\{-\frac{5+6x^3}{5x^3(5+3x^3)} + \frac{6}{25}\log\frac{5+3x^3}{x^3}\right\}.$$

$$\int \frac{x^4}{(x^2+1)^5}\,dx = \frac{3x}{128\Phi} - \frac{3x}{64\Phi^3} - \frac{x^3}{8\Phi^4} + \frac{3}{128}\,\text{arc tg}\,x.$$

$$\int \frac{x^4}{(x^2-1)^5}\,dx = \frac{1}{128}\left\{\frac{3x^7-11x^5-11x^3+3x}{(x^2-1)^4} + \frac{3}{2}\log\frac{1-x}{1+x}\right\}.$$

$$\int \frac{x^5}{(x^2+1)^6}\,dx = \frac{1}{4\Phi^4} - \frac{1}{6\Phi^3} - \frac{1}{10\Phi^5}.$$

$$\int \frac{dx}{x^7(x^6-2)} = \frac{1}{12x^6} - \frac{1}{24}\log\frac{x^6}{x^6-2}.$$

$$\int\frac{dx}{x^3(1-x^4)^2}=\frac{2x^4-1}{4x^4\Phi}+\frac{1}{2}\log\frac{x^4}{\Phi}.$$

$$\int\frac{dx}{(a+bx^4)^5}=\frac{1}{16a}\left\{\frac{x}{\Phi^4}+\frac{5x}{4a\Phi^3}+\frac{55x}{32a^2\Phi^2}+\frac{385x}{128a^3\Phi}+\frac{1155}{128a^3}\int\frac{dx}{\Phi}\right\}.$$

$$\int U\sqrt{x}\,dx$$

$$\int\frac{dx}{(x+1)\sqrt{x}}=\text{arc tg}\,\frac{2\sqrt{x}}{1-x}=\text{arc sin}\,\frac{2\sqrt{x}}{1+x}=-\text{arc sin}\,\frac{1-x}{1+x}=2\,\text{arc tg}\sqrt{x}.$$

$$\int\frac{dx}{(x-a)\sqrt{x}}=\frac{1}{\sqrt{a}}\log\frac{\sqrt{x}-\sqrt{a}}{\sqrt{x}+\sqrt{a}}=-\frac{2}{\sqrt{a}}\log\frac{\sqrt{x}+\sqrt{a}}{\sqrt{x-a}}=\frac{2}{\sqrt{a}}\log\frac{\sqrt{x}-\sqrt{a}}{\sqrt{x-a}}.$$

$$\int\frac{dx}{(ax+b)\sqrt{x}}=\frac{2}{\sqrt{ab}}\,\text{arc tg}\sqrt{\frac{ax}{b}}.$$

$$\int\frac{dx}{(ax-b)\sqrt{x}}=\frac{1}{\sqrt{ab}}\log\frac{2\sqrt{abx}-b-ax}{ax-b}.$$

$$\int\frac{dx}{(ax^2+b)\sqrt{x}}=\frac{k}{2b\sqrt{2}}\log\frac{x+k\sqrt{2x}+k^2}{x-k\sqrt{2x}+k^2}+\frac{k}{b\sqrt{2}}\,\text{arc tg}\,\frac{k\sqrt{2x}}{k^2-x}.\qquad k=\left(\frac{b}{a}\right)^{\frac{1}{4}}.$$

$$\int\frac{dx}{(ax^2-b)\sqrt{x}}=-\frac{k}{2b}\log\frac{k+\sqrt{x}}{k-\sqrt{x}}-\frac{k}{b}\,\text{arc tg}\,\frac{\sqrt{x}}{k}.\qquad k=\left(\frac{b}{a}\right)^{\frac{1}{4}}.$$

$$\int\frac{dx}{(x-a)^2\sqrt{x}}=-\frac{\sqrt{x}}{a(x-a)}+\frac{1}{a\sqrt{a}}\log\frac{\sqrt{x}+\sqrt{a}}{\sqrt{x-a}}.$$

$$\int\frac{dx}{(ax+b)^2\sqrt{x}}=\frac{\sqrt{x}}{b(ax+b)}+\frac{1}{2b}\int\frac{dx}{(ax+b)\sqrt{x}}.$$

$$\int\frac{dx}{(ax^2+b)^2\sqrt{x}}=\frac{\sqrt{x}}{2b(ax^2+b)}+\frac{3}{4b}\int\frac{dx}{(ax^2+b)\sqrt{x}}.$$

$$\int\frac{\sqrt{x}}{x-a}\,dx=2\sqrt{x}-\sqrt{a}\log\frac{\sqrt{x}+\sqrt{a}}{\sqrt{x}-\sqrt{a}}.$$

$$\int\frac{\sqrt{x}}{ax+b}\,dx=\frac{2}{a}\sqrt{x}-\frac{b}{a}\int\frac{dx}{(ax+b)\sqrt{x}}.$$

$$\int\frac{\sqrt{x}}{ax^2+b}\,dx=2\int\frac{z^2}{az^4+b}\,dz.\qquad x=z^2.$$

$$=\frac{1}{2ak\sqrt{2}}\log\frac{x-k\sqrt{2x}+k^2}{x+k\sqrt{2x}+k^2}+\frac{1}{ak\sqrt{2}}\,\text{arc tg}\,\frac{k\sqrt{2x}}{k^2-x}.\qquad k=\left(\frac{b}{a}\right)^{\frac{1}{4}}.$$

$$\int\frac{\sqrt{x}}{ax^2-b}\,dx=\frac{1}{2ak}\log\frac{k-\sqrt{x}}{k+\sqrt{x}}+\frac{1}{ak}\,\text{arc tg}\,\frac{\sqrt{x}}{k}.\qquad k=\left(\frac{b}{a}\right)^{\frac{1}{4}}.$$

$$\int\frac{\sqrt{x}}{(x-a)^2}\,dx=-\frac{\sqrt{x}}{x-a}+\frac{1}{2\sqrt{a}}\log\frac{\sqrt{x}-\sqrt{a}}{\sqrt{x}+\sqrt{a}}.$$

$$\int\frac{\sqrt{x}}{(ax+b)^2}\,dx=-\frac{\sqrt{x}}{a(ax+b)}+\frac{1}{a\sqrt{ab}}\,\text{arc tg}\sqrt{\frac{ax}{b}}.$$

$$\int\frac{\sqrt{x}}{(ax^2+b)^2}\,dx=\frac{x\sqrt{x}}{2b(ax^2+b)}+\frac{1}{4b}\int\frac{\sqrt{x}}{ax^2+b}\,dx.$$

$$\int U\sqrt{ax+b}\,dx$$

$$\int\sqrt{ax+b}\,dx = \frac{2}{3a}(ax+b)^{\frac{3}{2}}.$$

$$\int x\sqrt{ax+b}\,dx = \frac{2(3ax-2b)}{15a^2}(ax+b)^{\frac{3}{2}}.$$

$$\int x^2\sqrt{ax+b}\,dx = \frac{2}{a^3}\left\{\frac{\Phi^2}{7}-\frac{2b}{5}\Phi+\frac{b^2}{3}\right\}\Phi^{\frac{3}{2}}.$$

$$\int x^3\sqrt{ax+b}\,dx = \frac{2}{a^4}\left\{\frac{\Phi^3}{9}-\frac{3b}{7}\Phi^2+\frac{3b^2}{5}\Phi-\frac{b^3}{3}\right\}\Phi^{\frac{3}{2}}.$$

$$\int\frac{dx}{\sqrt{ax+b}} = \frac{2}{a}\sqrt{ax+b}.$$

$$\int\frac{dx}{x\sqrt{x-a}} = \frac{2}{\sqrt{a}}\,\text{arc sin}\sqrt{\frac{x-a}{x}} = \frac{1}{\sqrt{a}}\,\text{arc sin}\,\frac{x-2a}{x}.$$

$$\int\frac{dx}{x\sqrt{2ax-a^2}} = \frac{1}{a}\,\text{arc sin}\,\frac{x-a}{x}.$$

$$\int\frac{dx}{x\sqrt{b^2-a^2x}} = \frac{1}{b}\log\frac{\sqrt{b^2-a^2x}-b}{\sqrt{b^2-a^2x}+b}.$$

$$\int\frac{dx}{x\sqrt{ax+b}} = \frac{1}{\sqrt{b}}\log\frac{\sqrt{\Phi}-\sqrt{b}}{\sqrt{\Phi}+\sqrt{b}} = \frac{2}{\sqrt{b}}\log\frac{\sqrt{\Phi}-\sqrt{b}}{\sqrt{x}} = -\frac{2}{\sqrt{b}}\log\frac{\sqrt{\Phi}+\sqrt{b}}{\sqrt{x}}.$$

$$\int\frac{dx}{x\sqrt{ax-b}} = \frac{2}{\sqrt{b}}\,\text{arc tg}\sqrt{\frac{ax-b}{b}} = \frac{2}{\sqrt{b}}\,\text{arc sin}\sqrt{\frac{ax-b}{ax}} = \frac{2}{\sqrt{b}}\,\text{arc cos}\sqrt{\frac{b}{ax}} = \frac{1}{\sqrt{b}}\,\text{arc cos}\,\frac{2b-ax}{ax}.$$

$$\int\frac{dx}{(1-x)\sqrt{1+x}} = \frac{1}{\sqrt{2}}\log\frac{\sqrt{2}+\sqrt{1+x}}{\sqrt{2}-\sqrt{1+x}}.$$

$$\int\frac{dx}{(x-1)\sqrt{2-x}} = \log\frac{\sqrt{2-x}-1}{\sqrt{2-x}+1}.$$

$$\int\frac{dx}{(x-1)\sqrt{x-2}} = 2\,\text{arc tg}\sqrt{x-2}.$$

$$\int\frac{dx}{(x+a)\sqrt{2ax+a^2}} = \frac{1}{a}\,\text{arc sin}\,\frac{x}{x+a}.$$

$$\int\frac{dx}{(x+1)\sqrt{ax+b}} = \frac{2}{\sqrt{a-b}}\,\text{arc tg}\sqrt{\frac{ax+b}{a-b}}. \qquad a>b$$

$$= \frac{1}{b-a}\log\frac{\sqrt{ax+b}-\sqrt{b-a}}{\sqrt{ax+b}+\sqrt{b-a}}. \qquad a<b.$$

$$\int\frac{dx}{(ax+2b)\sqrt{ax+b}} = \frac{2}{a\sqrt{b}}\,\text{arc tg}\sqrt{\frac{ax+b}{b}}.$$

$$\int\frac{dx}{(ax+b)\sqrt{b-ax}} = \frac{1}{a\sqrt{2b}}\log\frac{\sqrt{b-ax}-\sqrt{2b}}{\sqrt{b-ax}+\sqrt{2b}}.$$

$$\int\frac{dx}{(mx+n)\sqrt{ax+b}} = \frac{1}{\sqrt{m(bm-an)}}\log\frac{\sqrt{m(ax+b)}-\sqrt{bm-an}}{\sqrt{m(ax+b)}+\sqrt{bm-an}}. \qquad bm-an>0.$$

$$= \frac{2}{\sqrt{m(an-bm)}} \operatorname{arc\,tg} \sqrt{\frac{m(ax+b)}{an-bm}} \qquad an - bm > 0.$$

$$\int \frac{dx}{(n-mx)\sqrt{b-ax}} = \frac{1}{\sqrt{m}\sqrt{bm-an}} \log \frac{\sqrt{m(b-ax)} + \sqrt{bm-an}}{\sqrt{m(b-ax)} - \sqrt{bm-an}}. \qquad bm - an > 0.$$

$$= -\frac{2}{\sqrt{m}\sqrt{an-bm}} \operatorname{arc\,tg} \sqrt{\frac{m(b-ax)}{an-bm}}. \qquad an - bm > 0.$$

$$\int \frac{dx}{x^2\sqrt{x+1}} = -\frac{\sqrt{x^2+1}}{x} + \frac{1}{2} \log \frac{\sqrt{x+1}+1}{\sqrt{x+1}-1}.$$

$$\int \frac{dx}{x^2\sqrt{x-1}} = \frac{\sqrt{x-1}}{x} + \operatorname{arc\,tg} \sqrt{x-1}.$$

$$\int \frac{dx}{x^2\sqrt{ax+b}} = -\frac{\sqrt{ax+b}}{bx} - \frac{a}{2b\sqrt{b}} \log \frac{\sqrt{ax+b}-\sqrt{b}}{\sqrt{ax+b}+\sqrt{b}}.$$

$$\int \frac{dx}{x^2\sqrt{ax-b}} = \frac{\sqrt{ax-b}}{bx} + \frac{a}{b\sqrt{b}} \operatorname{arc\,tg} \sqrt{\frac{ax-b}{b}}.$$

$$\int \frac{dx}{x(mx+n)\sqrt{ax+b}} = \frac{1}{n}\int \frac{dx}{x\sqrt{ax+b}} - \frac{m}{n}\int \frac{dx}{(mx+n)\sqrt{ax+b}}.$$

$$\int \frac{dx}{(mx+n)^2\sqrt{ax+b}} = \frac{1}{\sqrt{m}}\int \frac{dz}{z^2\sqrt{az+bm-an}}. \qquad mx + n = z.$$

$$\int \frac{dx}{x^3\sqrt{x-1}} = \frac{3x+2}{4x^2}\sqrt{x-1} + \frac{3}{4} \operatorname{arc\,tg} \sqrt{x-1} = \frac{3x+2}{4x^2}\sqrt{x-1} - \frac{3}{4} \operatorname{arc\,sin} \frac{1}{\sqrt{x}}.$$

$$\int \frac{dx}{x^3\sqrt{ax+b}} = \frac{3ax-2b}{4b^2x^2}\sqrt{ax+b} + \frac{3a^2}{8b^2}\int \frac{dx}{x\sqrt{ax+b}}.$$

$$\int \frac{dx}{x^2(mx+n)\sqrt{ax+b}} = \frac{1}{n}\int \frac{dx}{x^2\sqrt{\Phi}} - \frac{m}{n^2}\int \frac{dx}{x\sqrt{\Phi}} + \frac{m^2}{n^2}\int \frac{dx}{(mx+n)\sqrt{\Phi}}.$$

$$\int \frac{dx}{x^4\sqrt{x-1}} = \left\{ \frac{1}{x^2} + \frac{5}{4x} + \frac{15}{8} \right\} \frac{\sqrt{x-1}}{3x} + \frac{5}{8} \operatorname{arc\,tg} \sqrt{x-1}.$$

$$\int \frac{dx}{x^4\sqrt{ax+b}} = \left\{ -\frac{1}{3bx^3} + \frac{5a}{12b^2x^2} - \frac{5a^2}{8b^3x} \right\} \sqrt{\Phi} - \frac{5a^3}{16b^3}\int \frac{dx}{x\sqrt{\Phi}}.$$

$$\int \frac{dx}{(x^4+x^3)\sqrt{x+1}} = \left\{ \frac{15}{4} + \frac{5}{4x} - \frac{1}{2x^2} \right\} \frac{1}{\sqrt{x+1}} + \frac{15}{8} \log \frac{\sqrt{x+1}-1}{\sqrt{x+1}+1}.$$

$$\int \frac{x}{\sqrt{a-x}}\,dx = -\frac{2}{3}(x+2a)\sqrt{a-x}.$$

$$\int \frac{x}{\sqrt{ax+b}}\,dx = \frac{2ax-4b}{3a^2}\sqrt{ax+b}.$$

$$\int \frac{x}{(x+2)\sqrt{2x+4}}\,dx = \frac{2x+8}{\sqrt{2x+4}}.$$

$$\int \frac{x}{(mx+n)\sqrt{ax+b}}\,dx = \frac{1}{m}\int \frac{dx}{\sqrt{\Phi}} - \frac{n}{m}\int \frac{dx}{(mx+n)\sqrt{\Phi}}.$$

$$\int \frac{mx+n}{\sqrt{ax+b}}\,dx = \frac{2}{3a^2}(3an - 2bm + amx)\sqrt{ax+b}.$$

$$\int\frac{x^2}{\sqrt{x+1}}\,dx = \frac{2}{15}(3x^2-4x+8)\sqrt{x+1}\,.$$

$$\int\frac{x^2}{\sqrt{ax+b}}\,dx = \left\{\frac{\Phi^2}{5}-\frac{2b\Phi}{3}+b^2\right\}\frac{2\sqrt{\Phi}}{a^3}$$

$$\int\frac{x^3}{\sqrt{x-1}}\,dx = \left\{\frac{\Phi^3}{7}+\frac{3\Phi^2}{5}+x\right\}2\sqrt{\Phi}\,.$$

$$\int\frac{x^3}{\sqrt{1-x}}\,dx = -\frac{2}{35}(5x^3+6x^2+8x+16)\sqrt{1-x}\,.$$

$$\int\frac{x^3}{\sqrt{ax+b}}\,dx = \frac{2}{35a^4}(5a^3x^3-6a^2bx^2+8ab^2x-16b^3)\sqrt{\Phi}$$

$$\int\frac{\sqrt{x+1}}{x}\,dx = 2\sqrt{x+1}+\log\frac{\sqrt{x+1}-1}{\sqrt{x+1}+1}\,.$$

$$\int\frac{\sqrt{ax+b}}{x}\,dx = 2\sqrt{\Phi}+\sqrt{b}\log\frac{\sqrt{\Phi}-\sqrt{b}}{\sqrt{\Phi}+\sqrt{b}}\,.$$

$$\int\frac{\sqrt{ax-b}}{x}\,dx = 2\sqrt{\Phi}-2\sqrt{b}\,\text{arc tg}\sqrt{\frac{ax-b}{b}}\,.$$

$$\int\frac{\sqrt{x+1}}{x+5}\,dx = 2\sqrt{x+1}-4\,\text{arc tg}\frac{\sqrt{x+1}}{2}\,.$$

$$\int\frac{\sqrt{ax+b}}{mx+n}\,dx = \frac{2}{m}\sqrt{\Phi}-\frac{2}{m}\sqrt{\frac{an-bm}{m}}\,\text{arc tg}\sqrt{\frac{m\Phi}{an-bm}}\,.$$

$$\int\frac{\sqrt{ax+b}}{x^2}\,dx = -\frac{\sqrt{\Phi}}{x}+\frac{a}{2\sqrt{b}}\log\frac{\sqrt{\Phi}-\sqrt{b}}{\sqrt{\Phi}+\sqrt{b}}$$

$$\int\frac{\sqrt{ax-b}}{x^2}\,dx = -\frac{\sqrt{\Phi}}{x}+\frac{a}{\sqrt{b}}\,\text{arc tg}\sqrt{\frac{ax-b}{b}}\,.$$

$$\int\frac{\sqrt{ax+b}}{x^3}\,dx = -\frac{\Phi^{\frac{3}{2}}}{2bx^2}+\frac{a\sqrt{\Phi}}{4bx}-\frac{a^2}{8b}\int\frac{dx}{x\sqrt{\Phi}}$$

$\int U\sqrt{\frac{mx+n}{ax+b}}\,dx$

$$\int\sqrt{\frac{x}{x+1}}\,dx = \sqrt{x^2+x}-\log\left(\sqrt{x+1}+\sqrt{x}\right) = \sqrt{x^2+x}-\frac{1}{2}\log\left(x+\frac{1}{2}+\sqrt{x^2+x}\right).$$

$$\int\sqrt{\frac{x}{x+a}}\,dx = \sqrt{x^2+ax}-a\log\left(\sqrt{x+a}+\sqrt{x}\right).$$

$$\int\sqrt{\frac{x}{ax+b}}\,dx = \frac{1}{a}\sqrt{ax^2+bx}-\frac{b}{a\sqrt{a}}\log\left(\sqrt{ax}+\sqrt{ax+b}\right).$$

$$\int\sqrt{\frac{x}{2a-x}}\,dx = -\sqrt{2ax-x^2}-2a\,\text{arc sin}\sqrt{\frac{2a-x}{2a}}$$

$$\int\sqrt{\frac{x}{b-ax}}\,dx = -\frac{1}{a}\sqrt{bx-ax^2}-\frac{b}{a\sqrt{a}}\,\text{arc sin}\sqrt{\frac{b-ax}{b}}$$

$$\int\sqrt{\frac{x+4}{x}}\,dx = \sqrt{x^2+4x}+2\log\left(x+2+\sqrt{x^2+4x}\right).$$

$$\int\sqrt{\frac{x-a}{x}}\,dx = \sqrt{x^2-ax} - a\log\left(\sqrt{x}+\sqrt{x-a}\right).$$

$$\int\sqrt{\frac{ax+b}{x}}\,dx = \sqrt{ax^2+bx} + \frac{b}{\sqrt{a}}\log\left(\sqrt{ax}+\sqrt{ax+b}\right).$$

$$\int\sqrt{\frac{b-x}{x}}\,dx = \sqrt{bx-x^2} + b\arcsin\sqrt{\frac{x}{b}}.$$

$$\int\sqrt{\frac{b-ax}{x}}\,dx = \sqrt{bx-ax^2} - \frac{b}{\sqrt{a}}\arcsin\sqrt{\frac{b-ax}{b}}.$$

$$\int\sqrt{\frac{x+1}{x-1}}\,dx = \sqrt{x^2-1} + 2\log\left(\sqrt{x-1}+\sqrt{x+1}\right).$$

$$\int\sqrt{\frac{1-x}{1+x}}\,dx = \sqrt{1-x^2} + \arcsin x.$$

$$\int\sqrt{\frac{x+3}{x+1}}\,dx = \sqrt{(x+3)(x+1)} + \log\left(x+2+\sqrt{(x+3)(x+1)}\right).$$

$$\int\sqrt{\frac{a-x}{a+x}}\,dx = \sqrt{a^2-x^2} + a\arcsin\frac{x}{a} = \int\frac{a-x}{\sqrt{a^2-x^2}}\,dx = a\int\frac{dx}{\sqrt{a^2-x^2}} - \int\frac{x}{\sqrt{a^2-x^2}}\,dx.$$

$$\int\sqrt{\frac{x-a}{x-b}}\,dx = \sqrt{(x-a)(x-b)} - \frac{1}{2}(a-b)\log\frac{\sqrt{x-b}+\sqrt{x-a}}{\sqrt{x-b}-\sqrt{x-a}}.$$

$$\int\sqrt{\frac{a-x}{x-b}}\,dx = \sqrt{(a-x)(x-b)} - (a-b)\,\mathrm{arc\,tg}\sqrt{\frac{a-x}{x-b}}.$$

$$\int\sqrt{\frac{b-ax}{b+ax}}\,dx = \frac{1}{a}\sqrt{b^2-a^2x^2} + \frac{b}{a}\arcsin\frac{ax}{b}.$$

$$\int\sqrt{\frac{ax+b}{mx+n}}\,dx = \frac{1}{m}\sqrt{amx^2+(an+bm)x+bn} + \frac{bm-an}{m\sqrt{am}}\log\left\{\sqrt{amx+an}+\sqrt{amx+bm}\right\}.$$

$$\int\sqrt{\frac{ax+b}{n-mx}}\,dx = -\frac{1}{m}\sqrt{bn+(an-bm)x-amx^2} + \frac{an+bm}{m\sqrt{am}}\arcsin\sqrt{\frac{bm+amx}{an+bm}}.$$

$$\int\sqrt{\frac{b-ax}{mx+n}}\,dx = \frac{1}{m}\sqrt{bn+(bm-an)x-amx^2} - \frac{an+bm}{m\sqrt{am}}\arcsin\sqrt{\frac{bm-amx}{an+bm}}.$$

$$\int x\sqrt{\frac{x}{a-x}}\,dx = -\frac{1}{4}(3a+2x)\sqrt{ax-x^2} + \frac{3a^2}{4}\,\mathrm{arc\,tg}\sqrt{\frac{x}{a-x}}.$$

$$\int x\sqrt{\frac{x}{2a-x}}\,dx = -\frac{1}{2}(3a+x)\sqrt{2ax-x^2} + \frac{3a^2}{2}\arcsin\frac{x-a}{a}.$$

$$= -\frac{1}{2}(3a+x)\sqrt{2ax-x^2} + \frac{3a^2}{2}\,\mathrm{arc\,sin\,vers}\,\frac{x}{a}.$$

$$\int x\sqrt{\frac{a+x}{a-x}}\,dx = -\frac{2a+x}{2}\sqrt{a^2-x^2} + \frac{a^2}{2}\arcsin\frac{x}{a}.$$

$$\int x\sqrt{\frac{x+a}{x-a}}\,dx = \frac{2a+x}{2}\sqrt{x^2-a^2} + \frac{a^2}{2}\log\left(x+\sqrt{x^2-a^2}\right).$$

$$\int x\sqrt{\frac{a-x}{a+x}}\,dx = \frac{x-2a}{2}\sqrt{a^2-x^2} - a^2\arcsin\sqrt{\frac{x+a}{2a}} = \frac{x-2a}{2}\sqrt{a^2-x^2} - \frac{a^2}{2}\arcsin\frac{x}{a}.$$

$$\int x\sqrt{\frac{ax+b}{mx+n}}\,dx = a\int\frac{x^2}{\sqrt{(ax+b)(mx+n)}}\,dx + b\int\frac{x}{\sqrt{(ax+b)(mx+n)}}\,dx.$$

$$\int\frac{1}{x}\sqrt{\frac{x+a}{x-a}}\,dx = \log\left(x+\sqrt{x^2-a^2}\right) + \text{arc cos}\,\frac{a}{x} = \log\left(x+\sqrt{x^2-a^2}\right) + \text{arc séc}\,\frac{x}{a}.$$

$$\int\frac{1}{x}\sqrt{\frac{x+a}{a-x}}\,dx = \frac{b}{a}\log\frac{x}{a+\sqrt{a^2-x^2}} + \text{arc sin}\,\frac{x}{a}.$$

$$\int\frac{1}{x}\sqrt{\frac{ax+b}{mx+x}}\,dx = a\int\frac{dx}{\sqrt{(mx+n)(ax+b)}} + b\int\frac{dx}{x\sqrt{(mx+n)(ax+b)}}.$$

$$\int x^2\sqrt{\frac{a-x}{b-x}}\,dx = -b^2\int\sqrt{\frac{z+a-b}{z}}\,dz + 2b\int\sqrt{z^2+(a-b)z}\,dz - \int z\sqrt{z^2+(a-b)z}\,dz. \quad b-x=z.$$

$$\int x^2\sqrt{\frac{ax+b}{mx+n}}\,dx = \frac{1}{m^3\sqrt{m}}\left\{n^2\int\sqrt{\frac{az+k}{z}}\,dz - 2n\int\sqrt{az^2+kz}\,dz + \int z\sqrt{az^2+kz}\,dz.\right\} \cdot \quad mx+n=z. \quad k=bm-an.$$

$X\sqrt{ax^2+b}\,dx$

$$\int\sqrt{x^2+a^2}\,dx = \frac{x}{2}\sqrt{x^2+a^2} + \frac{a^2}{2}\log\left(x+\sqrt{x^2+a^2}\right).$$

$$\int\sqrt{x^2-a^2}\,dx = \frac{x}{2}\sqrt{x^2-a^2} - \frac{a^2}{2}\log\left(x+\sqrt{x^2-a^2}\right) = \frac{(x+\sqrt{x^2-a^2})^4-a^4}{8\,(x+\sqrt{x^2-a^2})^2} - \frac{a^2}{2}\log\left(x+\sqrt{x^2-a^2}\right).$$

$$\int\sqrt{a^2-x^2}\,dx = \frac{x}{2}\sqrt{a^2-x^2} + \frac{a^2}{2}\,\text{arc sin}\,\frac{x}{a}.$$

$$\int\sqrt{ax^2+b}\,dx = \frac{x}{2}\sqrt{ax^2+b} + \frac{b}{2}\int\frac{dx}{\sqrt{ax^2+b}} = \frac{x}{2}\sqrt{ax^2+b} + \frac{b}{2\sqrt{a}}\log\left(x\sqrt{a}+\sqrt{ax^2+b}\right).$$

$$\int\sqrt{b-ax^2}\,dx = \frac{x}{2}\sqrt{b-ax^2} + \frac{b}{2\sqrt{a}}\,\text{arc sin}\,x\sqrt{\frac{a}{b}}.$$

$$\int x\sqrt{ax^2+b}\,dx = \frac{1}{3a}\left(ax^2+b\right)^{\frac{3}{2}}.$$

$$\int x^2\sqrt{x^2+1}\,dx = \frac{x\,(2x^2+1)}{8}\sqrt{x^2+1} + \frac{1}{8}\log\left(\sqrt{x^2+1}-x\right).$$

$$\int x^2\sqrt{ax^2+b}\,dx = \frac{x\,(2ax^2+b)}{8a}\sqrt{ax^2+b} - \frac{b^2}{8a\sqrt{a}}\log\left(x\sqrt{a}+\sqrt{ax^2+b}\right)$$

$$= \frac{x\,(2ax^2+b)}{8a}\sqrt{ax^2+b} - \frac{b^2}{16a\sqrt{a}}\log\left(2ax^2+b+2x\sqrt{a}\sqrt{ax^2+b}\right).$$

$$\int x^2\sqrt{a^2-x^2}\,dx = \frac{x\,(2x^2-a^2)}{8}\sqrt{a^2-x^2} + \frac{a^4}{8}\,\text{arc sin}\,\frac{x}{a}.$$

$$\int x^2\sqrt{b-ax^2}\,dx = \frac{x\,(2ax^2-b)}{8a}\sqrt{b-ax^2} + \frac{b^2}{8a\sqrt{a}}\,\text{arc sin}\,x\sqrt{\frac{a}{b}}.$$

$$\int x^3\sqrt{ax^2+b}\,dx = \frac{1}{15a^2}\left(3a^2x^4+abx^2-2b^2\right)\sqrt{ax^2+b}.$$

$$\int \frac{dx}{(mx + n)\sqrt{ax^2 + b}}$$

$$\int \frac{dx}{\sqrt{x^2 - 1}} = \log\left(x + \sqrt{x^2 - 1}\right) = -2 \log\left(\sqrt{x + 1} - \sqrt{x - 1}\right).$$

$$\int \frac{dx}{\sqrt{1 - x^2}} = \text{arc sin } x.$$

$$\int \frac{dx}{\sqrt{x^2 + a}} = \log\left(x + \sqrt{x^2 + a}\right) = \frac{1}{2} \log \frac{x + \sqrt{x^2 + a}}{x - \sqrt{x^2 + a}} = \log \frac{x - \sqrt{a} + \sqrt{x^2 + a}}{x + \sqrt{a} - \sqrt{x^2 + a}}.$$

$$\int \frac{dx}{\sqrt{x^2 - a^2}} = \log\left(x + \sqrt{x^2 - a^2}\right) = \log\left\{\frac{x}{a} + \sqrt{\frac{x^2}{a^2} - 1}\right\} = -2 \log\left(\sqrt{x + a} - \sqrt{x - a}\right).$$

$$\int \frac{dx}{\sqrt{a^2 - x^2}} = \text{arc sin } \frac{x}{a} = -\text{arc sin } \frac{\sqrt{a^2 - x^2}}{a} = \frac{1}{2} \text{arc sin } \frac{2x\sqrt{a^2 - x^2}}{a^2} = -\text{arc cos } \frac{x}{a}.$$

$$= \text{arc tg } \frac{x}{\sqrt{a^2 - x^2}} = -\text{arc tg } \frac{\sqrt{a^2 - x^2}}{x} = 2 \text{ arc tg } \frac{a - \sqrt{a^2 - x^2}}{x} = 2 \text{ arc tg } \sqrt{\frac{a + x}{a - x}}.$$

$$\int \frac{dx}{\sqrt{1 - a^2x^2}} = \frac{1}{a} \text{arc sin } ax = -\frac{1}{a} \text{arc sin } \sqrt{1 - a^2x^2} = \frac{2}{a} \text{arc sin } \sqrt{\frac{ax + 1}{2}}.$$

$$\int \frac{dx}{\sqrt{ax^2 + b}} = \frac{1}{\sqrt{a}} \log\left(x\sqrt{a} + \sqrt{ax^2 + b}\right) = \frac{1}{2\sqrt{a}} \log \frac{\sqrt{ax^2 + b} + x\sqrt{a}}{\sqrt{ax^2 + b} - x\sqrt{a}} = -\frac{1}{\sqrt{a}} \log\left(\sqrt{ax^2 + b} - x\sqrt{a}\right).$$

$$\int \frac{dx}{\sqrt{2b^2 - a^2x^2}} = \frac{1}{2a} \text{arc sin } \frac{a^2x^2 - b^2}{b^2} = \frac{1}{a} \text{arc sin } \frac{ax}{b\sqrt{2}}.$$

$$\int \frac{dx}{\sqrt{b - ax^2}} = \frac{1}{\sqrt{a}} \text{arc sin } x\sqrt{\frac{a}{b}} = \frac{1}{\sqrt{a}} \text{arc cos } \sqrt{\frac{b - ax^2}{b}} = \frac{2}{a} \text{arc tg } \sqrt{\frac{b + ax}{b - ax}}.$$

$$\int \frac{dx}{x\sqrt{x^2 + a^2}} = \frac{1}{a} \log \frac{\sqrt{x^2 + a^2} - a}{x} = \frac{1}{a} \log \frac{x}{\sqrt{x^2 + a^2} + a}.$$

$$\int \frac{dx}{x\sqrt{x^2 - a^2}} = \frac{1}{a} \text{arc cos } \frac{a}{x} = \frac{1}{a} \text{arc tg } \frac{\sqrt{x^2 - a^2}}{a} = \frac{1}{a} \text{arc séc } \frac{x}{a}.$$

$$\int \frac{dx}{x\sqrt{1 - x^2}} = \log \frac{\sqrt{1 - x^2} - 1}{x} = -\log\left\{\frac{1}{x} + \sqrt{\frac{1}{x^2} - 1}\right\} = \log \frac{\sqrt{1 + x} - \sqrt{1 - x}}{\sqrt{1 + x} + \sqrt{1 - x}} = \log \text{tg } \frac{\alpha}{2}. \qquad x = \sin \alpha.$$

$$\int \frac{dx}{x\sqrt{a^2 - x^2}} = \frac{1}{a} \log \frac{x}{a + \sqrt{a^2 - x^2}}.$$

$$\int \frac{dx}{x\sqrt{2ax^2 - a^2}} = \frac{1}{a} \text{arc sin } \frac{\sqrt{2x^2 - a}}{x\sqrt{2}} = \frac{1}{2a} \text{arc sin } \frac{x^2 - a}{x^2}.$$

$$\int \frac{dx}{x\sqrt{ax^2 + b}} = \frac{1}{\sqrt{b}} \log \frac{\sqrt{ax^2 + b} - \sqrt{b}}{x} = \frac{1}{\sqrt{b}} \log \frac{x}{\sqrt{ax^2 + b} + \sqrt{b}} = \frac{1}{2\sqrt{b}} \log \frac{\sqrt{ax^2 + b} - \sqrt{b}}{\sqrt{ax^2 + b} + \sqrt{b}}.$$

$$\int \frac{dx}{x\sqrt{ax^2 - b}} = \frac{1}{\sqrt{b}} \text{arc tg } \frac{\sqrt{ax^2 - b}}{\sqrt{b}} = \frac{1}{\sqrt{b}} \text{arc sin } \frac{\sqrt{ax^2 - b}}{x\sqrt{a}}.$$

$$\int \frac{dx}{x\sqrt{b^2 - a^2x^2}} = \frac{1}{b} \log \frac{x}{\sqrt{b^2 - a^2x^2} + b} = \frac{1}{b} \log \frac{\sqrt{b^2 - a^2x^2} - b}{x} = \frac{1}{b} \log \frac{\sqrt{ax + b} - \sqrt{b - ax}}{\sqrt{ax + b} + \sqrt{b - ax}}.$$

$$\int \frac{dx}{(x + 1)\sqrt{x^2 + 1}} = \frac{1}{\sqrt{2}} \log \frac{x \quad 1 + \sqrt{2}\ \sqrt{x^2 + 1}}{x + 1}.$$

$$\int\frac{dx}{(x+a)\sqrt{x^2+1}}=\frac{1}{\sqrt{1+a^2}}\log\frac{x+a}{1-ax\sqrt{1+a^2}\sqrt{x^2+1}}=\frac{1}{\sqrt{1+a^2}}\log\frac{ax-1+\sqrt{1+a^2}\sqrt{x^2+1}}{x+a}.$$

$$\int\frac{dx}{(ax+b)\sqrt{x^2+1}}=\frac{1}{\sqrt{a^2+b^2}}\log\frac{ax+b}{a-bx+\sqrt{a^2+b^2}\sqrt{x^2+1}}=\frac{1}{\sqrt{a^2+b^2}}\log\frac{b+ax+\sqrt{a^2+b^2}-a\sqrt{x^2+1}}{b+ax-\sqrt{a^2+b^2}-a\sqrt{x^2+1}}.$$

$$=-2\int\frac{dz}{b+2az-bz^2}. \qquad 2zx=1-z^2.$$

$$\int\frac{dx}{(x+a)\sqrt{x^2-a^2}}=\frac{1}{a}\sqrt{\frac{x-a}{x+a}}.$$

$$\int\frac{dx}{(ax+b)\sqrt{a^2x^2-b^2}}=\frac{1}{ab}\sqrt{\frac{ax-b}{ax+b}}.$$

$$\int\frac{dx}{(x+a)\sqrt{x^2-1}}=\frac{1}{\sqrt{a^2-1}}\log\frac{1+ax+\sqrt{(a^2-1)(x^2-1)}}{x+a}. \qquad a^2>1.$$

$$=\frac{1}{\sqrt{1-a^2}}\text{arc cos}\frac{ax+1}{x+a}. \qquad a^2<1.$$

$$\int\frac{dx}{(x+b)\sqrt{x^2-a^2}}=-2\int\frac{dz}{(a+b)z^2+a-b}. \qquad \sqrt{x^2-a^2}=(x-a)z.$$

$$\int\frac{dx}{(r^2x+pq)\sqrt{r^2x^2-q^2}}=-\frac{1}{qr\sqrt{r^2-p^2}}\text{arc sin}\frac{prx+qr}{r^2x+pq}.$$

$$\int\frac{dx}{(x+a)\sqrt{1-x^2}}=-\frac{1}{\sqrt{1-a^2}}\log\frac{1+ax+\sqrt{(1-a^2)(1-x^2)}}{x+a}. \qquad a^2<1$$

$$=\frac{1}{\sqrt{a^2-1}}\text{arc sin}\frac{ax+1}{x+a}. \qquad a^2>1.$$

$$\int\frac{dx}{(x+a)\sqrt{a^2-x^2}}=-\frac{1}{a}\sqrt{\frac{a-x}{a+x}}.$$

$$\int\frac{dx}{(x-1)\sqrt{1-x^2}}=-\sqrt{\frac{1+x}{1-x}}.$$

$$\int\frac{dx}{(x+a)\sqrt{b^2-x^2}}=2\int\frac{dz}{(a+b)z^2+a-b}. \qquad \sqrt{b^2-x^2}=(b-x)z.$$

$$=\frac{1}{\sqrt{b^2-a^2}}\log\frac{\sqrt{(a+b)(b+x)}-\sqrt{(b-a)(b-x)}}{\sqrt{(a+b)(b+x)}+\sqrt{(b-a)(b-x)}}. \qquad a^2<b^2.$$

$$=\frac{2}{\sqrt{a^2-b^2}}\text{arc tg}\sqrt{\frac{a+b}{a-b}\,\frac{b+x}{b-x}}. \qquad a^2>b^2.$$

$$\int\frac{dx}{(x-a)\sqrt{1-x^2}}=2\int\frac{dz}{(1-a)z^2-a-1} \qquad x(z^2+1)=z^2-1.$$

$$=-2\int\frac{dz}{az^2+2z+a} \qquad x(z^2+1)=-2z.$$

$$=\frac{1}{\sqrt{1-a^2}}\log\frac{1-ax-\sqrt{1-a^2}\sqrt{1-x^2}}{x-a}=-\frac{1}{\sqrt{1-a^2}}\log\frac{1-ax+\sqrt{1-a^2}\sqrt{1-x^2}}{x-a} \qquad a^2<1.$$

$$=\frac{1}{\sqrt{a^2-1}}\text{arc sin}\frac{1-ax}{x-a}=\frac{2}{\sqrt{a^2-1}}\text{arc tg}\sqrt{\frac{a+1}{a-1}\cdot\frac{1-x}{1+x}}. \qquad a^2>1.$$

$$\int \frac{dx}{(ax+b)\sqrt{1-x^2}} = \frac{1}{\sqrt{a^2-b^2}} \log \frac{\sqrt{(a+b)(1+x)} - \sqrt{(a-b)(1-x)}}{\sqrt{(a+b)(1+x)} + \sqrt{(a-b)(1-x)}}. \qquad a > b.$$

$$= \frac{1}{\sqrt{b^2-a^2}} \text{arc sin} \frac{a+bx}{b+ax}. \qquad a < b.$$

$$\int \frac{dx}{(b-ax)\sqrt{b^2-a^2x^2}} = \frac{1}{ab}\sqrt{\frac{b+ax}{b-ax}}.$$

$$\int \frac{dx}{(mx+n)\sqrt{b^2-a^2x^2}} = \frac{1}{\sqrt{b^2m^2-a^2n^2}} \log \frac{\sqrt{(an+bm)(b+ax)} - \sqrt{(bm-an)(b-ax)}}{\sqrt{(an+bm)(b+ax)} + \sqrt{(bm-an)(b-ax)}}. \qquad b^2m^2 - a^2n^2 > 0.$$

$$= \frac{2}{\sqrt{a^2n^2-b^2m^2}} \text{arc tg} \sqrt{\frac{a+b}{a-b}\cdot\frac{b+x}{b-x}}. \qquad a^2n^2 - b^2m^2 > 0.$$

$$\int \frac{dx}{(r^2x+pq)\sqrt{q^2-r^2x^2}} = \frac{1}{qr\sqrt{p^2-r^2}} \text{arc sin} \frac{prx+qr}{r^2x+pq}.$$

$$\int \frac{dx}{(mx+n)\sqrt{ax^2+b}} = \frac{1}{\sqrt{bm^2+an^2}} \log \frac{bm-anx-\sqrt{bm^2+an^2}\sqrt{ax^2+b}}{mx+n}. \qquad bm^2+an^2 > 0.$$

$$= -\frac{1}{\sqrt{bm^2+an^2}} \log \frac{bm-anx+\sqrt{bm^2+an^2}\sqrt{ax^2+b}}{mx+n}. \qquad bm^2+an^2 > 0.$$

$$= -\frac{1}{\sqrt{-(bm^2+an^2)}} \text{arc sin} \frac{bm-anx}{(mx+n)\sqrt{-ab}}. \qquad bm^2+an^2 < 0.$$

$\int \frac{dx}{X^2\sqrt{ax^2+b}}$

$$\int \frac{dx}{x^2\sqrt{x^2+1}} = -\frac{\sqrt{x^2+1}}{x}.$$

$$\int \frac{dx}{(x^2+a^2)\sqrt{x^2+1}} = \frac{1}{a\sqrt{a^2-1}} \log \frac{a\sqrt{x^2+1}+x\sqrt{a^2-1}}{\sqrt{x^2+a^2}}. \qquad a^2 > 1.$$

$$= \frac{1}{a\sqrt{1-a^2}} \text{arc tg} \frac{x\sqrt{1-a^2}}{a\sqrt{1+x^2}} \qquad a^2 < 1.$$

$$\int \frac{dx}{(1-x^2)\sqrt{x^2+1}} = \frac{1}{\sqrt{2}} \log \frac{x\sqrt{2}+\sqrt{x^2+1}}{\sqrt{1-x^2}} = \frac{1}{2\sqrt{2}} \log \frac{\sqrt{x^2+1}+x\sqrt{2}}{\sqrt{x^2+1}-x\sqrt{2}}.$$

$$\int \frac{dx}{(x^2-a^2)\sqrt{x^2+1}} = \frac{1}{a\sqrt{a^2+1}} \log \frac{a\sqrt{x^2+1}-x\sqrt{a^2+1}}{\sqrt{a^2-x^2}} = \frac{1}{2a\sqrt{a^2+1}} \log \frac{x\sqrt{a^2+1}-a\sqrt{x^2+1}}{x\sqrt{a^2+1}+a\sqrt{x^2+1}}.$$

$$\int \frac{dx}{(1+2x^2)\sqrt{x^2+1}} = \text{arc tg} \frac{x}{\sqrt{x^2+1}}.$$

$$\int \frac{dx}{(x+a)^2\sqrt{x^2+1}} = -\frac{\sqrt{x^2+1}}{(1+a^2)(x+a)} + \frac{a}{(a^2+1)^{\frac{3}{2}}} \log \frac{x+a}{1-ax+\sqrt{(a^2+1)(x^2+1)}}.$$

$$\int \frac{dx}{(x^2+b^2)\sqrt{x^2+a^2}} = \frac{1}{2b\sqrt{b^2-a^2}} \log \frac{x\sqrt{b^2-a^2}+b\sqrt{x^2+a^2}}{x\sqrt{b^2-a^2}-b\sqrt{x^2+a^2}}. \qquad b^2 > a^2.$$

$$= \frac{1}{b\sqrt{a^2-b^2}} \text{arc tg} \frac{x\sqrt{a^2-b^2}}{b\sqrt{x^2+a^2}}. \qquad b^2 < a^2.$$

$$\int \frac{dx}{(a^2-x^2)\sqrt{x^2+a^2}} = \frac{1}{a^2\sqrt{2}} \log \frac{x\sqrt{2}+\sqrt{x^2+a^2}}{\sqrt{a^2-x^2}}.$$

$$\int \frac{dx}{x^2 \sqrt{ax^2+b}} = -\frac{\sqrt{ax^2+b}}{bx}.$$

$$\int \frac{dx}{(1+x^2)\sqrt{ax^2+b}} = \frac{1}{\sqrt{b-a}} \operatorname{arc\,sin} \frac{x\sqrt{b-a}}{\sqrt{b}\sqrt{ax^2+b}}.$$

$$\int \frac{dx}{(x^2+4)\sqrt{9x^2+1}} = \frac{1}{4\sqrt{35}} \log \frac{2\sqrt{9x^2+1}+x\sqrt{35}}{2\sqrt{9x^2+1}-x\sqrt{35}}.$$

$$\int \frac{dx}{(a^2-x^2)\sqrt{a^2-b^2-x^2}} = \frac{1}{ab} \operatorname{arc\,tg} \frac{bx}{a\sqrt{a^2-b^2-x^2}}.$$

$$\int \frac{dx}{(x^2-b)\sqrt{ax^2+b}} = -\frac{1}{2b\sqrt{a+1}} \log \frac{x\sqrt{a+1}+\sqrt{ax^2+b}}{x\sqrt{a+1}-\sqrt{ax^2+b}}.$$

$$\int \frac{dx}{(n-mx^2)\sqrt{ax^2+b}} = \frac{1}{2\sqrt{n(bm+an)}} \log \frac{x\sqrt{bm+an}+\sqrt{n(ax^2+b)}}{x\sqrt{bm+an}-\sqrt{n(ax^2+b)}}. \qquad bm+an>0.$$

$$= \frac{1}{\sqrt{-n(bm+an)}} \operatorname{arc\,tg} \frac{x\sqrt{-(bm+an)}}{\sqrt{n(ax^2+b)}}. \qquad bm+an<0.$$

$$\int \frac{dx}{(mx^2+n)\sqrt{ax^2+b}} = \frac{1}{2\sqrt{n(an-bm)}} \log \frac{\sqrt{n(ax^2+b)}+x\sqrt{an-bm}}{\sqrt{n(ax^2+b)}-x\sqrt{an-bm}}. \qquad an-bm>0.$$

$$= \frac{1}{2\sqrt{n(an-bm)}} \log \frac{x\sqrt{an-bm}+\sqrt{n(ax^2+b)}}{x\sqrt{an-bm}-\sqrt{n(ax^2+b)}}. \qquad an-bm>0.$$

$$= \frac{1}{\sqrt{n(bm-an)}} \operatorname{arc\,tg} \frac{x\sqrt{bm-an}}{\sqrt{n(ax^2+b)}}. \qquad bm-an>0.$$

$$= \frac{x}{n\sqrt{ax^2+b}}. \qquad an=bm.$$

$$\int \frac{dx}{x(mx+n)\sqrt{ax^2+b}} = \frac{1}{n}\int \frac{dx}{x\sqrt{ax^2+b}} - \frac{m}{n}\int \frac{dx}{(mx+n)\sqrt{ax^2+b}}.$$

$$\int \frac{dx}{(m^2x^2-an^2x^2-n^2b)\sqrt{ax^2+b}} = -\frac{1}{2mnb} \log \frac{mx+n\sqrt{ax^2+b}}{mx-n\sqrt{ax^2+b}}.$$

$\int \frac{dx}{X^2\sqrt{ax^2-b}}$

$$\int \frac{dx}{x^2\sqrt{x^2-1}} = \frac{\sqrt{x^2-1}}{x}.$$

$$\int \frac{dx}{(x^2+1)\sqrt{x^2-1}} = \frac{1}{\sqrt{2}} \log \frac{\sqrt{x^2+1}}{x\sqrt{2}-\sqrt{x^2-1}} = \frac{1}{\sqrt{2}} \log \frac{x\sqrt{2}+\sqrt{x^2-1}}{\sqrt{x^2+1}}.$$

$$\int \frac{dx}{(x^2+4)\sqrt{x^2-1}} = \frac{1}{4\sqrt{5}} \log \frac{x\sqrt{5}+2\sqrt{x^2-1}}{x\sqrt{5}-2\sqrt{x^2-1}}.$$

$$\int \frac{dx}{(x^2+a^2)\sqrt{x^2-1}} = -\frac{1}{a\sqrt{a^2+1}} \log \frac{x\sqrt{a^2+1}-a\sqrt{x^2-1}}{\sqrt{x^2+a^2}}.$$

$$\int \frac{dx}{(x^2-a^2)\sqrt{x^2-1}} = \frac{1}{a\sqrt{a^2-1}} \log \frac{x\sqrt{a^2-1}-a\sqrt{x^2-1}}{\sqrt{x^2-a^2}}. \qquad a^2>1.$$

$$= \frac{1}{a\sqrt{1-a^2}} \operatorname{arc\,sin} \frac{a\sqrt{x^2-1}}{\sqrt{x^2-a^2}}. \qquad a^2<1.$$

$$\int \frac{dx}{(x^2+b^2)\sqrt{x^2-a^2}} = \frac{1}{2b\sqrt{a^2+b^2}} \log \frac{x\sqrt{a^2+b^2}+b\sqrt{x^2-a^2}}{x\sqrt{a^2+b^2}-b\sqrt{x^2-a^2}}.$$

$$\int \frac{dx}{x^2\sqrt{ax^2-b}} = \frac{\sqrt{ax^2-b}}{bx}.$$

$$\int \frac{dx}{(mx^2+n)\sqrt{ax^2-b}} = \frac{1}{2\sqrt{n(bm+an)}} \log \frac{x\sqrt{bm+an}+\sqrt{n(ax^2-b)}}{x\sqrt{bm+an}-\sqrt{n(ax^2-b)}}. \qquad bm+an>0.$$

$$= \frac{1}{\sqrt{-n(bm+an)}} \operatorname{arc\,tg} \frac{x\sqrt{-(bm+an)}}{\sqrt{n(ax^2-b)}}. \qquad bm+an<0.$$

$$\int \frac{dx}{(n-mx^2)\sqrt{ax^2-b}} = \frac{1}{2\sqrt{n(an-bm)}} \log \frac{x\sqrt{an-bm}+\sqrt{n(ax^2-b)}}{x\sqrt{an-bm}-\sqrt{n(ax^2-b)}}. \qquad an-bm>0.$$

$$= \frac{1}{\sqrt{n(bm-an)}} \operatorname{arc\,tg} \frac{x\sqrt{bm-an}}{\sqrt{n(ax^2-b)}}. \qquad an-bm<0.$$

$\int \frac{dx}{X^2\sqrt{b-ax^2}}$

$$\int \frac{dx}{x^2\sqrt{1-x^2}} = -\frac{\sqrt{1-x^2}}{x}.$$

$$\int \frac{dx}{(x^2+1)\sqrt{1-x^2}} = \frac{1}{\sqrt{2}} \operatorname{arc\,tg} \frac{x\sqrt{2}}{\sqrt{1-x^2}} = \frac{1}{\sqrt{2}} \arccos \sqrt{\frac{1-x^2}{1+x^2}}.$$

$$\int \frac{dx}{(x^2+4)\sqrt{1-x^2}} = \frac{1}{2\sqrt{5}} \operatorname{arc\,cotg} \frac{x\sqrt{5}}{2\sqrt{1-x^2}}$$

$$\int \frac{dx}{(x^2-1)\sqrt{1-x^2}} = -\frac{x}{\sqrt{1-x^2}}.$$

$$\int \frac{dx}{(x^2+a^2)\sqrt{1-x^2}} = \frac{1}{a\sqrt{1+a^2}} \operatorname{arc\,tg} \frac{x\sqrt{1+a^2}}{a\sqrt{1-x^2}}.$$

$$\int \frac{dx}{(x^2-a^2)\sqrt{1-x^2}} = \frac{1}{a\sqrt{1-a^2}} \log \frac{a\sqrt{1-x^2}-x\sqrt{1-a^2}}{\sqrt{x^2-a^2}} = \frac{1}{a\sqrt{1-a^2}} \log \frac{a\sqrt{1-x^2}-x\sqrt{1-a^2}}{\sqrt{a^2-x^2}}. \quad a^2<1.$$

$$= -\frac{1}{a\sqrt{a^2-1}} \operatorname{arc\,tg} \frac{x\sqrt{a^2-1}}{a\sqrt{1-x^2}}. \qquad a^2>1.$$

$$\int \frac{dx}{(1-ax^2)\sqrt{1-x^2}} = \frac{1}{\sqrt{a-1}} \log \frac{x\sqrt{a-1}+\sqrt{1-x^2}}{\sqrt{1-ax^2}}. \qquad a>1.$$

$$= \frac{1}{\sqrt{1-a}} \arccos \sqrt{\frac{1-x^2}{1-ax^2}}. \qquad a<1.$$

$$\int \frac{dx}{(1+a^2x^2)\sqrt{1-x^2}} = \frac{1}{\sqrt{a^2+1}} \operatorname{arc\,tg} \frac{x\sqrt{a^2+1}}{\sqrt{1-x^2}} = \frac{1}{\sqrt{a^2+1}} \arccos \sqrt{\frac{1-x^2}{1+a^2x^2}}$$

$$\int \frac{dx}{(a+bx^2)\sqrt{1-x^2}} = \frac{1}{2na} \log \frac{\sqrt{1-x^2}+nx}{\sqrt{1-x^2}-nx}. \qquad n^2 = -\frac{a+b}{a}.$$

$$= \frac{1}{ma} \operatorname{arc\,tg} \frac{mx}{\sqrt{1-x^2}}. \qquad m^2 = \frac{a+b}{a}.$$

$$\int \frac{dx}{(x+a)^2\sqrt{1-x^2}} = \frac{\sqrt{1-x^2}}{(a^2-1)(x+a)} + \frac{a}{(a^2-1)^{\frac{3}{2}}} \arcsin \frac{1+ax}{a+x} = \frac{\sqrt{1-x^2}}{(a^2-1)(x+a)} - \frac{2a}{(a^2-1)^{\frac{3}{2}}} \operatorname{arc\,tg} \sqrt{\frac{a-1}{a+1}\cdot\frac{1-x}{1+x}}.$$

$$\int\frac{dx}{x^2\sqrt{a^2-x^2}}=-\frac{\sqrt{a^2-x^2}}{a^2x}.$$

$$\int\frac{dx}{(x^2+a^2)\sqrt{a^2-x^2}}=\frac{1}{a^2\sqrt{2}}\operatorname{arc\,tg}\frac{x\sqrt{2}}{\sqrt{a^2-x^2}}.$$

$$\int\frac{dx}{(x^2+b^2)\sqrt{a^2-x^2}}=\frac{1}{b\sqrt{a^2+b^2}}\operatorname{arc\,tg}\frac{x\sqrt{a^2+b^2}}{b\sqrt{a^2-x^2}}.$$

$$\int\frac{dx}{(mx^2+n)\sqrt{b-ax^2}}=\frac{1}{2\sqrt{-n(bm+an)}}\log\frac{x\sqrt{-(bm+an)}+\sqrt{n(b-ax^2)}}{x\sqrt{-(bm+an)}-\sqrt{n(b-ax^2)}}. \qquad bm+an<0.$$

$$=\frac{1}{\sqrt{n(bm+an)}}\operatorname{arc\,tg}\frac{x\sqrt{bm+an}}{\sqrt{n(b-ax^2)}}. \qquad bm+an>0.$$

$$\int\frac{dx}{(n-mx^2)\sqrt{b-ax^2}}=\frac{1}{2\sqrt{n(bm-an)}}\log\frac{x\sqrt{bm-an}+\sqrt{n(b-ax^2)}}{x\sqrt{bm-an}-\sqrt{n(b-ax^2)}}. \qquad bm-an>0.$$

$$=\frac{1}{\sqrt{n(an-bm)}}\operatorname{arc\,tg}\frac{x\sqrt{an-bm}}{\sqrt{n(b-ax^2)}}. \qquad an-bm>0.$$

$\int\frac{dx}{X\sqrt{ax^2+b}}$

$$\int\frac{dx}{x^3\sqrt{x^2+1}}=-\frac{\sqrt{x^2+1}}{2x^2}-\frac{1}{2}\log\frac{x}{1+\sqrt{x^2+1}}$$

$$\int\frac{dx}{x^4\sqrt{x^2+1}}=\frac{2x^2-1}{3x^3}\sqrt{x^2+1}.$$

$$\int\frac{dx}{x^5\sqrt{1+x^2}}=\frac{3x^2-2}{8x^4}\sqrt{1+x^2}-\frac{3}{8}\log\frac{1+\sqrt{1+x^2}}{x}.$$

$$\int\frac{dx}{x^3\sqrt{ax^2+b}}=-\frac{\sqrt{ax^2+b}}{2bx^2}-\frac{a}{2b}\int\frac{dx}{x\sqrt{ax^2+b}}.$$

$$\int\frac{dx}{x^2(mx+n)\sqrt{ax^2+b}}=\frac{1}{n}\int\frac{dx}{x^2\sqrt{\Phi}}-\frac{m}{n^2}\int\frac{dx}{x\sqrt{\Phi}}+\frac{m^2}{n^2}\int\frac{dx}{(mx+n)\sqrt{\Phi}}.$$

$$\int\frac{dx}{x^4\sqrt{ax^2+b}}=\frac{2ax^2-b}{3b^2x^3}\sqrt{ax^2+b}=-\frac{1}{b^2}\int(z^2-a)\,dz. \qquad ax^2+b=x^2z^2.$$

$$\int\frac{dx}{(n^4-m^4x^4)\sqrt{ax^2+b}}=\frac{1}{4n^3\sqrt{bm^2+an^2}}\log\frac{x\sqrt{bm^2+an^2}+n\sqrt{\Phi}}{x\sqrt{bm^2+an^2}-n\sqrt{\Phi}}+\frac{1}{4n^3\sqrt{an^2-bm^2}}\log\frac{x\sqrt{an^2-bm^2}+n\sqrt{\Phi}}{x\sqrt{an^2-bm^2}-n\sqrt{\Phi}}. \qquad an^2>bm^2.$$

$$=\frac{1}{4n^3\sqrt{bm^2+an^2}}\log\frac{x\sqrt{bm^2+an^2}+n\sqrt{\Phi}}{x\sqrt{bm^2+an^2}-n\sqrt{\Phi}}+\frac{1}{2n^3\sqrt{bm^2-an^2}}\operatorname{arc\,tg}\frac{x\sqrt{bm^2-an^2}}{n\sqrt{\Phi}}. \qquad an^2<bm^2.$$

$$\int\frac{dx}{(1+nx^2)^2\sqrt{a+bx^2}}=\frac{nx\sqrt{a+bx^2}}{2(an-b)(1+nx^2)}+\frac{an-2b}{2(an-b)}\int\frac{dx}{(1+nx^2)\sqrt{a+bx^2}}.$$

$$\int\frac{dx}{x^p(mx+n)\sqrt{ax^2+b}}=\frac{1}{n}\int\frac{dx}{x^p\sqrt{\Phi}}-\frac{m}{n^2}\int\frac{dx}{x^{p-1}\sqrt{\Phi}}+\frac{m^2}{n^3}\int\frac{dx}{x^{p-2}\sqrt{\Phi}}-\dots\pm\frac{m^{p-1}}{n^p}\int\frac{dx}{x\sqrt{\Phi}}\mp\frac{m^p}{n^p}\int\frac{dx}{(mx+n)\sqrt{\Phi}}.$$

$$\int\frac{dx}{x^3\sqrt{x^2-1}}=\frac{\sqrt{x^2-1}}{2x^2}+\frac{1}{2}\operatorname{arc\,séc}x.$$

$$\int\frac{dx}{x^6\sqrt{x^2-1}}=\frac{8x^4+4x^2+3}{15x^5}\sqrt{x^2-1}.$$

$$\int\frac{dx}{x^4\sqrt{x^2-a^2}}=\frac{2x^2+a^2}{3a^4x^2}\sqrt{x^2-a^2}\,.$$

$$\int\frac{dx}{x^4\sqrt{7x^2-9}}=\frac{1}{9}\left\{\frac{1}{3x^3}+\frac{14}{27x}\right\}\sqrt{7x^2-9}\,.$$

$$\int\frac{dx}{(n^4-m^4x^4)\sqrt{ax^2-b)}}=\frac{1}{4n^3\sqrt{an^2-bm^2}}\log\frac{x\sqrt{an^2-bm^2}+n\sqrt{\Phi}}{x\sqrt{an^2-bm^2}-n\sqrt{\Phi}}+\frac{1}{4n^3\sqrt{bm^2+an^2}}\log\frac{x\sqrt{bm^2+an^2}+n\sqrt{\Phi}}{x\sqrt{bm^2+an^2}-n\sqrt{\Phi}}\,.$$

$$=\frac{1}{4n^3\sqrt{bm^2+an^2}}\log\frac{x\sqrt{bm^2+an^2}+n\sqrt{\Phi}}{x\sqrt{bm^2+an^2}-n\sqrt{\Phi}}+\frac{1}{2n^3\sqrt{bm^2-an^2}}\operatorname{arc\,tg}\frac{x\sqrt{bm^2-an^2}}{n\sqrt{\Phi}}\,.$$

$$\int\frac{dx}{(mx+n)^2\sqrt{1-x^2}}=-\frac{m\sqrt{1-x^2}}{(m^2-n^2)(mx+n)}-\frac{n}{m^2-n^2}\int\frac{dx}{(mx+n)\sqrt{1-x^2}}\,.$$

$$\int\frac{dx}{x^3\sqrt{1-x^2}}=-\frac{\sqrt{1-x^2}}{2x^2}+\frac{1}{2}\int\frac{dx}{x\sqrt{1-x^2}}\,.$$

$$\int\frac{dx}{x^4\sqrt{1-x^2}}=-\frac{\sqrt{1-x^2}}{3x^3}+\frac{2}{3}\int\frac{dx}{x^2\sqrt{1-x^2}}\,.$$

$$\int\frac{dx}{(x^2+a)^2\sqrt{1-x^2}}=\frac{x\sqrt{1-x^2}}{(a^2+a)(x^2+a)}+\frac{2a+1}{2(a^2+a)^{\frac{3}{2}}}\operatorname{arc\,tg}\frac{x\sqrt{1+a}}{\sqrt{a(1-x^2)}}\,.$$

$$\int\frac{dx}{x^7\sqrt{1-x^2}}=-\left\{\frac{1}{x^4}+\frac{5}{4x^2}+\frac{15}{8}\right\}\frac{\sqrt{1-x^2}}{6x^2}+\frac{5}{16}\log\frac{1-\sqrt{1-x^2}}{x}\,.$$

$$\int\frac{dx}{x^3\sqrt{a^2-x^2}}=-\frac{\sqrt{a^2-x^2}}{2a^2x^2}+\frac{1}{2a^3}\log\frac{a-\sqrt{a^2-x^2}}{x}\,.$$

$$\int\frac{dx}{(n^4-m^4x^4)\sqrt{b-ax^2}}=\frac{1}{4n^3\sqrt{bm^2-an^2}}\log\frac{x\sqrt{bm^2-an^2}+n\sqrt{\Phi}}{x\sqrt{bm^2-an^2}-n\sqrt{\Phi}}+\frac{1}{2n^3\sqrt{bm^2+an^2}}\operatorname{arc\,tg}\frac{x\sqrt{bm^2+an^2}}{n\sqrt{\Phi}}\,.$$

$$=\frac{1}{2n^3\sqrt{an^2-bm^2}}\operatorname{arc\,tg}\frac{x\sqrt{an^2-bm^2}}{n\sqrt{\Phi}}+\frac{1}{2n^3\sqrt{bm^2+an^2}}\operatorname{arc\,tg}\frac{x\sqrt{bm^2+an^2}}{n\sqrt{\Phi}}\,.$$

$\int\frac{X}{\sqrt{ax^2+b}}\,dx$

$$\int\frac{x}{\sqrt{ax^2+b}}\,dx=\frac{1}{a}\sqrt{ax^2+b}\,.$$

$$\int\frac{x+1}{\sqrt{1-x^2}}\,dx=-\sqrt{1-x^2}+\operatorname{arc\,sin}x.$$

$$\int\frac{x^2}{\sqrt{x^2+a^2}}\,dx=\frac{x}{2}\sqrt{x^2+a^2}-\frac{a^2}{2}\log\left(x+\sqrt{x^2+a^2}\right)\,.$$

$$\int\frac{x^2}{\sqrt{x^2-a^2}}\,dx=\frac{x}{2}\sqrt{x^2-a^2}+\frac{a^2}{2}\log\left(x+\sqrt{x^2-a^2}\right)\,.$$

$$\int\frac{x^2}{\sqrt{a^2-x^2}}\,dx=-\frac{x}{2}\sqrt{a^2-x^2}+\frac{a^2}{2}\operatorname{arc\,sin}\frac{x}{a}\,.$$

$$\int\frac{x^2}{\sqrt{ax^2+b}}\,dx=\frac{x\sqrt{ax^2+b}}{2a}-\frac{b}{2a}\int\frac{dx}{\sqrt{ax^2+b}}=\frac{x\sqrt{ax^2+b}}{2a}-\frac{b}{2a\sqrt{a}}\log\left((x\sqrt{a}+\sqrt{ax^2+b}\right)\,.$$

$$\int\frac{x^2}{\sqrt{b-ax^2}}\,dx=-\frac{x\sqrt{b-ax^2}}{2a}-\frac{b}{2a\sqrt{a}}\operatorname{arc\,sin}\sqrt{\frac{b-ax^2}{b}}=-\frac{x\sqrt{b-ax^2}}{2a}+\frac{b}{2a\sqrt{a}}\operatorname{arc\,sin}x\sqrt{\frac{a}{b}}\,.$$

$$\int\frac{2x^2-1}{\sqrt{1-x^2}}\,dx = \frac{1}{2}\left(x-\sqrt{1-x^2}\right)^2.$$

$$\int\frac{2x^2-a^2}{\sqrt{x^2-a^2}}\,dx = x\sqrt{x^2-a^2}.$$

$$\int\frac{(x+a)^2}{\sqrt{a^2-x^2}}\,dx = -\left(\frac{x}{2}+2a\right)\sqrt{a^2-x^2}+\frac{3a^2}{2}\ \text{arc sin}\,\frac{x}{a}.$$

$$\int\frac{x^3}{\sqrt{1-x^2}}\,dx = -\frac{x^2}{3}\sqrt{1-x^2}+\frac{2}{3}\int\frac{x}{\sqrt{1-x^2}}\,dx = -\frac{x^2+2}{3}\sqrt{1-x^2}$$

$$\int\frac{x^3}{\sqrt{a^2-x^2}}\,dx = -\frac{x^2+2a^2}{3}\sqrt{a^2-x^2}.$$

$$\int\frac{x^3}{\sqrt{ax^2+b}}\,dx = \frac{ax^2-2b}{3a^2}\sqrt{ax^2+b}.$$

$$\int\frac{x^4}{\sqrt{a^2-x^2}}\,dx = -\frac{3a^2+2x^2}{8}\,x\sqrt{a^2-x^2}+\frac{3a^4}{8}\ \text{arc sin}\,\frac{x}{a}.$$

$$\int\frac{x^5}{\sqrt{a^2-x^2}}\,dx = -\frac{1}{15}\left(3x^4+4a^2x^2+8a^4\right)\sqrt{a^2-x^2}.$$

$$\int\frac{x^5}{\sqrt{ax^2+b}}\,dx = \left\{\frac{\Phi^2}{5}-\frac{2b\Phi}{3}+b^2\right\}\frac{\sqrt{\Phi}}{a^3}.$$

$$\int\frac{x^6}{\sqrt{1+x^2}}\,dx = \left\{x^5-\frac{5x^3}{4}+\frac{15x}{8}\right\}\frac{\sqrt{1+x^2}}{6}.$$

$$\int\frac{x^6}{\sqrt{1-x^2}}\,dx = -\left\{x^5+\frac{5x^3}{4}+\frac{15x}{8}\right\}\frac{\sqrt{1-x^2}}{6}+\frac{5}{16}\ \text{arc sin}\,x.$$

$\int\frac{\sqrt{ax^2+b}}{X}\,dx$

$$\int\frac{\sqrt{x^2+a^2}}{x}\,dx = \sqrt{x^2+a^2}+a\log\frac{x}{a+\sqrt{x^2+a^2}}.$$

$$\int\frac{\sqrt{x^2-1}}{x}\,dx = \int\frac{x}{\sqrt{x^2-1}}\,dx-\int\frac{dx}{x\sqrt{x^2-1}} = \sqrt{x^2-1}+\text{arc sin}\,\frac{1}{x} = \sqrt{x^2-1}-\text{arc tg}\sqrt{x^2-1} = \sqrt{x^2-1}-\text{arc séc}\,x.$$

$$\int\frac{\sqrt{x^2-a^2}}{x}\,dx = \sqrt{x^2-a^2}+a\ \text{arc sin}\,\frac{a}{x}.$$

$$\int\frac{\sqrt{ax^2-1}}{x}\,dx = \sqrt{ax^2-1}-\text{arc tg}\sqrt{ax^2-1}.$$

$$\int\frac{\sqrt{ax^2+b}}{x}\,dx = \sqrt{ax^2+b}+b\int\frac{dx}{x\sqrt{ax^2+b}}.$$

$$\int\frac{\sqrt{a^2-x^2}}{x^2}\,dx = -\frac{\sqrt{a^2-x^2}}{x}-\text{arc sin}\,\frac{x}{a}.$$

$$\int\frac{\sqrt{ax^2+b}}{x^2}\,dx = -\frac{\sqrt{ax^2+b}}{x}+a\int\frac{dx}{\sqrt{ax^2+b}}.$$

$$\int\frac{\sqrt{x^2+1}}{x^2+2}\,dx = \frac{1}{2\sqrt{2}}\log\frac{\sqrt{2(1+x^2)}-x}{\sqrt{2(1+x^2)}+x}+\log\left(x+\sqrt{x^2+1}\right).$$

$$\int \frac{\sqrt{x^2+1}}{2x^2+1}\,dx = \frac{1}{4}\log\frac{x+\sqrt{x^2+1}}{x-\sqrt{x^2+1}} + \frac{1}{2}\,\text{arc tg}\,\frac{x}{\sqrt{x^2+1}}.$$

$$\int \frac{\sqrt{x^2+1}}{1-x^2}\,dx = \frac{1}{2\sqrt{2}}\log\frac{\sqrt{\Phi}+x\sqrt{2}}{\sqrt{\Phi}-x\sqrt{2}} + \frac{1}{2}\log\frac{(\sqrt{\Phi}-x)(\sqrt{\Phi}+x\sqrt{2})}{(\sqrt{\Phi}+x)(\sqrt{\Phi}-x\sqrt{2})}.$$

$$\int \frac{\sqrt{1-x^2}}{1+x^2}\,dx = \sqrt{2}\,\text{arc tg}\,\frac{x\sqrt{2}}{\sqrt{1-x^2}} - \text{arc sin}\,x.$$

$$\int \frac{\sqrt{x^2-4}}{x^2-3}\,dx = \log(x+\sqrt{x^2-4}) - \frac{1}{\sqrt{3}}\,\text{arc tg}\,\frac{\sqrt{3(x^2-4)}}{x}.$$

$$\int \frac{\sqrt{ax^2+b}}{mx^2+n}\,dx = \frac{a}{m}\int\frac{dx}{\sqrt{ax^2+b}} + \frac{bm-an}{m}\int\frac{dx}{(mx^2+n)\sqrt{ax^2+b}}.$$

$$\int \frac{\sqrt{ax^2+b}}{x^3}\,dx = -\frac{\sqrt{ax^2+b}}{2x^2} + \frac{a}{2}\int\frac{dx}{x\sqrt{ax^2+b}}.$$

$$\int \frac{\sqrt{ax^2+b}}{x^4}\,dx = -\frac{(ax^2+b)^{\frac{3}{2}}}{3bx^3}.$$

$\int \frac{Y}{X\sqrt{ax^2+b}}\,dx$

$$\int \frac{x}{(x-1)\sqrt{x^2-1}}\,dx = -\sqrt{\frac{x+1}{x-1}} + \log(x+\sqrt{x^2-1}).$$

$$\int \frac{x}{(mx+n)\sqrt{ax^2+b}}\,dx = \frac{1}{m}\int\frac{dx}{\sqrt{ax^2+b}} - \frac{n}{m}\int\frac{dx}{(mx+n)\sqrt{ax^2+b}}.$$

$$= \frac{1}{\sqrt{bm^2+an^2}}\log\frac{bm-anx-\sqrt{bm^2+an^2}\sqrt{ax^2+b}}{mx+n}. \qquad bm^2+an^2>0.$$

$$= \frac{1}{\sqrt{-(bm^2+an^2)}}\,\text{arc tg}\,\frac{bm-anx}{\sqrt{-(bm^2+an^2)}\sqrt{ax^2+b}} \qquad bm^2+an^2<0.$$

$$\int \frac{x}{(x^2+3)\sqrt{5+2x^2}}\,dx = \text{arc tg}\sqrt{5+2x^2}.$$

$$\int \frac{x}{(x^2+a^2)\sqrt{x^2-1}}\,dx = \frac{1}{\sqrt{a^2+1}}\,\text{arc tg}\,\frac{\sqrt{x^2-1}}{\sqrt{a^2+1}}.$$

$$\int \frac{x}{(x^2+a^2)\sqrt{1-x^2}}\,dx = \frac{1}{2\sqrt{a^2+1}}\log\frac{\sqrt{a^2+1}-\sqrt{1-x^2}}{\sqrt{a^2+1}+\sqrt{1-x^2}}$$

$$\int \frac{x}{(x^2-b^2)\sqrt{a^2-x^2}}\,dx = \text{arc sin}\sqrt{\frac{x^2-b^2}{a^2-b^2}}.$$

$$\int \frac{x}{(mx^2+n)\sqrt{ax^2+b}}\,dx = \frac{1}{\sqrt{bm^2-amn}}\log\frac{m\sqrt{ax^2+b}-\sqrt{bm^2-amn}}{\sqrt{mx^2+n}}. \qquad bm-an>0.$$

$$= \frac{1}{\sqrt{amn-bm^2}}\,\text{arc tg}\,\frac{m\sqrt{ax^2+b}}{\sqrt{amn-bm^2}}. \qquad bm-an<0.$$

$$\int \frac{l+mx}{\{(x-\alpha)^2+\beta^2\}\sqrt{ax^2+2bx+c}}\,dx = \int\frac{2\sqrt{a}\,\{m(az^2-c)+2l(b-az)\}}{\{az^2-c-2\alpha(b-az)\}^2+4\beta^2(b-az)^2}\,dz. \qquad x=\frac{az^2-c}{2(b-az)}.$$

$$= \int\frac{2\sqrt{c}\,\{l(a-cz^2)+2m(cz-b)\}}{\{\alpha(a-cz^2)-2(cz-b)\}^2+\beta^2(a-cz^2)^2}\,dz \qquad x=\frac{2(cz-b)}{a-cz^2}.$$

$$\int \frac{x^2}{(mx+n)\sqrt{ax^2+b}}\,dx = \frac{1}{m}\int \frac{x}{\sqrt{\Phi}}\,dx - \frac{n}{m^2}\int \frac{dx}{\sqrt{\Phi}} + \frac{n^2}{m^2}\int \frac{dx}{(mx+n)\sqrt{\Phi}}.$$

$$\int \frac{x^2}{(1-x^2)\sqrt{x^2+1}}\,dx = \frac{1}{2}\log\frac{(\sqrt{x^2+1}-x)\,(\sqrt{x^2+1}+x\sqrt{2})}{(\sqrt{x^2+1}+x)\,(\sqrt{x^2+1}-x\sqrt{2})}$$

$$\int \frac{x^2}{(1+x^2)\sqrt{1-x^2}}\,dx = \text{arc sin}\, x - \frac{1}{\sqrt{2}}\,\text{arc tg}\,\frac{x\sqrt{2}}{\sqrt{1-x^2}}.$$

$$\int \frac{x^2}{(mx^2+n)\sqrt{ax^2+b}}\,dx = \frac{1}{m}\int \frac{dx}{\sqrt{ax^2+b}} - \frac{n}{m}\int \frac{dx}{(mx+n)\sqrt{ax^2+b}}.$$

$$\int \frac{x^2}{(a^4-b^4x^4)\sqrt{\alpha+\gamma x^2}}\,dx = \frac{1}{4ab^2\sqrt{\eta}}\log\frac{x\sqrt{\eta}+a\sqrt{\Phi}}{x\sqrt{\eta}-a\sqrt{\Phi}} - \frac{1}{2ab^2\sqrt{\mu}}\,\text{arc tg}\,\frac{x\sqrt{\mu}}{a\sqrt{\Phi}}. \qquad \eta = \alpha b^2 + \gamma a^2. \quad \mu = \alpha b^2 - \gamma a^2.$$

$$\int \frac{x^2}{(a^4-b^4x^4)\sqrt{\gamma x^2-\alpha}}\,dx = \frac{1}{4ab^2\sqrt{\eta}}\log\frac{x\sqrt{\eta}-a\sqrt{\Phi}}{x\sqrt{\eta}+a\sqrt{\Phi}} + \frac{1}{2ab^2\sqrt{\mu}}\,\text{arc tg}\,\frac{x\sqrt{\mu}}{a\sqrt{\Phi}}. \qquad » \qquad »$$

$$\int \frac{x^2}{(a^4-b^4x^4)\sqrt{\alpha-\gamma x^2}}\,dx = \frac{1}{4ab^2\sqrt{\mu}}\log\frac{x\sqrt{\mu}+a\sqrt{\Phi}}{x\sqrt{\mu}-a\sqrt{\Phi}} - \frac{1}{2ab^2\sqrt{\eta}}\,\text{arc tg}\,\frac{x\sqrt{\eta}}{a\sqrt{\Phi}}. \qquad » \qquad »$$

$$\int \frac{x^2}{(a^4-b^4x^4)\sqrt{\alpha+\gamma x^2}}\,dx = \frac{1}{4ab^2\sqrt{\eta}}\log\frac{x\sqrt{\eta}+a\sqrt{\Phi}}{x\sqrt{\eta}-a\sqrt{\Phi}} + \frac{1}{4ab^2\sqrt{-\mu}}\log\frac{x\sqrt{-\mu}-a\sqrt{\Phi}}{x\sqrt{-\mu}+a\sqrt{\Phi}}. \qquad » \qquad »$$

$$\int \frac{x^2}{(a^4-b^4x^4)\sqrt{\gamma x^2-\alpha}}\,dx = \frac{1}{4ab^2\sqrt{-\mu}}\log\frac{x\sqrt{-\mu}+a\sqrt{\Phi}}{x\sqrt{-\mu}-a\sqrt{\Phi}} + \frac{1}{4ab^2\sqrt{\eta}}\log\frac{x\sqrt{\eta}-a\sqrt{\Phi}}{x\sqrt{\eta}+a\sqrt{\Phi}}. \qquad » \qquad »$$

$$\int \frac{x^2}{(a^4-b^4x^4)\sqrt{\alpha-\gamma x^2}}\,dx = \frac{1}{2ab^2\sqrt{-\mu}}\,\text{arc tg}\,\frac{x\sqrt{-\mu}}{a\sqrt{\Phi}} - \frac{1}{2ab^2\sqrt{\eta}}\,\text{arc tg}\,\frac{x\sqrt{\eta}}{a\sqrt{\Phi}}. \qquad » \qquad »$$

$$\int \frac{x^2+5}{(1+x^2)^2\sqrt{1-x^2}}\,dx = \frac{x\sqrt{1-x^2}}{1+x^2} + 2\sqrt{2}\,\text{arc tg}\,\frac{x\sqrt{2}}{\sqrt{1-x^2}}$$

$$\int \frac{1-3x^2}{(1+x^2)^2\sqrt{1-x^2}}\,dx = \frac{x\sqrt{1-x^2}}{1+x^2}$$

$$\int \frac{x^3}{(mx^2+n)\sqrt{ax^2+b}}\,dx = \frac{1}{m}\int \frac{x}{\sqrt{\Phi}}\,dx - \frac{n}{m}\int \frac{x}{(mx+n)\sqrt{\Phi}}\,dx.$$

$$\int \frac{x^p}{(mx+n)\sqrt{ax^2+b}}\,dx = \frac{1}{m}\int \frac{x^{p-1}}{\sqrt{\Phi}}\,dx - \frac{n}{m^2}\int \frac{x^{p-2}}{\sqrt{\Phi}}\,dx + \frac{n^2}{m^3}\int \frac{x^{p-3}}{\sqrt{\Phi}}\,dx - \cdots \pm \frac{n^{p-1}}{m^p}\int \frac{dx}{\sqrt{\Phi}} \mp \frac{n^p}{m^p}\int \frac{dx}{(mx+n)\sqrt{\Phi}}.$$

$\int \frac{X\sqrt{ax^2+b}}{Y}\,dx$

$$\int \frac{x\sqrt{2x^2+5}}{x^2+3}\,dx = \sqrt{2x^2+5} - \text{arc tg}\,\sqrt{2x^2+5}.$$

$$\int \frac{x\sqrt{ax^2+b}}{mx^2+n}\,dx = \frac{a}{m}\int \frac{x}{\sqrt{\Phi}}\,dx + \frac{bm-an}{m}\int \frac{x}{(mx^2+n)\sqrt{\Phi}}\,dx.$$

$$\int \frac{x^2\sqrt{ax^2+b}}{mx^2+n}\,dx = \frac{a}{m}\int \frac{x^2}{\sqrt{\Phi}}\,dx + \frac{bm^2-an}{m^2}\int \frac{dx}{\sqrt{\Phi}} - \frac{bmn-an^2}{m^2}\int \frac{dx}{(mx^2+n)\sqrt{\Phi}}$$

$$\int \frac{x\sqrt{ax^2+b}}{ax^2+b+1}\,dx = \frac{\sqrt{ax^2+b}}{a} - \frac{1}{a}\,\text{arc tg}\,\sqrt{ax^2+b}.$$

$$\int U \sqrt{\frac{mx^2+n}{ax^2+b}}\,dx$$

$$\int \sqrt{\frac{ax^2+b}{mx^2+n}}\,dx = \int \frac{ax^2+b}{\sqrt{(ax^2+b)(mx^2+n)}}\,dx, \quad \text{elliptic integral.}$$

$$\int x \sqrt{\frac{a-bx^2}{m-nx^2}}\,dx = -\frac{1}{n\sqrt{n}} \int \sqrt{an-bm+bz^2}\,dz. \qquad m-nx^2=z^2.$$

$$\int x^3 \sqrt{\frac{a^2-x^2}{a^2+x^2}}\,dx = \frac{x^2-2a^2}{4}\sqrt{a^4-x^4} - \frac{a^4}{4}\arcsin\frac{x^2}{a^2}.$$

$$\int \frac{1}{x} \sqrt{\frac{a^2-x^2}{x^2-b^2}}\,dx = \operatorname{arc\,tg}\sqrt{\frac{a^2-x^2}{x^2-b^2}} - \frac{a}{b}\operatorname{arc\,tg}\frac{b}{a}\sqrt{\frac{a^2-x^2}{x^2-b^2}}.$$

$$\int U \sqrt{ax^2+bx}\,dx$$

$$\int \sqrt{x^2-ax}\,dx = \frac{2x-a}{4}\sqrt{x^2-ax} - \frac{a^2}{8}\log\left(2x-a+2\sqrt{x^2-ax}\right).$$

$$\int \sqrt{2ax-x^2}\,dx = \frac{x-a}{2}\sqrt{2ax-x^2} + \frac{a^2}{2}\arcsin\frac{x-a}{a} = -\frac{a-x}{2}\sqrt{2ax-x^2} - \frac{a^2}{2}\arcsin\frac{a-x}{a}$$

$$\int \sqrt{ax^2+bx}\,dx = \frac{2ax+b}{4a}\sqrt{ax^2+bx} - \frac{b^2}{8a\sqrt{a}}\log\left(2ax+b+2\sqrt{a}\sqrt{ax^2+bx}\right).$$

$$\int \sqrt{bx-ax^2}\,dx = \frac{2ax-b}{4a}\sqrt{bx-ax^2} + \frac{b^2}{8a\sqrt{a}}\arcsin\frac{2ax-b}{b}.$$

$$\int x\sqrt{2ax-x^2}\,dx = \left\{\frac{x^2}{3} - \frac{ax}{6} - \frac{a^2}{2}\right\}\sqrt{2ax-x^2} + \frac{a^3}{2}\arcsin\frac{x-a}{a}.$$

$$\int x\sqrt{ax^2+bx}\,dx = \frac{1}{3a}\left(ax^2+bx\right)^{\frac{3}{2}} - \frac{b}{2a}\int\sqrt{ax^2+bx}\,dx.$$

$$\int x^2\sqrt{ax^2+bx}\,dx = \left\{\frac{x}{4a} - \frac{5b}{24a^2}\right\}\left(ax^2+bx\right)^{\frac{3}{2}} + \frac{5b^2}{16a^2}\int\sqrt{ax^2+bx}\,dx.$$

$$\int x^3\sqrt{2ax-x^2}\,dx = -\left\{\frac{x^2}{5} + \frac{7}{20}ax + \frac{7a^2}{12}\right\}\left(2ax-x^2\right)^{\frac{3}{2}} + \frac{7}{4}a^3\int\sqrt{2ax-x^2}\,dx.$$

$$\int x^m\sqrt{ax^2+bx}\,dx = \frac{x^{m+1}}{m+1}\sqrt{\Phi} - \frac{b}{2(m+1)}\int\frac{x^{m+1}}{\sqrt{\Phi}}\,dx - \frac{a}{m+1}\int\frac{x^{m+2}}{\sqrt{\Phi}}\,dx.$$

$$\int\frac{dx}{\sqrt{x^2+bx}} = \log\left(x+\frac{b}{2}+\sqrt{x^2+bx}\right).$$

$$\int\frac{dx}{\sqrt{x-x^2}} = \arcsin(2x-1) = \arcsin 2\sqrt{x-x^2}.$$

$$\int\frac{dx}{\sqrt{ax-x^2}} = \arcsin\frac{2x-a}{a} = 2\arcsin\sqrt{\frac{x}{a}} = \arccos\frac{a-2x}{a}.$$

$$\int\frac{dx}{\sqrt{x-ax^2}} = -\frac{2}{\sqrt{a}}\arcsin\sqrt{1-ax} = -\frac{1}{\sqrt{a}}\arcsin(1-2ax).$$

$$\int\frac{dx}{\sqrt{ax^2+bx}} = \frac{2}{\sqrt{a}}\log\left(\sqrt{ax}+\sqrt{ax+b}\right) = \frac{1}{\sqrt{a}}\log\frac{\sqrt{\Phi}+x\sqrt{a}}{\sqrt{\Phi}-x\sqrt{a}} = \frac{1}{\sqrt{a}}\log\left\{2ax+b+2\sqrt{a}\sqrt{\Phi}\right\}.$$

$$\int \frac{dx}{\sqrt{bx-ax^2}} = \frac{1}{\sqrt{a}} \arcsin \frac{2ax-b}{b} = \frac{2}{\sqrt{a}} \arcsin \sqrt{\frac{ax}{b}} = \frac{2}{\sqrt{a}} \operatorname{arc\,tg} \sqrt{\frac{ax}{b-ax}} = \frac{2}{\sqrt{a}} \arccos \sqrt{\frac{b-ax}{b}}.$$

$$\int \frac{dx}{x\sqrt{ax^2+bx}} = -\frac{2}{bx}\sqrt{ax^2+bx}.$$

$$\int \frac{dx}{(x+1)\sqrt{8x-x^2}} = \frac{2}{3} \operatorname{arc\,tg} 3\sqrt{\frac{x}{8-x}}.$$

$$\int \frac{dx}{(3x+2)\sqrt{3x^2+4x}} = -\frac{1}{2\sqrt{3}} \operatorname{arc\,tg} \frac{2}{\sqrt{3}\sqrt{3x^2+4x}}.$$

$$\int \frac{dx}{(mx+n)\sqrt{ax^2+bx}} = \frac{1}{\sqrt{an^2-bmn}} \log \frac{(bm-2an)x-bn-2\sqrt{(an^2-bmn)\Phi}}{mx+n}. \qquad an^2-bmn>0.$$

$$= \frac{1}{\sqrt{bmn-an^2}} \operatorname{arc\,tg} \frac{(bm-2an)x-bn}{2\sqrt{(bmn-an^2)\Phi}}. \qquad bmn > an^2.$$

$$\int \frac{dx}{x^2\sqrt{2ax-x^2}} = -\frac{a+x}{3a^2x^2}\sqrt{2ax-x^2}.$$

$$\int \frac{dx}{x^2\sqrt{ax^2+bx}} = -\frac{2}{3bx}\left(\frac{1}{x}-\frac{2a}{b}\right)\sqrt{ax^2+bx}.$$

$$\int \frac{dx}{x^3\sqrt{2ax-x^2}} = -\frac{3a^2+2ax+2x^2}{15a^3x^3}\sqrt{2ax-x^2}.$$

$$\int \frac{dx}{x^3\sqrt{ax^2+bx}} = -\frac{2}{5bx}\left\{\frac{1}{x^2}-\frac{4a}{3bx}+\frac{8a^2}{3b^2}\right\}\sqrt{ax^2+bx}.$$

$$\int \frac{x}{\sqrt{2ax-x^2}}\,dx = -\sqrt{2ax-x^2} + a \arcsin \frac{x-a}{a} = \int \frac{z+a}{\sqrt{a^2-z^2}}\,dz. \qquad x = z+a.$$

$$\int \frac{x}{\sqrt{ax^2+bx}}\,dx = \frac{1}{a}\sqrt{ax^2+bx} - \frac{b}{2a}\int \frac{dx}{\sqrt{ax^2+bx}}.$$

$$\int \frac{2x+1}{\sqrt{x^2+x}}\,dx = 2\sqrt{x^2+x}.$$

$$\int \frac{a-x}{\sqrt{2ax-x^2}}\,dx = \sqrt{2ax-x^2}.$$

$$\int \frac{x^2}{\sqrt{2ax-x^2}}\,dx = -\frac{3a+x}{2}\sqrt{2ax-x^2} + \frac{3}{2}a^2 \arcsin \frac{x-a}{a} = -\frac{3a+x}{2}\sqrt{2ax-x^2} + 3a^2 \arcsin \sqrt{\frac{x}{2a}}.$$

$$\int \frac{x^2}{\sqrt{ax^2+bx}}\,dx = \left\{\frac{x}{2a}-\frac{3b}{4a^2}\right\}\sqrt{ax^2+bx} + \frac{3b^2}{8a^2}\int \frac{dx}{\sqrt{ax^2+bx}}.$$

$$\int \frac{x^3}{\sqrt{2ax-x^2}}\,dx = -\frac{1}{6}(2x^2+5ax+15a^2)\sqrt{2ax-x^2} + \frac{5}{2}a^3 \arcsin \frac{x-a}{a}.$$

$$\int \frac{x^n}{\sqrt{2ax-x^2}}\,dx = -\frac{x^{n-1}\sqrt{\Phi}}{n} + \frac{a(2n-1)}{n}\int \frac{x^{n-1}}{\sqrt{\Phi}}\,dx.$$

$$\int \frac{\sqrt{ax^2+bx}}{x}\,dx = \sqrt{ax^2+bx} + \frac{b}{2}\int \frac{dx}{\sqrt{ax^2+bx}}.$$

$$\int \frac{\sqrt{ax^2+bx}}{x^2}\,dx = -\frac{\sqrt{ax^2+bx}}{x} + \frac{b}{2}\int \frac{dx}{x\sqrt{ax^2+bx}} + a\int \frac{dx}{\sqrt{ax^2+bx}}.$$

$$\int \frac{\sqrt{ax^2+bx}}{x^3}\,dx = -\frac{1}{2x^2}\sqrt{ax^2+bx} + \frac{b}{4}\int \frac{dx}{x^2\sqrt{ax^2+bx}} + \frac{a}{2}\int \frac{dx}{x\sqrt{ax^2+bx}}.$$

$$\int \frac{\sqrt{ax^2+bx}}{x^n}\,dx = -\frac{\sqrt{ax^2+bx}}{(n-1)\,x^{n-1}} + \frac{b}{2\,(n-1)}\int \frac{dx}{x^{n-1}\sqrt{ax^2+bx}} + \frac{a}{n-1}\int \frac{dx}{x^{n-2}\sqrt{ax^2+bx}}$$

$\int X\sqrt{ax^2+bx+c}\,dx$

$$\int \sqrt{x^2+x+1}\,dx = \frac{2x+1}{4}\sqrt{\Phi} + \frac{3}{8}\log\left(2x+1+2\sqrt{\Phi}\right).$$

$$\int \sqrt{b^2-(x-a)^2}\,dx = \frac{x-a}{2}\sqrt{\Phi} + \frac{b^2}{2}\,\text{arc sin}\,\frac{x-a}{b}.$$

$$\int \sqrt{(x-a)(b-x)}\,dx = \frac{2x-a-b}{4}\sqrt{\Phi} + \frac{(b-a)^2}{8}\,\text{arc sin}\,\frac{2x-a-b}{b-a}.$$

$$\int \sqrt{ax^2+bx+c}\,dx = \frac{2ax+b}{4a}\sqrt{\Phi} + \frac{4ac-b^2}{8a}\int \frac{dx}{\sqrt{\Phi}} = \frac{2ax+b}{4a}\sqrt{\Phi} + \frac{4ac-b^2}{8a\sqrt{a}}\log\left(2ax+b+2\sqrt{a\Phi}\right).$$

$$\int \sqrt{c+bx-ax^2}\,dx = \frac{2ax-b}{4a}\sqrt{\Phi} + \frac{4ac+b^2}{8a\sqrt{a}}\,\text{arc sin}\,\frac{2ax-b}{\sqrt{4ac+b^2}}.$$

$$\int x\sqrt{x^2+x+1}\,dx = \frac{1}{3}\Phi^{\frac{3}{2}} - \frac{2x+1}{8}\sqrt{\Phi} - \frac{3}{16}\log\left(1+2x+2\sqrt{\Phi}\right).$$

$$\int x\sqrt{b^2-(x-a)^2}\,dx = -\frac{1}{3}\Phi^{\frac{3}{2}} + \frac{a}{2}(x-a)\sqrt{\Phi} + \frac{ab^2}{2}\,\text{arc sin}\,\frac{x-a}{b}.$$

$$\int x\sqrt{x^2-3ax+2a^2}\,dx = \int (z+a)\sqrt{z^2-az}\,dz. \qquad x = z+a.$$

$$\int x\sqrt{ax^2+bx+c}\,dx = \frac{1}{3a}\Phi^{\frac{3}{2}} - \frac{b}{2a}\int \sqrt{\Phi}\,dx.$$

$$\int x^2\sqrt{x^2+x+1}\,dx = \frac{6x-5}{24}\Phi^{\frac{3}{2}} + \frac{2x+1}{64}\sqrt{\Phi} + \frac{3}{128}\log\left(1+2x+2\sqrt{\Phi}\right).$$

$$\int x^2\sqrt{5x^2+2x+1}\,dx = \frac{1}{20}\left(x-\frac{1}{3}\right)\Phi^{\frac{3}{2}}.$$

$$\int x^2\sqrt{ax^2+bx+c}\,dx = \frac{6ax-5b}{24a^2}\Phi^{\frac{3}{2}} - \frac{4ac-5b^2}{16a^2}\int \sqrt{\Phi}\,dx.$$

$$\int x^m\sqrt{ax^2+bx+c}\,dx = \frac{x^{m+1}}{m+1}\sqrt{\Phi} - \frac{b}{2\,(m+1)}\int \frac{x^{m+1}}{\sqrt{\Phi}}\,dx - \frac{a}{m+1}\int \frac{x^{m+2}}{\sqrt{\Phi}}\,dx.$$

$\int \frac{dx}{\sqrt{ax^2+bx+c}}$

$$\int \frac{dx}{\sqrt{x^2+x+1}} = \log\left(1+2x+2\sqrt{\Phi}\right).$$

$$\int \frac{dx}{\sqrt{x^2-x+1}} = \log\left(2x-1+2\sqrt{\Phi}\right)$$

$$\int \frac{dx}{\sqrt{x^2-x-1}} = \log\left(2x-1+2\sqrt{\Phi}\right).$$

$$\int \frac{dx}{\sqrt{1 + x - x^2}} = \text{arc sin} \frac{2x - 1}{\sqrt{5}} .$$

$$\int \frac{dx}{\sqrt{1 - x - x^2}} = \text{arc sin} \frac{2x + 1}{\sqrt{5}} = -2 \text{ arc tg} \frac{\sqrt{\Phi} - 1}{x} .$$

$$\int \frac{dx}{\sqrt{2x^2 + x + 1}} = \frac{1}{\sqrt{2}} \log \left(4x + 1 + 2\sqrt{2\Phi}\right) .$$

$$\int \frac{dx}{\sqrt{2 + 2x + 4x^2}} = \frac{1}{2} \log \left(4x + 1 + 2\sqrt{\Phi}\right) .$$

$$\int \frac{dx}{\sqrt{1 + 2x - x^2}} = \text{arc sin} \frac{x - 1}{\sqrt{2}}$$

$$\int \frac{dx}{\sqrt{1 + x - 2x^2}} = \frac{1}{\sqrt{2}} \text{ arc sin} \frac{4x - 1}{3} .$$

$$\int \frac{dx}{\sqrt{1 - 2x - x^2}} = - \text{arc cos} \frac{x + 1}{\sqrt{2}} = - \frac{1}{2} \text{ arc cos} (x^2 + 2x).$$

$$\int \frac{dx}{\sqrt{x^2 - 2x \cos \alpha + 1}} = \log \left(x - \cos \alpha + \sqrt{\Phi}\right) .$$

$$\int \frac{dx}{\sqrt{1 - (ax + b)^2}} = \frac{1}{a} \text{ arc sin} (ax + b).$$

$$\int \frac{dx}{\sqrt{b^2 - (x - a)^2}} = \frac{x - a}{2} \sqrt{\Phi} + \frac{b^2}{2} \text{ arc sin} \frac{x - a}{b} .$$

$$\int \frac{dx}{\sqrt{(x - 1)(5 - x)}} = 2 \text{ arc sin} \frac{\sqrt{x - 1}}{2} = -2 \text{ arc sin} \frac{\sqrt{5 - x}}{2} .$$

$$\int \frac{dx}{\sqrt{(x - 1)(x - 2)}} = \log \frac{\sqrt{\Phi} + x - 1}{\sqrt{\Phi} - x + 1} .$$

$$\int \frac{dx}{\sqrt{(x - a)(x - b)}} = 2 \log \left\{\sqrt{x - a} + \sqrt{x - b}\right\} = -2 \log \left\{\sqrt{a - x} + \sqrt{b - x}\right\} = \log \frac{\sqrt{\Phi} + x - a}{\sqrt{\Phi} - x + a} .$$

$$\int \frac{dx}{\sqrt{(x - a)(b - x)}} = 2 \text{ arc tg} \sqrt{\frac{x - a}{b - x}} = -2 \text{ arc tg} \sqrt{\frac{b - x}{x - a}} = -2 \text{ arc sin} \sqrt{\frac{b - x}{b - a}} .$$

$$\int \frac{dx}{\sqrt{(ax + b)(n - ax)}} = \frac{n}{a} \text{ arc tg} \sqrt{\frac{ax + b}{n - ax}} = \frac{n}{a} \text{ arc sin} \sqrt{\frac{ax + b}{b + n}} .$$

$$\int \frac{dx}{\sqrt{(ax + b)(mx + n)}} = 2 \int \frac{dz}{m - az^2} . \qquad \sqrt{(ax + b)(mx + n)} = z\,(ax + b).$$

$$= \frac{1}{\sqrt{am}} \log \frac{\sqrt{m(ax + b)} + \sqrt{a(mx + n)}}{\sqrt{m(ax + b)} - \sqrt{a(mx + n)}} = \frac{2}{\sqrt{am}} \log \left\{\sqrt{m(ax + b)} + \sqrt{a(mx + n)}\right\} .$$

$$\int \frac{dx}{\sqrt{(b - ax)(mx + n)}} = \frac{2}{\sqrt{am}} \text{ arc tg} \sqrt{\frac{a(mx + n)}{m(b - ax)}} .$$

$$\int \frac{dx}{\sqrt{ax^2 + bx + c}} = \frac{1}{\sqrt{a}} \log \left(b + 2ax + 2\sqrt{a\Phi}\right) = - \frac{1}{\sqrt{a}} \log \left(b + 2ax - 2\sqrt{a\Phi}\right) .$$

$$\int \frac{dx}{\sqrt{c+bx-ax^2}} = \frac{1}{\sqrt{a}} \text{arc sin} \frac{2ax-b}{\sqrt{b^2+4ac}} = \frac{1}{\sqrt{a}} \text{arc tg} \frac{2ax-b}{2\sqrt{a\Phi}} = -\frac{2}{\sqrt{a}} \text{arc tg} \frac{\sqrt{c}+\sqrt{\Phi}}{x\sqrt{a}} = \frac{2}{\sqrt{a}} \text{arc tg} \sqrt{\frac{\sqrt{b^2+4ac}+2ax-b}{\sqrt{b^2+4ac}-2ax+b}}$$

$$= -\frac{2}{\sqrt{a}} \text{arc tg} \sqrt{\frac{\alpha-x}{x-\beta}} = -\frac{2}{\sqrt{a}} \text{arc sin} \sqrt{\frac{\alpha-x}{\alpha-\beta}}.$$ α, β being the roots of $+ bx - ax^2 = 0$.

$\int \frac{dx}{X\sqrt{ax^2+bx+c}}$

$$\int \frac{dx}{x\sqrt{x^2+x+1}} = \log \frac{x}{2+x+2\sqrt{\Phi}} = \log \frac{2+x-2\sqrt{\Phi}}{x}.$$

$$\int \frac{dx}{x\sqrt{x^2+x-1}} = \text{arc sin} \frac{x-2}{x\sqrt{5}}.$$

$$\int \frac{dx}{x\sqrt{3x^2+4x-4}} = \frac{1}{2} \text{arc sin} \frac{x-2}{2x}.$$

$$\int \frac{dx}{x\sqrt{x^2+2x+3}} = \frac{1}{\sqrt{3}} \log \frac{x+\sqrt{3}-\sqrt{\Phi}}{x-\sqrt{3}-\sqrt{\Phi}} = \frac{1}{\sqrt{3}} \log \frac{x}{x+3+\sqrt{3\Phi}}$$

$$\int \frac{dx}{x\sqrt{4x^2-4x-1}} = \text{arc cos} \frac{2x+1}{2x\sqrt{2}}.$$

$$\int \frac{dx}{x\sqrt{7x^2+6x-1}} = \text{arc sin} \frac{3x-1}{4x}.$$

$$\int \frac{dx}{x\sqrt{(mx+n)(ax+b)}} = -2\int \frac{dz}{n-bz^2}. \qquad mx+n = z^2(ax+b).$$

$$\int \frac{dx}{x\sqrt{x^2+2ax-a^2}} = \frac{1}{a} \text{arc sin} \frac{x-a}{x\sqrt{2}}.$$

$$\int \frac{dx}{x\sqrt{a^2+2ax-x^2}} = \frac{1}{a} \log \frac{x\sqrt{2}}{x+a+\sqrt{\Phi}}.$$

$$\int \frac{dx}{x\sqrt{ax^2+bx+c}} = -\frac{1}{\sqrt{c}} \log \frac{2c+bx+2\sqrt{c\Phi}}{x} = \frac{1}{\sqrt{c}} \log \frac{2c+bx-2\sqrt{c\Phi}}{x} = \frac{1}{\sqrt{c}} \log \frac{x\sqrt{a}-\sqrt{\Phi}+\sqrt{c}}{x\sqrt{a}-\sqrt{\Phi}-\sqrt{c}}.$$

$$\int \frac{dx}{x\sqrt{ax^2+bx-c}} = \frac{1}{\sqrt{c}} \text{arc sin} \frac{bx-2c}{x\sqrt{b^2+4ac}} = \frac{1}{\sqrt{c}} \text{arc tg} \frac{bx-2c}{2\sqrt{c\Phi}}.$$

$$\int \frac{dx}{(x+1)\sqrt{x^2+x+1}} = \log \frac{x+\sqrt{\Phi}}{x+2+\sqrt{\Phi}} = \log \frac{1-x-2\sqrt{\Phi}}{x+1}.$$

$$\int \frac{dx}{(x+1)\sqrt{1+x-x^2}} = \text{arc tg} \frac{3x+1}{2\sqrt{\Phi}} = -2 \text{arc tg} \frac{x+1+\sqrt{\Phi}}{x} = -2 \text{arc tg} \frac{\sqrt{\Phi}-x-1}{x}.$$

$$\int \frac{dx}{(x+1)\sqrt{1-x-x^2}} = \log \frac{x+3-2\sqrt{\Phi}}{x+1} = \log \frac{x+1}{x+3+2\sqrt{\Phi}}.$$

$$\int \frac{dx}{(x-2)\sqrt{x^2-4x+1}} = -\frac{1}{\sqrt{3}} \text{arc sin} \frac{\sqrt{3}}{x-2}.$$

$$\int \frac{dx}{(x-2)\sqrt{4x-x^2-3}} = \log \frac{1-\sqrt{\Phi}}{x-2}.$$

$$\int \frac{dx}{(x+4)\sqrt{x^2+3x-4}} = \frac{2}{5}\sqrt{\frac{x-1}{x+4}}.$$

$$\int\frac{dx}{(x+3)\sqrt{(6-x)(x-1)}}=\frac{1}{3}\operatorname{arc\,tg}\frac{3}{2}\sqrt{\frac{x-1}{6-x}}.$$

$$\int\frac{dx}{(x+5)\sqrt{(x+1)(x+4)}}=\frac{1}{2}\log\frac{2\sqrt{x+4}+\sqrt{x+1}}{2\sqrt{x+4}-\sqrt{x+1}}.$$

$$\int\frac{dx}{(x-a)\sqrt{(x-a)(x-b)}}=\frac{2}{b-a}\sqrt{\frac{x-b}{x-a}}.$$

$$\int\frac{dx}{(x-a)\sqrt{(x-b)(c-x)}}=\frac{2}{\sqrt{(b-a)(c-a)}}\operatorname{arc\,tg}\sqrt{\frac{(x-b)(c-a)}{(c-x)(b-a)}}.$$

$$\int\frac{dx}{(ax+b)\sqrt{ax^2+2bx+c}}=\frac{1}{ak}\log\frac{\sqrt{\Phi}-k}{ax+b}.\qquad k^2=\frac{ac-b^2}{a}.$$

$$=\frac{1}{ak}\operatorname{arc\,tg}\frac{\sqrt{\Phi}}{k}.\qquad k^2=\frac{b^2-ac}{a}.$$

$$\int\frac{dx}{(bx+c)\sqrt{ax^2+2bx+c}}=\frac{1}{ck}\log\frac{bx+c}{\sqrt{\Phi}-k}.\qquad k^2=\frac{ac-b^2}{c}.$$

$$=\frac{1}{ck}\operatorname{arc\,tg}\frac{kx}{\sqrt{\Phi}}.\qquad k^2=\frac{b^2-ac}{c}.$$

$$\int\frac{dx}{(x+n)\sqrt{ax^2+bx+c}}=-\frac{1}{\sqrt{an^2+c-bn}}\log\left\{\frac{\sqrt{\Phi}+\sqrt{an^2+c-bn}}{x+n}+\frac{b-2an}{2\sqrt{an^2+c-bn}}\right\}.\quad an^2+c-bn>0.$$

$$=\frac{1}{\sqrt{bn-an^2-c}}\operatorname{arc\,sin}\frac{(b-2an)x+2(c-bn+an^2)}{(x+n)\sqrt{b^2-4ac}}.\qquad an^2+c-bn<0.$$

$$=-\frac{2\sqrt{\Phi}}{(b-2an)(x+n)}.\qquad an^2+c-bn=0.$$

$$\int\frac{dx}{(x-n)\sqrt{ax^2+bx+c}}=-\int\frac{dz}{\sqrt{(an^2+bn+c)z^2+(2an+b)z+a}}.\qquad x-n=\frac{1}{z}.$$

$$=-\frac{1}{\sqrt{an^2+bn+c}}\log\left\{\frac{\sqrt{\Phi}+\sqrt{an^2+bn+c}}{x-n}+\frac{b+2an}{2\sqrt{an^2+bn+c}}\right\}\cdot\quad an^2+bn+c>0.$$

$$=\frac{1}{\sqrt{-(an^2+bn+c)}}\operatorname{arc\,sin}\frac{2c+bn+(b+2an)x}{(x-n)\sqrt{b^2-4ac}}.\qquad an^2+bn+c<0.$$

$$\int\frac{dx}{(mx+n)\sqrt{n^2+\frac{n}{m}(m^2+p^2-a)x-ax^2}}=\frac{2}{np}\operatorname{arc\,tg}\frac{px}{mx+n+\sqrt{\Phi}}.$$

$$\int\frac{dx}{(mx+n)\sqrt{ax^2+bx+c}}=-\int\frac{dz}{\sqrt{a(1-nz)^2+bm(1-nz)z+cm^2z^2}}.\qquad mx+n=\frac{1}{z}.$$

$$=\frac{1}{\sqrt{an^2+cm^2-bmn}}\log\frac{2cm-bn+(bm-2an)x-2\sqrt{(an^2+cm^2-bmn)\Phi}}{mx+n}.\quad an^2+cm^2-bmn>0$$

$$=\frac{1}{\sqrt{bmn-an^2-cm^2}}\operatorname{arc\,tg}\frac{2cm-bn+(bm-2an)x}{2\sqrt{(bmn-an^2-cm^2)\Phi}}.\quad an^2+cm^2-bmn<0$$

$$\int\frac{dx}{x^2\sqrt{x^2+x+1}}=-\frac{\sqrt{\Phi}}{x}-\frac{1}{2}\log\frac{2+x-2\sqrt{\Phi}}{x}$$

$$\int\frac{dx}{x^2\sqrt{ax^2+bx+c}}=-\frac{\sqrt{\Phi}}{cx}-\frac{b}{2c}\int\frac{dx}{x\sqrt{\Phi}}.$$

$$\int \frac{dx}{(x^2+m^2)\sqrt{a+2bx+cx^2}} = (\alpha-\beta)\int \frac{z+1}{\{(\alpha^2+m^2)z^2+\beta^2+m^2\}\sqrt{(a+2b\alpha+c\alpha^2)z^2+a+2b\beta+c\beta^2}}\,dz. \qquad x=\frac{\alpha z+\beta}{z+1}.$$

α, β being given by $c\alpha\beta + b(\alpha+\beta)+a=0. \qquad \alpha\beta+m^2=0.$

$$\int \frac{dx}{(x+m)^2\sqrt{ax^2+bx+c}} = -\int \frac{z}{\sqrt{a(1-mz)^2+b(1-mz)z+cz^2}}\,dz. \qquad x+m=\frac{1}{z}.$$

$$\int \frac{dx}{x^3\sqrt{x^2+x+1}} = \left\{\frac{3}{4x}-\frac{1}{2x^2}\right\}\sqrt{\Phi}-\frac{1}{8}\log\frac{2+x-2\sqrt{\Phi}}{x}.$$

$$\int \frac{dx}{x^3\sqrt{ax^2+bx+c}} = \frac{3bx-2c}{4c^2x^2}\sqrt{\Phi}+\frac{3b^2-4ac}{8c^2}\int\frac{dx}{x\sqrt{\Phi}}.$$

$$\int \frac{dx}{x^m\sqrt{ax^2+bx+c}} = -\int\frac{z^{m-1}}{\sqrt{cz^2-bz+a}}\,dz. \qquad xz=1.$$

$$= -\frac{\sqrt{\Phi}}{c(m-1)x^{m-1}}-\frac{b(2m-3)}{2c(m-1)}\int\frac{dx}{x^{m-1}\sqrt{\Phi}}-\frac{a(m-2)}{c(m-1)}\int\frac{dx}{x^{m-2}\sqrt{\Phi}}.$$

$$\int \frac{dx}{(x+h)^m\sqrt{ax^2+bx+c}} = -\int\frac{z^{m-1}}{\sqrt{(ah^2-bh+c)z^2-(b-2ah)z+a}}\,dz. \qquad zx=1-hz.$$

$$\int \frac{dx}{(x-h)^m\sqrt{ax^2+bx+c}} = -\int\frac{z^{m-1}}{\sqrt{(ah^2+bh+c)z^2+(2ah+b)z+a}}\,dz. \qquad zx=1+hz.$$

$$\int \frac{dx}{x^p(mx+n)\sqrt{ax^2+bx+c}} = \frac{1}{n}\int\frac{dx}{x^p\sqrt{\Phi}}-\frac{m}{n^2}\int\frac{dx}{x^{p-1}\sqrt{\Phi}}+\frac{m^2}{n^3}\int\frac{dx}{x^{p-2}\sqrt{\Phi}}-\dots\pm\frac{m^{p-1}}{n^p}\int\frac{dx}{x\sqrt{\Phi}}\mp\frac{m^p}{n^p}\int\frac{dx}{(mx+n)\sqrt{\Phi}}.$$

$\int U\sqrt{ax^2+bx+c}\,dx$

$$\int \frac{x}{\sqrt{x^2+x+1}}\,dx = \sqrt{\Phi}-\frac{1}{2}\log\left(1+2x+2\sqrt{\Phi}\right).$$

$$\int \frac{x}{\sqrt{1-x-x^2}}\,dx = -\sqrt{\Phi}+\text{arc tg}\frac{\sqrt{\Phi}-1}{x} = -\frac{z+2}{z^2+1}-\text{arc tg}\, z. \qquad \sqrt{1-x-x^2}=1-zx.$$

$$\int \frac{x}{\sqrt{(ax+b)(mx+n)}}\,dx = \frac{\sqrt{\Phi}}{am}-\frac{an+bm}{am\sqrt{am}}\log\left\{\sqrt{m(ax+b)}+\sqrt{a(mx+n)}\right\}.$$

$$\int \frac{x}{\sqrt{(ax+b)(n-mx)}}\,dx = -\frac{\sqrt{\Phi}}{am}+\frac{an-bm}{2am\sqrt{am}}\text{arc sin}\frac{2amx+bm-an}{an+bm}.$$

$$\int \frac{x}{\sqrt{ax^2+bx+c}}\,dx = \frac{\sqrt{\Phi}}{a}-\frac{b}{2a}\int\frac{dx}{\sqrt{\Phi}} = \frac{\sqrt{\Phi}}{a}-\frac{b}{2a\sqrt{a}}\log\left(2ax+b+2\sqrt{a\Phi}\right).$$

$$\int \frac{x}{\sqrt{c+bx-ax^2}}\,dx = -\frac{\sqrt{\Phi}}{a}+\frac{b}{2a\sqrt{a}}\text{arc sin}\frac{2ax-b}{\sqrt{b^2+4ac}}.$$

$$\int \frac{x}{(mx+n)\sqrt{ax^2+bx+c}}\,dx = \frac{1}{m}\int\frac{dx}{\sqrt{\Phi}}-\frac{n}{m}\int\frac{dx}{(mx+n)\sqrt{\Phi}}.$$

$$\int \frac{mx+n}{\sqrt{ax^2+bx+c}}\,dx = \frac{m}{a}\sqrt{\Phi}+\frac{2an-bm}{2a}\int\frac{dx}{\sqrt{\Phi}}.$$

$$\int \frac{mx+n}{\{(x-\alpha)^2+\beta^2\}\sqrt{ax^2+2bx+c}}\,dx = 2\sqrt{a}\int\frac{m(az^2-c)+2n(b-az)}{\{az^2-c-2\alpha(b-az)\}^2+4\beta^2(b-az)^2}\,dz. \qquad x=\frac{az^2-c}{2(b-az)}$$

$$\int \frac{6bcx + ac - 3b^2}{(x^2 + \beta^2)^2 \sqrt{a + 2bx + cx^2}}\, dx = \frac{2c^2 (cx - b) \sqrt{\Phi}}{(ac - b^2)(x^2 + \beta^2)}. \qquad \text{where } ac - 3b^2 = 2c^2\beta^2.$$

$$\int \frac{x^2}{\sqrt{x^2 + x + 1}}\, dx = \frac{2x - 3}{4} \sqrt{\Phi} - \frac{1}{8} \log \left(1 + 2x + 2\sqrt{\Phi}\right).$$

$$\int \frac{x^2}{\sqrt{(ax + b)(mx + n)}}\, dx = \frac{2amx - 3(an + bm)}{4a^2m^2} \sqrt{\Phi} + \frac{3(an + bm)^2 - 4abmn}{8a^2m^2} \int \frac{dx}{\sqrt{\Phi}}.$$

$$\int \frac{x^2}{\sqrt{(ax + b)(n - mx)}}\, dx = \frac{3(bm - an) - 2amx}{4a^2m^2} \sqrt{\Phi} + \frac{3(a^2n^2 + b^2m^2) - 2abmn}{8a^2m^2 \sqrt{am}} \arcsin \frac{2amx + bm - an}{an + bm}.$$

$$\int \frac{x^2}{\sqrt{ax^2 + bx + c}}\, dx = \frac{2ax - 3b}{4a^2} \sqrt{\Phi} + \frac{3b^2 - 4ac}{8a^2} \int \frac{dx}{\sqrt{\Phi}}.$$

$$\int \frac{x^2}{(mx + n)\sqrt{ax^2 + bx + c}}\, dx = \frac{1}{m} \int \frac{x}{\sqrt{\Phi}}\, dx - \frac{n}{m^2} \int \frac{dx}{\sqrt{\Phi}} + \frac{n^2}{m^2} \int \frac{dx}{(mx + n)\sqrt{\Phi}}.$$

$$\int \frac{mx^2 + nx + p}{\sqrt{ax^2 + bx + c}}\, dx = \frac{2amx + 4an - 3bm}{4a^2} \sqrt{\Phi} + \frac{4a(2ap - mc) - b(4an - 3bm)}{8a^2} \int \frac{dx}{\sqrt{\Phi}}.$$

$$\int \frac{x^3}{\sqrt{x^2 + x + 1}}\, dx = \frac{8x^2 - 10x - 1}{24} \sqrt{\Phi} + \frac{7}{16} \log \left(1 + 2x + 2\sqrt{\Phi}\right).$$

$$\int \frac{x^3}{\sqrt{1 - x - x^2}}\, dx = -\frac{8x^2 - 10x + 31}{24} \sqrt{\Phi} - \frac{17}{16} \arcsin \frac{2x + 1}{\sqrt{5}}.$$

$$\int \frac{x^3}{\sqrt{ax^2 + bx + c}}\, dx = \left\{ \frac{x^2}{3a} - \frac{5bx}{12a^2} + \frac{5b^2}{8a^3} - \frac{2c}{3a^2} \right\} \sqrt{\Phi} + \left\{ \frac{3bc}{4a^2} - \frac{5b^3}{16a^3} \right\} \int \frac{dx}{\sqrt{\Phi}}.$$

$$\int \frac{x^n}{\sqrt{ax^2 + bx + c}}\, dx = \frac{x^{n-1}}{an} \sqrt{\Phi} - \frac{b(2n - 1)}{2an} \int \frac{x^{n-1}}{\sqrt{\Phi}}\, dx - \frac{c(n - 1)}{an} \int \frac{x^{n-2}}{\sqrt{\Phi}}\, dx.$$

$$\int \frac{9x^3 - 3x^2 + 2}{\sqrt{1 - 2x + 3x^2}}\, dx = \frac{3x^2 + x - 1}{3} \sqrt{\Phi} + \frac{4}{3\sqrt{3}} \log \left(3x - 1 + \sqrt{3\Phi}\right).$$

$$\int \frac{\sqrt{ax^2 + bx + c}}{x}\, dx = \sqrt{\Phi} + \frac{b}{2} \int \frac{dx}{\sqrt{\Phi}} + c \int \frac{dx}{x\sqrt{\Phi}}.$$

$$\int \frac{\sqrt{ax^2 + bx + c}}{x^2}\, dx = -\frac{\sqrt{\Phi}}{x} + \frac{b}{2} \int \frac{dx}{x\sqrt{\Phi}} + a \int \frac{dx}{\sqrt{\Phi}}.$$

$$\int \frac{\sqrt{ax^2 + bx + c}}{x^3}\, dx = -\frac{bx + 2c}{4cx^2} \sqrt{\Phi} + \frac{4ac - b^2}{8c} \int \frac{dx}{x\sqrt{\Phi}}.$$

$$\int \frac{\sqrt{ax^2 + bx + c}}{x^n}\, dx = -\frac{\sqrt{\Phi}}{(n - 1)x^{n-1}} + \frac{b}{2(n - 1)} \int \frac{dx}{x^{n-1}\sqrt{\Phi}} + \frac{a}{n - 1} \int \frac{dx}{x^{n-2}\sqrt{\Phi}}.$$

$\int U \sqrt{X^3}\, dx$

$$\int x^2 \sqrt{a + bx^3}\, dx = \frac{2}{9b} (a + bx^3)^{\frac{3}{2}}. \qquad (*)$$

$$\int \frac{dx}{x\sqrt{a^2 + x^3}} = \frac{1}{3a} \log \frac{\sqrt{a^2 + x^3} - a}{\sqrt{a^2 + x^3} + a}.$$

(*) $\int \sqrt{1 + x^3}\, dx$, $\int \frac{dx}{\sqrt{1 + x^3}}$ are elliptic integrals.

$$\int \frac{x+1}{(x-2)\sqrt{1+x^3}}\,dx = -\frac{2}{3}\log\left\{\left(\frac{x+1}{x-2}\right)^{\frac{3}{2}} + \sqrt{\left(\frac{x+1}{x-2}\right)^3 - 1}\right\}.$$

$$\int \frac{x+1}{(x-1)\sqrt{x^3+x}}\,dx = \sqrt{2}\log\frac{x-1}{\sqrt{x^2+1}+\sqrt{2x}}.$$

$$\int \frac{x-1}{(x+2)\sqrt{x^3-1}}\,dx = \frac{2}{3}\arccos\sqrt{1-\left(\frac{x-1}{x+2}\right)^3}.$$

$$\int \frac{x-a}{(x-b)\sqrt{(x-a)\{3x^2-3(a+b)x+a^2+ab+b^2\}}}\,dx = \frac{1}{\sqrt{a-b}}\int\sqrt{\frac{z}{1-z^3}}\,dz. \qquad x-a = z(x-b).$$

$$\int \frac{x^2}{\sqrt{a+bx^3}}\,dx = \frac{2}{3b}\sqrt{a+bx^3}.$$

$$\int \frac{x^2-x+4}{(x+1)(x-2)\sqrt{x^3-\mathrm{A}^2x^2-3x-2}}\,dx = \frac{2}{\mathrm{A}}\arccos\frac{\mathrm{A}x}{(x+1)\sqrt{x-2}}.$$

$$\int \frac{amx^2-anx-2bn}{(mx+n)(ax+b)\sqrt{(ax+b)(mx+n)^2-p^2x^2}}\,dx = \frac{2}{p}\arccos\frac{px}{(mx+n)\sqrt{ax+b}}.$$

$$\int\sqrt{\frac{x}{a+x^3}}\,dx = \frac{2}{3}\log\left(x^{\frac{3}{2}}+\sqrt{a+x^3}\right).$$

$$\int\sqrt{\frac{x}{x^3-a}}\,dx = \frac{2}{3}\log\left(x^{\frac{3}{2}}+\sqrt{x^3-a}\right).$$

$$\int\sqrt{\frac{x}{a-x^3}}\,dx = \frac{2}{3}\arcsin\sqrt{\frac{x^3}{a}}.$$

$$\int\sqrt{\frac{x}{(1-x)^3}}\,dx = 2\sqrt{\frac{x}{1-x}} - 2\,\mathrm{arc\,tg}\sqrt{\frac{x}{1-x}}.$$

$$\int\sqrt{\frac{3mx+n}{mx^3+nx^2+\frac{n^2}{3m}x+p}}\,dx = \frac{2}{\sqrt{3}}\log\left\{\sqrt{mx^3+nx^2+\frac{n^2}{3m}x+\frac{n^3}{27m^2}}+\sqrt{mx^3+nx^2+\frac{n^2}{3m}x+p}\right\}.$$

$\int \mathrm{U}\sqrt{\mathrm{X}^4}\,dx$

$$\int\frac{dx}{x\sqrt{a^2+x^4}} = \frac{1}{2a}\log\frac{x^2}{a+\sqrt{a^2+x^4}}$$

$$\int\frac{dx}{x\sqrt{x^4-1}} = \frac{1}{2}\arccos\frac{1}{x^2} = \frac{1}{2}\,\mathrm{arc\,séc}\,x^2.$$

$$\int\frac{dx}{x\sqrt{a^2-x^4}} = \frac{1}{2a}\log\frac{x^2}{a+\sqrt{a^2-x^4}}.$$

$$\int\frac{dx}{x\sqrt{(a+bx^2)(a+cx^2)}} = -\frac{1}{a}\log\left\{\sqrt{\frac{a}{x^2}+b}+\sqrt{\frac{a}{x^2}+c}\right\}.$$

$$\int\frac{dx}{x\sqrt{(x^2-a^2)(b^2-x^2)}} = \frac{1}{ab}\arcsin\frac{b}{x}\sqrt{\frac{x^2-a^2}{b^2-a^2}} = \frac{1}{ab}\,\mathrm{arc\,tg}\,\frac{b}{a}\sqrt{\frac{x^2-a^2}{b^2-x^2}}.$$

$$\int\frac{dx}{x^3\sqrt{a+bx^4}} = -\frac{\sqrt{a+bx^4}}{2ax^2}.$$

$$\int\frac{x}{\sqrt{1-x^4}}\,dx = \frac{1}{2}\arccos\sqrt{1-x^4}$$

$$\int \frac{x}{\sqrt{a^4 - x^4}}\,dx = \frac{1}{2}\,\text{arc sin}\,\frac{x^2}{a^2}.$$

$$\int \frac{x}{\sqrt{a^2x^4 + b}}\,dx = \frac{1}{2a}\log\left(ax^2 + \sqrt{a^2x^4 + b}\right).$$

$$\int \frac{x}{\sqrt{(ax^2 + b)(ax^2 + c)}}\,dx = \frac{1}{a}\log\left\{\sqrt{ax^2 + b} + \sqrt{ax^2 + c}\right\}.$$

$$\int \frac{x}{\sqrt{(x^2 - a^2)(b^2 - x^2)}}\,dx = \text{arc sin}\sqrt{\frac{x^2 - a^2}{b^2 - a^2}} = \text{arc tg}\sqrt{\frac{x^2 - a^2}{b^2 - x^2}}.$$

$$\int \frac{x}{\sqrt{ax^4 + bx^2 + c}}\,dx = \frac{1}{2}\int \frac{dz}{\sqrt{az^2 + bz + c}}. \qquad x^2 = z.$$

$$\int \frac{x}{\sqrt{x^4 + 2x^3 - 3x^2 - ax + a}}\,dx = \frac{1}{3}\log\frac{(x+2)\sqrt{x-1} + \sqrt{x^3 + 3x^2 - a}}{(x+2)\sqrt{x-1} - \sqrt{x^3 + 3x^2 - a}} = \frac{2}{3}\log\left\{\sqrt{x^3 + 3x^2 - 4} + \sqrt{x^3 + 3x^2 - a}\right\}.$$

$$\int \frac{x}{\sqrt{mx^4 + \frac{2}{3}nx^3 - \frac{n^2}{3m}x + px - \frac{np}{3m}}}\,dx = \frac{2}{3\sqrt{m}}\log\left\{\sqrt{mx^3 + nx^2 - \frac{4n^3}{27m^2}} + \sqrt{mx^3 + nx^2 + p}\right\}.$$

$$\int \frac{x}{(a^2 + x^2)\sqrt{a^4 - x^4}}\,dx = -\frac{1}{2a^2}\sqrt{\frac{a^2 - x^2}{a^2 + x^2}}.$$

$$\int \frac{x}{(1 - x^3)\sqrt{x^4 - 2x}}\,dx = \frac{1}{3}\,\text{arc sin}\,\frac{1}{1 - x^3}.$$

$$\int \frac{3x + a}{\sqrt{(x^2 + ax)^2 + bx}}\,dx = \log\frac{\sqrt{\Phi} + x^2 + ax}{\sqrt{\Phi} - x^2 - ax}.$$

$$\int \frac{1 + x}{(1 - x)\sqrt{ax^4 + bx^3 + cx^2 + bx + a}}\,dx = -\int \frac{1 - \frac{1}{x^2}}{\left(x + \frac{1}{x} - 2\right)\sqrt{a\left(x^2 + \frac{1}{x^2}\right) + b\left(x + \frac{1}{x}\right) + c}}\,dx$$

$$= -\int \frac{dz}{(z - 2)\sqrt{az^2 + bz - 2a + c}}. \qquad x + \frac{1}{x} = z.$$

$$\int \frac{x^2}{(1 - x^4)\sqrt{1 + x^4}}\,dx = \frac{1}{4\sqrt{2}}\log\frac{x\sqrt{2} + \sqrt{1 + x^4}}{1 - x^2} - \frac{1}{4\sqrt{2}}\,\text{arc sin}\,\frac{x\sqrt{2}}{1 + x^2}.$$

$$\int \frac{x^2 + 1}{(1 - x^2)\sqrt{1 + x^4}}\,dx = \frac{1}{\sqrt{2}}\log\frac{x\sqrt{2} + \sqrt{1 + x^4}}{1 - x^2}.$$

$$\int \frac{x^2 + 1}{(x^2 - 1)\sqrt{1 + x^4}}\,dx = \frac{1}{\sqrt{2}}\log\frac{x^2 - 1}{x\sqrt{2} + \sqrt{1 + x^4}}.$$

$$\int \frac{x^2 + 1}{(x^2 + 2cx - 1)\sqrt{(x^2 + 2ax - 1)(x^2 + 2bx - 1)}}\,dx = \int \frac{dz}{(a - c)z^2 + c - b}. \qquad \frac{x^2 + 2bx - 1}{x^2 + 2ax - 1} = z^2.$$

$$\int \frac{3x^2 + 1}{(2x^4 - x^2 + 1)\sqrt{1 + x^4}}\,dx = -\frac{1}{\sqrt{2}}\,\text{arc sin}\,\frac{x\sqrt{2}}{2x^4 - x^2 + 1}.$$

$$\int \frac{x^2 - 1}{(x^2 + 2cx + 1)\sqrt{(x^2 + 2ax + 1)(x^2 + 2bx + 1)}}\,dx = \int \frac{dz}{(a - c)z^2 + c - b}. \qquad \frac{x^2 + 2bx + 1}{x^2 + 2ax + 1} = z^2.$$

$$\int \frac{1 - x^2}{(1 + x^2)\sqrt{1 + x^4}}\,dx = \frac{1}{\sqrt{2}}\,\text{arc sin}\,\frac{x\sqrt{2}}{1 + x^2} = -\frac{1}{\sqrt{2}}\,\text{arc tg}\,\frac{\sqrt{1 + x^4}}{x\sqrt{2}}.$$

$$\int \frac{1 - x^2}{(1 + x^2)\sqrt{x^4 + 2x^2\cos 2\alpha + 1}}\,dx = \frac{1}{2\sin\alpha}\,\text{arc sin}\,\frac{2x\sin\alpha}{1 + x^2}.$$

$$\int \frac{1 - cx^2}{(1 + cx^2)\sqrt{c^2x^4 + ax^2 + 1}}\,dx = \frac{1}{\sqrt{a - 2c}} \log \frac{x\sqrt{a - 2c} + \sqrt{\Phi}}{1 + cx^2}. \qquad a > 2c.$$

$$= -\frac{1}{\sqrt{2c - a}} \operatorname{arc\,cos} \frac{x\sqrt{2c - a}}{1 + cx^2}. \qquad a < 2c.$$

$$\int \frac{2x^2 + x - 1}{(x^2 - x + 2)\sqrt{x^4 - 1}}\,dx = \int \frac{2x - 1}{x^2 - x + 2}\sqrt{\frac{x + 1}{x - 1}}\,\frac{1}{\sqrt{x^2 + 1}}\,dx = \operatorname{arc\,sin} \frac{(x + 1)\sqrt{2(x^2 - 1)}}{2(x^2 - x + 2)}.$$

$$\int \frac{x^3}{\sqrt{ax^4 + b}}\,dx = \frac{1}{2a}\sqrt{ax^4 + b}.$$

$$\int \frac{x^4}{\sqrt{ax^4 + bx^2 + c}}\,dx = \frac{x\sqrt{\Phi}}{3a} - \frac{c}{3a}\int \frac{dx}{\sqrt{\Phi}} - \frac{2b}{3a}\int \frac{x^2}{\sqrt{\Phi}}\,dx.$$

$$\int \frac{x^2(x^2 + \sqrt{3})}{(2x^4 - x^2\sqrt{3} - 3)\sqrt{1 - x^4}}\,dx = \frac{1}{(108)^{\frac{1}{4}}} \operatorname{arc\,tg} \frac{3\sqrt{1 - x^4}}{(12)^{\frac{1}{4}}\,x^3}.$$

$$\int \frac{\sqrt{1 + x^4}}{x}\,dx = \int \frac{dx}{x\sqrt{\Phi}} + \int \frac{x^3}{\sqrt{\Phi}}\,dx = \frac{1}{2}\sqrt{\Phi} + \frac{1}{4} \log \frac{\sqrt{\Phi} - 1}{\sqrt{\Phi} + 1}.$$

$$\int \frac{\sqrt{a^2 - x^4}}{x^3}\,dx = -\frac{\sqrt{a^2 - x^4}}{2x^2} - \frac{1}{2} \operatorname{arc\,sin} \frac{x^2}{a}.$$

$$\int \frac{\sqrt{1 + x^4}}{1 - x^4}\,dx = \frac{1}{2\sqrt{2}} \log \frac{x\sqrt{2} + \sqrt{\Phi}}{x^2 - 1} + \frac{1}{2\sqrt{2}} \operatorname{arc\,sin} \frac{x\sqrt{2}}{1 + x^2}.$$

$$= \frac{1}{2\sqrt{2}} \log \frac{x\sqrt{2} + \sqrt{\Phi}}{1 - x^2} - \frac{1}{2\sqrt{2}} \operatorname{arc\,tg} \frac{\sqrt{\Phi}}{x\sqrt{2}}.$$

$$\int \frac{\sqrt{x^4 + 2x^2\cos 2\alpha + 1}}{1 - x^4}\,dx = \frac{1}{2}\cos\alpha \log \frac{2x\cos\alpha + \sqrt{\Phi}}{1 - x^2} + \frac{1}{2}\sin\alpha \operatorname{arc\,sin} \frac{2x\sin\alpha}{1 + x^2}.$$

$$\int \mathrm{U}\sqrt{x^n}\,dx$$

$$\int \frac{2x - 1}{(x + 1)\sqrt{(x + 1)^6 - a^2x^2}}\,dx = \int \frac{2x - 1}{(x + 1)^4}\,\frac{1}{\sqrt{1 - \frac{a^2x^2}{(x + 1)^6}}}\,dx = \frac{1}{a}\int d\theta. \qquad \frac{ax}{(x + 1)^3} = \cos\theta.$$

$$= \frac{1}{a} \operatorname{arc\,cos} \frac{ax}{(x + 1)^3}.$$

$$\int \frac{x^2}{\sqrt{a^6 + x^6}}\,dx = \frac{1}{3} \log \left\{ x^3 + \sqrt{a^6 + x^6} \right\}.$$

$$\int \frac{x^2}{\sqrt{a^2b^2 - x^6}}\,dx = \frac{1}{3} \operatorname{arc\,sin} \frac{x^3}{ab}.$$

$$\int \frac{x^3}{\sqrt{1 + x^8}}\,dx = \frac{1}{4} \log \left\{ x^4 + \sqrt{1 + x^8} \right\}.$$

$$\int \frac{x^{m-1}}{\sqrt{a^{2m} - x^{2m}}}\,dx = \frac{1}{m} \operatorname{arc\,sin} \left(\frac{x}{a}\right)^m.$$

$$\int \frac{x^{m-1}}{\sqrt{1 - (ax^m + b)^2}}\,dx = \frac{1}{ma} \operatorname{arc\,sin}(ax^m + b).$$

$$\int \frac{x^{m-1}}{\sqrt{(x^m + a)(x^m + b)}}\,dx = \frac{2}{m} \log \left\{ \sqrt{x^m + a} + \sqrt{x^m + b} \right\}.$$

$$\int \frac{x^{m-1}(x^m+1)}{(x^m-1)\sqrt{x^m(x^{2m}+1)}}\,dx = -\frac{\sqrt{2}}{m}\log\frac{\sqrt{x^{2m}+1}+\sqrt{2x^m}}{x^m-1}.$$

$\int U\, x^{\frac{3}{2}}\, dx$

$$\int \frac{dx}{(x-a)\,x^{\frac{3}{2}}} = \frac{2}{a\sqrt{x}} + \frac{1}{a\sqrt{a}}\log\frac{\sqrt{x}-\sqrt{a}}{\sqrt{x}+\sqrt{a}}.$$

$$\int \frac{dx}{(a+bx)\,x^{\frac{3}{2}}} = -\frac{2}{a\sqrt{x}} - \frac{b}{a}\int\frac{dx}{(a+bx)\sqrt{x}}.$$

$$\int \frac{dx}{(x-a)^2\,x^{\frac{3}{2}}} = \frac{3x-2a}{a^2(x-a)\sqrt{x}} + \frac{3}{2a^2\sqrt{a}}\log\frac{\sqrt{x}+\sqrt{a}}{\sqrt{x}-\sqrt{a}} = 2\int\frac{dz}{z^2(z^2-a)^2}. \qquad x = z^2.$$

$$\int \frac{dx}{(a+bx)^2\,x^{\frac{3}{2}}} = -\frac{2}{a(a+bx)\sqrt{x}} - \frac{3b}{a}\int\frac{dx}{(a+bx)^2\sqrt{x}}.$$

$$\int \frac{x^{\frac{3}{2}}}{x-a}\,dx = \frac{2}{3}x^{\frac{3}{2}} + 2a\sqrt{x} + a\sqrt{a}\log\frac{\sqrt{x}-\sqrt{a}}{\sqrt{x}+\sqrt{a}}.$$

$$\int \frac{x^{\frac{3}{2}}}{a+bx}\,dx = \frac{bx-3a}{3b^2}\,2\sqrt{x} + \frac{a^2}{b^2}\int\frac{dx}{(a+bx)\sqrt{x}}.$$

$$\int \frac{x^{\frac{3}{2}}}{a+bx^2}\,dx = \frac{2}{a}\sqrt{x} - \int\frac{dx}{(a+bx^2)\sqrt{x}}.$$

$$\int \frac{x^{\frac{3}{2}}}{(x-a)^2}\,dx = \frac{2x-3a}{x-a}\sqrt{x} + \frac{3}{2}\sqrt{a}\log\frac{\sqrt{x}-\sqrt{a}}{\sqrt{x}+\sqrt{a}}.$$

$$\int \frac{x^{\frac{3}{2}}}{(a+bx)^2}\,dx = \frac{2x^{\frac{3}{2}}}{b(a+bx)} - \frac{3a}{b}\int\frac{\sqrt{x}}{(a+bx)^2}\,dx.$$

$$\int \frac{x^{\frac{3}{2}}}{(a+bx^2)^2}\,dx = -\frac{\sqrt{x}}{2b(a+bx^2)} + \frac{1}{4b}\int\frac{dx}{(a+bx^2)\sqrt{x}}.$$

$\int U\,(a+bx)^{\frac{3}{2}}\, dx$

$$\int (a+bx)^{\frac{3}{2}}\,dx = \frac{2}{5b}(a+bx)^{\frac{5}{2}}.$$

$$\int x(a+bx)^{\frac{3}{2}}\,dx = \frac{2}{b^2}\left\{\frac{\Phi}{7} - \frac{a}{5}\right\}\Phi^{\frac{5}{2}}.$$

$$\int x^2(a+bx)^{\frac{3}{2}}\,dx = \frac{2}{b^3}\left\{\frac{\Phi^2}{9} - \frac{2}{7}a\Phi + \frac{a^2}{5}\right\}\Phi^{\frac{5}{2}}.$$

$$\int \frac{dx}{(a+bx)^{\frac{3}{2}}} = -\frac{2}{b\sqrt{a+bx}}.$$

$$\int \frac{dx}{x(a+bx)^{\frac{3}{2}}} = \frac{2}{a\sqrt{a+bx}} + \frac{1}{a}\int\frac{dx}{x\sqrt{a+bx}}.$$

$$\int \frac{dx}{x^2(a+bx)^{\frac{3}{2}}} = -\frac{a+3bx}{a^2x\sqrt{a+bx}} - \frac{3b}{2a^2}\int\frac{dx}{x\sqrt{a+bx}}.$$

$$\int \frac{x}{(a+bx)^{\frac{3}{2}}}\,dx = \frac{2(2a+bx)}{b^2\sqrt{a+bx}}.$$

$$\int \frac{x-2}{(x-1)^{\frac{3}{2}}}\, dx = \frac{2x}{\sqrt{x-1}}$$

$$\int \frac{x^2}{(a+bx)^{\frac{3}{2}}}\, dx = \left\{\frac{\Phi^2}{3} - 2a\Phi - a^2\right\}\frac{2}{b^3\sqrt{\Phi}}.$$

$$\int \frac{x^3}{(a+bx)^{\frac{3}{2}}}\, dx = \left\{\frac{\Phi^3}{5} - a\Phi^2 + 3a^2\Phi - a^3\right\}\frac{2}{b^4\sqrt{\Phi}}.$$

$$\int \frac{(a+bx)^{\frac{3}{2}}}{x}\, dx = \left\{\frac{2}{3}\Phi + 2a\right\}\sqrt{\Phi} + a^2\int\frac{dx}{x\sqrt{\Phi}}.$$

$$\int \frac{(a+bx)^{\frac{3}{2}}}{x^2}\, dx = -\frac{\Phi^{\frac{5}{2}}}{ax} + \frac{3b}{2a}\int\frac{\Phi^{\frac{3}{2}}}{x}\, dx = 2b\int\frac{z^4}{(z^2-a)^2}\, dz. \qquad a + bx = z^2.$$

$$\int \frac{(a+bx)^{\frac{3}{2}}}{x^3}\, dx = 2b^2\int\frac{z^4}{(z^2-a)^3}\, dz. \qquad a + bx = z^2.$$

$$\int \left(\frac{x-a}{x-b}\right)^{\frac{3}{2}} dx = \sqrt{(x-a)(x-b)} + 2(a-b)\sqrt{\frac{x-a}{x-b}} + \frac{3}{2}(a-b)\log\frac{\sqrt{x-b}-\sqrt{x-a}}{\sqrt{x-b}+\sqrt{x-a}}.$$

$\int U\,(a+bx^2)^{\frac{3}{2}}\, dx$

$$\int (a^2-x^2)^{\frac{3}{2}}\, dx = \frac{5a^2-2x^2}{8}\, x\sqrt{a^2-x^2} + \frac{3a^4}{8}\arcsin\frac{x}{a}.$$

$$\int (a+bx^2)^{\frac{3}{2}}\, dx = a\int\sqrt{\Phi}\, dx + b\int x^2\sqrt{\Phi}\, dx.$$

$$\int x\,(a+bx^2)^{\frac{3}{2}}\, dx = \frac{1}{5b}\Phi^{\frac{5}{2}}.$$

$$\int x^2(a+bx^2)^{\frac{3}{2}}\, dx = \frac{1}{6b}\, x\,\Phi^{\frac{5}{2}} - \frac{a}{6b}\int\Phi^{\frac{3}{2}}\, dx = \frac{x}{48b}(8b^2x^4 + 14abx^2 + 9a^2)\sqrt{\Phi} + \frac{a^3}{16b}\int\frac{dx}{\sqrt{\Phi}}.$$

$$\int x^3(1-x^2)^{\frac{3}{2}}\, dx = -\frac{1}{7}\left\{x^4 - \frac{8}{5}x^2 + \frac{1}{5}\right\}x^2\sqrt{\Phi} - \frac{2}{35}\sqrt{\Phi}.$$

$$\int x^4(1-x^2)^{\frac{3}{2}}\, dx = -\frac{1}{8}\left\{x^6 - \frac{3}{2}x^4 + \frac{1}{8}x^2 + \frac{3}{16}\right\}x\sqrt{\Phi} + \frac{3}{128}\arcsin x.$$

$$\int \frac{dx}{(a+bx^2)^{\frac{3}{2}}} = \frac{x}{a\sqrt{\Phi}}.$$

$$\int \frac{dx}{x\,(ax^2-1)^{\frac{3}{2}}} = -\frac{1}{\sqrt{\Phi}} - \operatorname{arc\,tg}\sqrt{\Phi}.$$

$$\int \frac{dx}{x(a+bx^2)^{\frac{3}{2}}} = \frac{1}{a\sqrt{\Phi}} + \frac{1}{a}\int\frac{dx}{x\sqrt{\Phi}}.$$

$$\int \frac{dx}{x^2(a+bx^2)^{\frac{3}{2}}} = -\frac{a+2bx^2}{a^2x\sqrt{\Phi}}.$$

$$\int \frac{dx}{x^3(1-x^2)^{\frac{3}{2}}} = \frac{3x^2-1}{2x^2\sqrt{1-x^2}} + \frac{3}{2}\log\frac{\sqrt{1-x^2}-1}{x}.$$

$$\int \frac{dx}{x^3(a+bx^2)^{\frac{3}{2}}} = \frac{1}{ax^2\sqrt{\Phi}} - \frac{3\sqrt{\Phi}}{2a^2x^2} + \frac{3b}{2a^2\sqrt{a}}\log\frac{\sqrt{a}+\sqrt{\Phi}}{x}.$$

$$\int \frac{dx}{x^4 (1 - x^2)^{\frac{3}{2}}} = - \frac{1 + 4x^2 - 8x^4}{3x^3 \sqrt{1 - x^2}}.$$

$$\int \frac{dx}{x^4 (a + bx^2)^{\frac{3}{2}}} = - \frac{\Phi^2 - 6bx^2\Phi - 3b^2x^4}{3a^3x^3 \sqrt{\Phi}}.$$

$$\int \frac{x}{(a + bx^2)^{\frac{3}{2}}} dx = - \frac{1}{b\sqrt{a + bx^2}}.$$

$$\int \frac{x^2}{(1 - x^2)^{\frac{3}{2}}} dx = \frac{x}{\sqrt{1 - x^2}} + \text{arc cos } x.$$

$$\int \frac{x^2}{(a^2 - x^2)^{\frac{3}{2}}} dx = \frac{x}{\sqrt{\Phi}} + \text{arc sin } \frac{\sqrt{\Phi}}{a}.$$

$$\int \frac{x^2}{(a + bx^2)^{\frac{3}{2}}} dx = - \frac{x}{b\sqrt{\Phi}} + \frac{1}{b} \int \frac{dx}{\sqrt{\Phi}}.$$

$$\int \frac{x^3}{(a + bx^2)^{\frac{3}{2}}} dx = \frac{2a + bx^2}{b^2 \sqrt{\Phi}}.$$

$$\int \frac{x^4}{(1 - x^2)^{\frac{3}{2}}} dx = \frac{3x - x^3}{2\sqrt{1 - x^2}} - \frac{3}{2} \text{arc sin } x.$$

$$\int \frac{x^4}{(a^2 - x^2)^{\frac{3}{2}}} dx = \frac{x^3}{\sqrt{a^2 - x^2}} - 3 \int \frac{x^2}{\sqrt{a^2 - x^2}} dx.$$

$$\int \frac{(a + bx^2)^{\frac{3}{2}}}{x} dx = \frac{1}{3} (4a + bx^2)\sqrt{\Phi} + a^2 \int \frac{dx}{x\sqrt{\Phi}}.$$

$$\int \frac{(a + bx^2)^{\frac{3}{2}}}{x^2} dx = - \frac{\Phi^{\frac{5}{2}}}{ax} + \frac{4b}{a} \int \Phi^{\frac{3}{2}} dx.$$

$$\int \frac{(a + bx^2)^{\frac{3}{2}}}{x^3} dx = \frac{2bx^2 - a}{2x^2} \sqrt{\Phi} + \frac{3ab}{2} \int \frac{dx}{x\sqrt{\Phi}}.$$

$$\int \frac{(a + bx^2)^{\frac{3}{2}}}{x^5} dx = - \frac{\Phi^{\frac{3}{2}}}{4x^4} + \frac{3b}{4} \int \frac{\sqrt{\Phi}}{x^3} dx.$$

$\int \mathbf{U} (ax^2 + bx)^{\frac{3}{2}} dx$

$$\int (ax^2 + bx)^{\frac{3}{2}} dx = \left\{ \frac{\Phi}{a} - \frac{3b^2}{8a^2} \right\} \frac{b + 2ax}{8} \sqrt{\Phi} + \frac{3b^4}{128a^2} \int \frac{dx}{\sqrt{\Phi}}.$$

$$\int x (ax^2 + bx)^{\frac{3}{2}} dx = \frac{\Phi^{\frac{5}{2}}}{5a} - \frac{b}{2a} \int \Phi^{\frac{3}{2}} dx.$$

$$\int \frac{dx}{(ax^2 + bx)^{\frac{3}{2}}} = - \frac{2(2ax + b)}{b^2 \sqrt{\Phi}}.$$

$$\int \frac{dx}{x (ax^2 + bx)^{\frac{3}{2}}} = - \frac{2}{3bx\sqrt{\Phi}} - \frac{4a}{3b} \int \frac{dx}{\Phi^{\frac{3}{2}}}.$$

$$\int \frac{dx}{x^2 (ax^2 + bx)^{\frac{3}{2}}} = \frac{2(2ax - b)}{5b^2x^2\sqrt{\Phi}} + \frac{8a^2}{5b^2} \int \frac{dx}{\Phi^{\frac{3}{2}}}.$$

$$\int \frac{dx}{x^3 (ax^2 + bx)^{\frac{3}{2}}} = \frac{2}{7bx\sqrt{\Phi}} \left\{ - \frac{1}{x^2} + \frac{8a}{5bx} - \frac{16a^2}{5b^2} \right\} - \frac{64a^3}{35b^3} \int \frac{dx}{\Phi^{\frac{3}{2}}}.$$

$$\int \frac{x}{(2ax - x^2)^{\frac{3}{2}}} dx = \frac{1}{a} \sqrt{\frac{x}{2a - x}}.$$

$$\int \frac{x}{(ax^2 + bx)^{\frac{3}{2}}} dx = \frac{2x}{b\sqrt{ax^2 + bx}}.$$

$$\int \frac{x^2}{(ax^2 + bx)^{\frac{3}{2}}} dx = -\frac{2x}{a\sqrt{\Phi}} + \frac{1}{a}\int \frac{dx}{\sqrt{\Phi}}.$$

$$\int \mathrm{U}\,(ax^2 + bx + c)^{\frac{3}{2}}\,dx$$

$$\int (ax^2 + bx + c)^{\frac{3}{2}} dx = \frac{2ax + b}{8a}\Phi^{\frac{3}{2}} + \frac{3(4ac - b^2)}{16a}\int \sqrt{\Phi}\,dx.$$

$$\int x\,(ax^2 + bx + c)^{\frac{3}{2}} dx = \frac{\Phi^{\frac{5}{2}}}{5a} - \frac{b}{2a}\int \Phi^{\frac{3}{2}} dx.$$

$$\int x^2\,(ax^2 + bx + c)^{\frac{3}{2}} dx = \left\{ x - \frac{7b}{10a} \right\} \frac{\Phi^{\frac{5}{2}}}{6a} + \frac{7b^2 - 4ac}{24a^2}\int \Phi^{\frac{3}{2}} dx.$$

$$\int x^3\,(ax^2 + bx + c)^{\frac{3}{2}} dx = \left\{ x^2 - \frac{3bx}{4a} + \frac{21b^2}{40a^2} - \frac{2c}{5a} \right\} \frac{\Phi^{\frac{5}{2}}}{7a} + \left\{ \frac{bc}{4a^2} - \frac{3b^3}{16a^3} \right\} \int \Phi^{\frac{3}{2}} dx.$$

$$\int \frac{dx}{(1 + x + x^2)^{\frac{3}{2}}} = \frac{2(2x + 1)}{3\sqrt{\Phi}}.$$

$$\int \frac{dx}{\{(x - a)(x - b)\}^{\frac{3}{2}}} = \frac{2(a + b - 2x)}{(a - b)^2\sqrt{\Phi}}.$$

$$\int \frac{dx}{(ax^2 + bx + c)^{\frac{3}{2}}} = \frac{2(2ax + b)}{(4ac - b^2)\sqrt{\Phi}}.$$

$$\int \frac{dx}{x\,(x^2 + x + 1)^{\frac{3}{2}}} = \frac{2(1 - x)}{3\sqrt{\Phi}} + \log \frac{2 + x - 2\sqrt{\Phi}}{x}.$$

$$\int \frac{dx}{x\,(ax^2 + bx + c)^{\frac{3}{2}}} = \frac{4ac - 2b^2 - 2abx}{c\,(4ac - b^2)\sqrt{\Phi}} + \frac{1}{c}\int \frac{dx}{x\sqrt{\Phi}}.$$

$$\int \frac{dx}{x^2\,(x^2 + x + 1)^{\frac{3}{2}}} = -\frac{5x^2 + 7x + 3}{3x\sqrt{\Phi}} - \frac{3}{2}\log \frac{2 + x - 2\sqrt{\Phi}}{x}.$$

$$\int \frac{dx}{x^2\,(ax^2 + bx + c)^{\frac{3}{2}}} = -\frac{4a(2ax + b)}{c\,(4ac - b^2)\sqrt{\Phi}} - \frac{1}{cx\sqrt{\Phi}} - \frac{3b}{2c}\int \frac{dx}{x\Phi^{\frac{3}{2}}}.$$

$$\int \frac{dx}{x^3\,(x^2 + x + 1)^{\frac{3}{2}}} = \frac{37x^3 + 23x^2 + 15x - 6}{12x^2\sqrt{\Phi}} + \frac{3}{8}\log \frac{2 + x - 2\sqrt{\Phi}}{x}.$$

$$\int \frac{dx}{x^3\,(ax^2 + bx + c)^{\frac{3}{2}}} = \left\{ \frac{5b}{2c} - \frac{1}{x} \right\} \left\{ \frac{1}{2cx\sqrt{\Phi}} + \frac{5ab\,(2ax + b)}{c^2\,(4ac - b^2)\sqrt{\Phi}} \right\} + \frac{3(5b^2 - 4ac)}{8c^2}\int \frac{dx}{x\Phi^{\frac{3}{2}}}.$$

$$\int \frac{x}{(x^2 + x + 1)^{\frac{3}{2}}} dx = -\frac{2(x + 2)}{3\sqrt{\Phi}}.$$

$$\int \frac{x}{(ax^2 + bx + c)^{\frac{3}{2}}} dx = -\frac{2(2c + bx)}{(4ac - b^2)\sqrt{\Phi}}.$$

$$\int \frac{mx + n}{(5x^2 + 6x + 2)^{\frac{3}{2}}} dx = \frac{3n - 2m + (5n - 3m)\,x}{\sqrt{5x^2 + 6x + 2}}.$$

$$\int \frac{mx+n}{\{(x-a)(x-b)\}^{\frac{3}{2}}}\,dx = \frac{2n(a+b-2x)+2m\{2ab-(a+b)x\}}{(a-b)^2\sqrt{\Phi}}.$$

$$\int \frac{x^2}{(x^2+x+1)^{\frac{3}{2}}}\,dx = -\frac{2(x-1)}{3\sqrt{\Phi}} + \log\left(1+2x+2\sqrt{\Phi}\right).$$

$$\int \frac{x^2}{(ax^2+bx+c)^{\frac{3}{2}}}\,dx = -\frac{(4ac-2b^2)x-2bc}{a(4ac-b^2)\sqrt{\Phi}} + \frac{1}{a}\int\frac{dx}{\sqrt{\Phi}}.$$

$$\int \frac{x^3}{(x^2+x+1)^{\frac{3}{2}}}\,dx = \frac{3x^2+7x+5}{3\sqrt{\Phi}} - \frac{3}{2}\log\left(1+2x+2\sqrt{\Phi}\right).$$

$$\int \frac{x^3}{(ax^2+bx+c)^{\frac{3}{2}}}\,dx = \left\{\frac{x^2}{a} + \frac{10ac-3b^2}{a^2(4ac-b^2)}bx + \frac{c(8ac-3b^2)}{a^2(4ac-b^2)}\right\}\frac{1}{\sqrt{\Phi}} - \frac{3b}{2a^2}\int\frac{dx}{\sqrt{\Phi}}$$

$$\int \frac{(ax^2+bx+c)^{\frac{3}{2}}}{x}\,dx = \frac{3c+\Phi}{3}\sqrt{\Phi} + \frac{b}{2}\int\sqrt{\Phi}\,dx + \frac{bc}{2}\int\frac{dx}{\sqrt{\Phi}} + c^2\int\frac{dx}{x\sqrt{\Phi}}.$$

$$\int \frac{(ax^2+bx+c)^{\frac{3}{2}}}{x^2}\,dx = \frac{3bx-2\Phi}{2x}\sqrt{\Phi} + \frac{3b^2}{4}\int\frac{dx}{\sqrt{\Phi}} + 3a\int\sqrt{\Phi}\,dx + \frac{3bc}{2}\int\frac{dx}{x\sqrt{\Phi}}.$$

$$\int \frac{(ax^2+bx+c)^{\frac{3}{2}}}{x^3}\,dx = -\left\{\frac{\Phi}{2x^2} + \frac{3b}{4x} - \frac{3a}{2}\right\}\sqrt{\Phi} + \left\{\frac{3b^2}{8} + \frac{3ac}{2}\right\}\int\frac{dx}{x\sqrt{\Phi}} + \frac{3ab}{2}\int\frac{dx}{\sqrt{\Phi}}.$$

—— $\int UX^{\frac{3}{2}}\,dx$ ——

$$\int \frac{x^2}{(a+bx^3)^{\frac{3}{2}}}\,dx = -\frac{2}{3b\sqrt{a+bx^3}}.$$

$$\int \frac{x^2}{(x^3-1)(x^3+1)^{\frac{3}{2}}}\,dx = \frac{1}{3\sqrt{x^3+1}} + \frac{1}{3\sqrt{2}}\log\frac{\sqrt{x^3+1}-\sqrt{2}}{\sqrt{x^3-1}}.$$

$$\int \frac{x}{(a+bx^4)^{\frac{3}{2}}}\,dx = \frac{x^2}{2a\sqrt{a+bx^4}}.$$

—— $\int UX^{\frac{5}{2}}\,dx$ ——

$$\int (a+bx)^{\frac{5}{2}}\,dx = \frac{2}{7b}(a+bx)^{\frac{7}{2}}.$$

$$\int \frac{dx}{(a+bx)^{\frac{5}{2}}} = -\frac{2}{3b(a+bx)^{\frac{3}{2}}}.$$

$$\int \frac{dx}{x(a+bx)^{\frac{5}{2}}} = \frac{2(4a+3bx)}{3a^2\Phi^{\frac{3}{2}}} + \frac{1}{a^2}\int\frac{dx}{x\sqrt{\Phi}}.$$

$$\int \frac{x}{(a+bx)^{\frac{5}{2}}}\,dx = -\frac{2(2a+3bx)}{3b^2(a+bx)^{\frac{3}{2}}}.$$

$$\int \frac{x^2}{(a+bx)^{\frac{5}{2}}}\,dx = \frac{2}{3b^3}\,\frac{3\Phi^2+6a\Phi-a^2}{\Phi^{\frac{3}{2}}}.$$

$$\int \frac{x^3}{(a+bx)^{\frac{5}{2}}}\,dx = \left\{\frac{\Phi^3}{3} - 3a\Phi^2 - 3a^2\Phi + \frac{a^3}{3}\right\}\frac{2}{b^4\Phi^{\frac{3}{2}}}.$$

$$\int \frac{(a+bx)^{\frac{5}{2}}}{x}\,dx = \left\{\frac{\Phi^2}{5} + \frac{a\Phi}{3} + a^2\right\}2\sqrt{\Phi} + a^3\int\frac{dx}{x\sqrt{\Phi}}.$$

$$\int (a^2 - x^2)^{\frac{5}{2}} dx = \frac{x}{48} \left\{ 8x^4 - 26a^2x^2 + 33a^4 \right\} \sqrt{\Phi} + \frac{5a^6}{16} \text{arc sin} \frac{x}{a} \cdot$$

$$\int \frac{dx}{(a^2 + x^2)^{\frac{5}{2}}} = \frac{x\,(2x^2 + 3a^2)}{3a^4\,(a^2 + x^2)^{\frac{3}{2}}} \cdot$$

$$\int \frac{dx}{(a + bx^2)^{\frac{5}{2}}} = \frac{x\,(3a + 2bx^2)}{3a^2\,(a + bx^2)^{\frac{3}{2}}} \cdot$$

$$\int \frac{dx}{x\,(a + bx^2)^{\frac{5}{2}}} = \frac{1}{a^2\sqrt{\Phi}} + \frac{1}{3a\Phi^{\frac{3}{2}}} + \frac{1}{2a^2\sqrt{a}} \log \frac{\sqrt{\Phi} - \sqrt{a}}{\sqrt{\Phi} + \sqrt{a}} \cdot$$

$$\int \frac{x^2}{(a + bx^2)^{\frac{5}{2}}} dx = \frac{x^3}{3a\,(a + bx^2)^{\frac{3}{2}}} \cdot$$

$$\int \frac{dx}{(ax^2 + bx)^{\frac{5}{2}}} = \frac{2}{3b^4} \frac{8a^2x^2 + 8abx - b^2}{(ax^2 + bx)^{\frac{3}{2}}} (2ax + b).$$

$$\int (a + bx + cx^2)^{\frac{5}{2}} dx = \frac{2cx + b}{12c} \Phi^{\frac{5}{2}} + \frac{5\,(4ac - b^2)}{24c} \int \Phi^{\frac{3}{2}} dx.$$

$$\int x^2\,(a + bx + cx^2)^{\frac{5}{2}} dx = \left(x - \frac{9b}{14c}\right) \frac{\Phi^{\frac{7}{2}}}{8c} + \frac{9b^2 - 4ac}{32c^2} \int \Phi^{\frac{5}{2}}\, dx.$$

$$\int \frac{dx}{(a + bx + cx^2)^{\frac{5}{2}}} = \frac{2\,(2cx + b)}{3\,(4ac - b^2)\sqrt{\Phi}} \left\{ \frac{1}{\Phi} + \frac{8c}{4ac - b^2} \right\} \cdot$$

$$\int \frac{x}{(a + bx + cx^2)^{\frac{5}{2}}} dx = - \frac{2\,(bx + 2a)}{3\,(4ac - b^2)\,\Phi^{\frac{3}{2}}} - \frac{8b\,(2cx + b)}{3\,(4ac - b^2)^2\sqrt{\Phi}} \cdot$$

$$\int \frac{x^2}{(a + bx + cx^2)^{\frac{5}{2}}} dx = \frac{(2b^2 - 4ac)\,x + 2ab}{3c\,(4ac - b^2)\,\Phi^{\frac{3}{2}}} + \frac{2\,(4ac + b^2)\,(2cx + b)}{3c\,(4ac - b^2)^2\sqrt{\Phi}} \cdot$$

$\int \text{U}\,\text{X}^{\frac{1}{3}} dx$

$$\int (a + bx)^{\frac{1}{3}} dx = \frac{3}{4b} (a + bx)^{\frac{4}{3}}.$$

$$\int x\,(3x + 7)^{\frac{1}{3}} dx = \frac{1}{28} (4x - 7)\,(3x + 7)^{\frac{4}{3}} \cdot$$

$$\int x\,(a + x)^{\frac{1}{3}} dx = \frac{3}{28} (4x - 3a)(a + x)^{\frac{4}{3}} \cdot$$

$$\int \frac{dx}{(a + bx)^{\frac{1}{3}}} = \frac{3}{2b} (a + bx)^{\frac{2}{3}} \cdot$$

$$\int \frac{dx}{x\,(a + bx)^{\frac{1}{3}}} = 3 \int \frac{z}{z^3 - a} dz. \qquad a + bx = z^3.$$

$$\int \frac{x}{(a - x)^{\frac{1}{3}}} dx = - \frac{3}{10} (2x + 3a)\,(a - x)^{\frac{2}{3}} \cdot$$

$$\int \frac{x}{(a + bx)^{\frac{1}{3}}} dx = \frac{3}{10b^2} (2bx - 3a)\,(a + bx)^{\frac{2}{3}} \cdot$$

$$\int \frac{2 + x}{(3 - x)^{\frac{1}{3}}} dx = - \frac{3}{5} \left(\frac{19}{2} + x\right) (3 - x)^{\frac{2}{3}} \cdot$$

$$\int \frac{x^2}{(1+x)^{\frac{1}{3}}}\,dx = \frac{3}{40}(5x^2 - 6x + 9)(1+x)^{\frac{2}{3}}.$$

$$\int \frac{x^2}{(a+bx)^{\frac{4}{3}}}\,dx = 3\left\{\frac{x^2}{8b} - \frac{3ax}{20b^2} + \frac{9a^2}{40b^3}\right\}(a+bx)^{\frac{2}{3}}.$$

$$\int \frac{x^3}{(a+bx)^{\frac{1}{3}}}\,dx = 3\left\{\frac{x^3}{11b} - \frac{9}{11.8}\cdot\frac{ax^2}{b^2} + \frac{9.6}{11.8.5}\cdot\frac{a^2x}{b^3} - \frac{9.6.3}{11.8.5.2}\,\frac{a^3}{b^4}\right\}(a+bx)^{\frac{2}{3}}.$$

$$\int\left(\frac{1+x}{x}\right)^{\frac{1}{3}} dx = -3\int\frac{z^3}{(z^3-1)^2}\,dz. \qquad \frac{1+x}{x} = z^3.$$

$$\int\left(\frac{x-a}{x-b}\right)^{\frac{1}{3}} dx = (x-a)^{\frac{1}{3}}(x-b)^{\frac{2}{3}} + \frac{1}{2}(a-b)\log\left\{(x-b)^{\frac{1}{3}} - (x-a)^{\frac{1}{3}}\right\} - \frac{a-b}{\sqrt{3}}\operatorname{arc\,tg}\frac{2(x-a)^{\frac{1}{3}} + (x-b)^{\frac{1}{3}}}{\sqrt{3}(x-b)^{\frac{1}{3}}}.$$

$$\int\frac{1}{x}\left(\frac{x-a}{x-b}\right)^{\frac{1}{3}} dx = 3\int\frac{dz}{1-z^3} - 3a\int\frac{dz}{a-bz^3}. \qquad x = \frac{a-bz^3}{1-z^3}.$$

$$\int x(a+bx^2)^{\frac{1}{3}}\,dx = \frac{3}{8b}(a+bx^2)^{\frac{4}{3}}.$$

$$\int x^3(a+x^2)^{\frac{1}{3}}\,dx = \frac{3}{56}(4x^2 - 3a)(a+x^2)^{\frac{4}{3}}.$$

$$\int x(a-bx^3)^{\frac{1}{3}}\,dx = -a\int\frac{z^3}{(z^3+b)^2}\,dz. \qquad a - bx^3 = z^3x^3.$$

$$\int x^4(a+bx^3)^{\frac{1}{3}}\,dx = -a^2\int\frac{dz}{(z^3-b)^2} - a^2b\int\frac{dz}{(z^3-b)^3}. \qquad a + bx^3 = z^3x^3.$$

$$\int x^5(1+x^3)^{\frac{1}{3}}\,dx = \frac{4x^3-3}{28}(1+x^3)^{\frac{4}{3}}.$$

$$\int x^3(a+bx^4)^{\frac{1}{3}}\,dx = \frac{3}{16b}(a+bx^4)^{\frac{4}{3}}.$$

$$\int\frac{dx}{(1+x)(1+3x+3x^2)^{\frac{1}{3}}} = \int\frac{dz}{(1-z^3)^{\frac{1}{3}}}. \qquad \frac{x}{1+x} = z.$$

$$\int\frac{dx}{(1-x^3)^{\frac{1}{3}}} = \frac{1}{2}\log\left\{x + (1-x^3)^{\frac{1}{3}}\right\} - \frac{1}{\sqrt{3}}\operatorname{arc\,tg}\frac{2(1-\quad)^{\frac{1}{3}} - x}{x\sqrt{3}}.$$

$$\int\frac{dx}{(1+x^3)^{\frac{1}{3}}} = -\frac{1}{6}\log\frac{(z-1)^2}{z^2+z+1} - \frac{1}{\sqrt{3}}\operatorname{arc\,tg}\frac{2z+1}{\sqrt{3}}. \qquad 1 + x^3 = z^3x^3.$$

$$\int\frac{dx}{(a+bx^3)^{\frac{1}{3}}} = -\int\frac{z}{z^3-b}\,dz. \qquad a + bx^3 = z^3x^3.$$

$$\int\frac{dx}{x^3(a+bx^3)^{\frac{1}{3}}} = -\frac{1}{2ax^2}(a+bx^3)^{\frac{2}{3}}.$$

$$\int\frac{x^3}{(a^2-x^2)^{\frac{1}{3}}}\,dx = -\frac{3}{20}(2x^2+3a^2)(a^2-x^2)^{\frac{2}{3}}.$$

$$\int\frac{x^5}{(7+4x^3)^{\frac{1}{3}}}\,dx = \frac{1}{160}(8x^3-21)(7+4x^3)^{\frac{2}{3}}.$$

$$\int (a+bx)^{\frac{2}{3}}\, dx = \frac{3}{5b}(a+bx)^{\frac{5}{3}}.$$

$$\int x^3 (a+x)^{\frac{2}{3}}\, dx = 3\left\{\frac{1}{14}\Phi^3 - \frac{3a}{11}\Phi^2 + \frac{3a^2}{8}\Phi - \frac{a^3}{5}\right\}\Phi^{\frac{5}{3}}.$$

$$\int x^5 (a+bx^2)^{\frac{2}{3}}\, dx = \frac{3}{2a^3}\left\{\frac{1}{11}\Phi^2 - \frac{a}{4}\Phi + \frac{b^2}{5}\right\}\Phi^{\frac{5}{3}}.$$

$$\int (a+bx^3)^{\frac{2}{3}}\, dx = \frac{az^2}{3(z^3-b)} - \frac{2a}{3}\int \frac{z}{z^3-b}\, dz. \qquad a+bx^3 = z^3x^3.$$

$$\int \frac{dx}{(x-a)\, x^{\frac{2}{3}}} = \frac{k^2}{2}\log\frac{(x^{\frac{1}{3}}-k)^2}{x^{\frac{2}{3}}+kx^{\frac{1}{3}}+k^2} - \frac{3}{k^2\sqrt{3}}\operatorname{arc\,tg}\frac{2x^{\frac{1}{3}}+k}{k\sqrt{3}}. \qquad k = a^{\frac{1}{3}}.$$

$$\int \frac{dx}{x(a+bx)^{\frac{2}{3}}} = 3\int \frac{dz}{z^3-a}. \qquad a+bx = z^3.$$

$$\int \frac{dx}{x(a+bx^3)^{\frac{2}{3}}} = \int \frac{dz}{z^3-a}. \qquad a+bx^3 = z^3.$$

$$\int \frac{dx}{x^2(a+bx^3)^{\frac{2}{3}}} = -\frac{1}{ax}(a+bx^3)^{\frac{1}{3}}.$$

$$\int \frac{x}{(a+bx)^{\frac{2}{3}}}\, dx = \frac{3}{4b^2}(bx-3a)(a+bx)^{\frac{1}{3}}.$$

$$\int \frac{1-x}{(1+2x-x^2)^{\frac{2}{3}}}\, dx = \frac{3}{2}(1+2x-x^2)^{\frac{1}{3}}.$$

$$\int \frac{x}{(1+x^3)^{\frac{2}{3}}}\, dx = -\frac{1}{2}\log(\Phi^{\frac{1}{3}}-x) + \frac{1}{\sqrt{3}}\operatorname{arc\,tg}\frac{2\Phi^{\frac{1}{3}}+x}{x\sqrt{3}}.$$

$$\int \frac{x}{(1-x^3)^{\frac{2}{3}}}\, dx = \frac{1}{6}\log\frac{z^2-z+1}{(z+1)^2} - \frac{1}{\sqrt{3}}\operatorname{arc\,tg}\frac{2z-1}{\sqrt{3}}. \qquad 1-x^3 = z^3x^3.$$

$$\int \frac{3x^2+2}{x^{\frac{2}{3}}}\, dx = \left(\frac{9}{7}x^2+6\right)x^{\frac{1}{3}}.$$

$$\int \frac{x^2}{(x+1)^{\frac{2}{3}}}\, dx = \frac{3}{14}(2x^2-3x+9)(1+x)^{\frac{1}{3}}.$$

$$\int \frac{x^2}{(a+bx^3)^{\frac{2}{3}}}\, dx = \frac{1}{b}(a+bx^3)^{\frac{1}{3}}.$$

$$\int \frac{(8+2x)^{\frac{2}{3}}}{x}\, dx = \frac{3}{2}\Phi^{\frac{2}{3}} + 6\log\frac{\Phi^{\frac{1}{3}}-2}{x^{\frac{1}{3}}} + 4\sqrt{3}\operatorname{arc\,tg}\frac{\Phi^{\frac{1}{3}}+1}{\sqrt{3}}.$$

$$\int \left(\frac{x-a}{x-b}\right)^{\frac{2}{3}} dx = (x-a)^{\frac{2}{3}}(x-b)^{\frac{1}{3}} + (a-b)\left\{(x-b)^{\frac{1}{3}} - (x-a)^{\frac{1}{3}}\right\} + \frac{2}{\sqrt{3}}(a-b)\operatorname{arc\,tg}\frac{2(x-a)^{\frac{1}{3}}+(x-b)^{\frac{1}{3}}}{(x-b)^{\frac{1}{3}}\sqrt{3}}.$$

$$\int U X^{\frac{m}{n}} dx$$

$$\int (a^2 - x^2)^{\frac{7}{2}} dx = \frac{x\sqrt{\Phi}}{8} \left\{ \Phi^3 + \frac{7}{6} a^2\Phi^2 + \frac{35}{24} a^4\Phi + \frac{105}{48} a^6 \right\} + \frac{105}{384} \operatorname{arc\,sin} \frac{x}{a} \cdot$$

$$\int \frac{x^3}{(a+bx)^{\frac{7}{2}}} dx = \frac{2\sqrt{\Phi}}{7b} \left\{ x^3 - \frac{6}{5}\frac{ax^2}{b} + \frac{8}{5}\frac{a^2x}{b^2} - \frac{16}{5}\frac{a^3}{b^3} \right\} \cdot$$

$$\int \frac{bx+c}{(a^2+x^2)^{m+\frac{1}{2}}} dx = -\frac{b}{(2m-1)\,\Phi^{m-\frac{1}{2}}} + c\int \frac{dx}{\Phi^{m+\frac{1}{2}}} \cdot$$

$$\int \frac{dx}{x^3(a+bx^3)^{\frac{4}{3}}} = -\frac{a+3bx^3}{2a^2x^2\,\Phi^{\frac{1}{3}}} \cdot$$

$$\int \frac{x^2}{(a+bx^3)^{\frac{4}{3}}} dx = -\frac{1}{b\,\Phi^{\frac{1}{3}}} \cdot$$

$$\int \frac{x^3}{(a+bx^3)^{\frac{4}{3}}} dx = -\frac{x}{b\,\Phi^{\frac{1}{3}}} - \frac{1}{2b\sqrt{b}} \log \frac{\Phi^{\frac{1}{3}} - x\sqrt{b}}{\Phi^{\frac{1}{3}} + x\sqrt{b}} \cdot$$

$$\int \frac{dx}{x^2(1+x^3)^{\frac{5}{3}}} = -\frac{3x^3+2a}{2a^2x\,\Phi^{\frac{2}{3}}} \cdot$$

$$\int x^2(1+x^4)^{\frac{1}{4}} dx = \frac{1}{4}x^3\Phi^{\frac{1}{4}} + \frac{1}{16}\log\frac{\Phi^{\frac{1}{4}}+x}{\Phi^{\frac{1}{4}}-x} + \frac{1}{8}\operatorname{arc\,tg}\frac{\Phi^{\frac{1}{4}}}{x} = -\int\frac{z^4}{(z^4-1)^2}dz. \qquad 1+x^4 = z^4x^4.$$

$$\int \frac{dx}{(x^4-1)^{\frac{1}{4}}} = \frac{1}{4}\log\frac{x+\Phi^{\frac{1}{4}}}{x-\Phi^{\frac{1}{4}}} - \frac{1}{2}\operatorname{arc\,tg}\frac{\Phi^{\frac{1}{4}}}{x} \cdot$$

$$\int \frac{dx}{(1-x^2)(2x^2-1)^{\frac{1}{4}}} = \frac{1}{4}\log\frac{\Phi^{\frac{1}{4}}-x}{\Phi^{\frac{1}{4}}+x} - \frac{1}{2}\operatorname{arc\,tg}\frac{x}{\Phi^{\frac{1}{4}}} \cdot$$

$$\int \frac{dx}{x(1+x)^{\frac{3}{4}}} = \log\frac{\Phi^{\frac{1}{4}}-1}{\Phi^{\frac{1}{4}}+1} - 2\operatorname{arc\,tg}\Phi^{\frac{1}{4}} \cdot$$

$$\int \frac{x^2}{(a+bx^4)^{\frac{3}{4}}} dx = -\frac{1}{4b^{\frac{3}{4}}}\log\frac{\Phi^{\frac{1}{4}} - b^{\frac{1}{4}}x}{\Phi^{\frac{1}{4}} + b^{\frac{1}{4}}x} + \frac{1}{2b^{\frac{3}{4}}}\operatorname{arc\,tg}\frac{\Phi^{\frac{1}{4}}}{b^{\frac{1}{4}}x} \cdot$$

$$\int \frac{x^7}{(9x^4-3)^{\frac{3}{4}}} dx = \frac{3x^4+4}{135}\Phi^{\frac{1}{4}} \cdot$$

$$\int \frac{x}{(a+bx)^{\frac{1}{5}}} dx = \frac{4bx-5a}{36b^2}\Phi^{\frac{4}{5}} \cdot$$

$$\int \frac{(1+x^2)^{\frac{2}{5}}}{x^3} dx = -\frac{z^3}{2(z^5-1)} + \frac{3}{2}\int\frac{z^2}{z^5-1}dz. \qquad (1+x^2)^{\frac{3}{5}} = z.$$

$$\int x^5(1-x^2)^{\frac{2}{7}} dx = -\frac{7}{2}\left\{\frac{1}{9}\Phi - \frac{1}{8}\Phi^2 + \frac{1}{23}\Phi^3\right\}\Phi^{\frac{2}{7}} \cdot$$

$$\int x^A \left(a+bx^B\right)^C dx$$

$$\int \sqrt{x}\,(1+\sqrt{x})^{\frac{2}{3}} dx = 6\int z^4(z^3-1)^2 dz. \qquad 1+\sqrt{x} = z^3.$$

$$\int x^{\frac{1}{8}}\left(1+x^{\frac{3}{2}}\right)^{\frac{1}{4}} dx = -\frac{2}{3}\int\frac{z^{\frac{1}{4}}}{(z-1)^2}dz. \qquad 1+x^{\frac{3}{2}} = zx^{\frac{3}{2}} \cdot$$

$$\int \frac{dx}{x^{\frac{5}{2}}(a+bx)^{\frac{3}{2}}} = -\frac{2}{3ax\sqrt{ax+bx^2}} - \frac{4b}{3a}\int \frac{dx}{(ax+bx^2)^{\frac{3}{2}}}.$$

$$\int \frac{dx}{\sqrt{x}\,(1+x^{\frac{1}{3}})} = 6\int \frac{z^2}{z^2+1}\,dz. \qquad x = z^6.$$

$$\int \frac{dx}{\sqrt{x}\,(a-\sqrt{x})^2} = \frac{1}{a}\cdot\frac{a+\sqrt{x}}{a-\sqrt{x}}.$$

$$\int \frac{dx}{\sqrt{x}\,(1+x^2)^{\frac{5}{4}}} = \frac{2\sqrt{x}}{(1+x^2)^{\frac{1}{4}}}.$$

$$\int \frac{dx}{x^{\frac{1}{4}}\left(1-x^{\frac{1}{6}}\right)^2} = 12\int \frac{z^8}{(1-z^2)^2}\,dz. \qquad x = z^{12}.$$

$$\int \frac{(\sqrt{x}+1)^2}{x^{\frac{3}{2}}}\,dx = 2\,\frac{x-1}{\sqrt{x}} + 2\log x.$$

$$\int \frac{(1-\sqrt{x})^{\frac{1}{3}}}{x^{\frac{3}{2}}}\,dx = -\frac{2\Phi^{\frac{1}{3}}}{\sqrt{x}} - \log\frac{1-\Phi^{\frac{1}{3}}}{x^{\frac{1}{6}}} + \frac{2}{\sqrt{3}}\,\text{arc tg}\,\frac{2\Phi^{\frac{1}{3}}+1}{\sqrt{3}}.$$

$$\int \frac{(a+bx^2)^{\frac{1}{3}}}{x^{\frac{11}{3}}}\,dx = -\frac{3}{8a}\left\{\frac{a+bx^2}{x^2}\right\}^{\frac{4}{3}}.$$

$$\int \frac{(\sqrt{x}+1)^{\frac{2}{3}}}{x^{\frac{7}{3}}}\,dx = -6\int z^4(z^3-1)\,dz. \qquad \sqrt{x}+1 = z^3\sqrt{x}.$$

$\int x^A \left(a+bx^n\right)^C dx$

$$\int x\sqrt{a+bx^n}\,dx = \frac{2x^{2-n}\,\Phi^{\frac{3}{2}}}{b(n+4)} - \frac{2a(2-n)}{b(n+4)}\int \frac{\sqrt{\Phi}}{x^{n-1}}\,dx.$$

$$\int x^2\sqrt{a+bx^n}\,dx = \frac{2x^{3-n}\,\Phi^{\frac{3}{2}}}{b(n+6)} - \frac{2a(3-n)}{b(n+6)}\int \frac{\sqrt{\Phi}}{x^{n-2}}\,dx.$$

$$\int \frac{dx}{1+x^n} = -\frac{1}{n}\left\{\cos\frac{\pi}{n}\log\left(1-2x\cos\frac{\pi}{n}+x^2\right) + \cos\frac{3\pi}{n}\log\left(1-2x\cos\frac{3\pi}{n}+x^2\right) + \cos\frac{5\pi}{n}\log\left(1-2x\cos\frac{5\pi}{n}+x^2\right) + \cdots\right\}$$
$$+\frac{2}{n}\left\{\sin\frac{\pi}{n}\,\text{arc tg}\,\frac{x\sin\frac{\pi}{n}}{1-x\cos\frac{\pi}{n}} + \sin\frac{3\pi}{n}\,\text{arc tg}\,\frac{x\sin\frac{3\pi}{n}}{1-x\cos\frac{3\pi}{n}} + \sin\frac{5\pi}{n}\,\text{arc tg}\,\frac{x\sin\frac{5\pi}{n}}{1-x\cos\frac{5\pi}{n}} + \cdots\right\}.$$

$$\int \frac{dx}{1-x^n} = -\frac{1}{n}\left\{\cos\frac{2\pi}{n}\log\left(1-2x\cos\frac{2\pi}{n}+x^2\right) + \cos\frac{4\pi}{n}\log\left(1-2x\cos\frac{4\pi}{n}+x^2\right) + \cos\frac{6\pi}{n}\log\left(1-2x\cos\frac{6\pi}{n}+x^2\right) + \cdots\right\}$$
$$+\frac{2}{n}\left\{\sin\frac{2\pi}{n}\,\text{arc tg}\,\frac{x\sin\frac{2\pi}{n}}{1-x\cos\frac{2\pi}{n}} + \sin\frac{4\pi}{n}\,\text{arc tg}\,\frac{x\sin\frac{4\pi}{n}}{1-x\cos\frac{4\pi}{n}} + \sin\frac{6\pi}{n}\,\text{arc tg}\,\frac{x\sin\frac{6\pi}{n}}{1-x\cos\frac{6\pi}{n}} + \cdots\right\} - \frac{1}{n}\log(1-x).$$

$$\int \frac{dx}{a+bx^n} = \frac{1}{a}\left(\frac{a}{b}\right)^{\frac{1}{n}}\int \frac{dz}{1+z^n}. \qquad x = \left(\frac{a}{b}\right)^{\frac{1}{n}} z.$$

$$\int \frac{dx}{x(x^n-a)} = \frac{1}{an}\log\frac{x^n-a}{x^n}.$$

$$\int \frac{dx}{x(a+bx^n)} = \frac{1}{an}\log\frac{x^n}{a+bx^n}.$$

$$\int \frac{dx}{x^2(a+bx^n)} = -\frac{1}{ax} - \frac{b}{a}\int \frac{x^{n-2}}{a+bx^n}\,dx.$$

$$\int \frac{dx}{x^3(a+bx^n)} = -\frac{1}{2ax^2} - \frac{b}{a}\int \frac{x^{n-3}}{a+bx^n}\,dx.$$

$$\int \frac{dx}{(a+bx^n)^2} = \frac{x}{an\Phi} + \frac{n-1}{an}\int \frac{dx}{\Phi}.$$

$$\int \frac{dx}{x.(a+bx^n)^2} = -\frac{1}{a^2}\int \frac{z^n-b}{z^{n+1}}\,dz. \qquad a+bx^n = x^n z^n.$$

$$\int \frac{dx}{\sqrt{a+bx^n}} = -\frac{2x\sqrt{\Phi}}{an} + \frac{n+2}{an}\int \sqrt{\Phi}\,dx.$$

$$\int \frac{dx}{x\sqrt{x^n-1}} = \frac{2}{n}\,\text{arc cos}\,\frac{1}{\sqrt{x^n}}.$$

$$\int \frac{dx}{x\sqrt{x^n-a}} = \frac{1}{n\sqrt{a}}\,\text{arc sin}\,\frac{x^n-2a}{x^n} = \frac{2}{n}\,\text{arc séc}\,\sqrt{x^n}.$$

$$\int \frac{dx}{x\sqrt{a+bx^n}} = \frac{1}{n\sqrt{a}}\log\frac{\sqrt{\Phi}-\sqrt{a}}{\sqrt{\Phi}+\sqrt{a}} = \frac{2}{n\sqrt{a}}\log\frac{\sqrt{\Phi}-\sqrt{a}}{x^{\frac{n}{2}}} = -\frac{2}{n\sqrt{a}}\log\frac{\sqrt{\Phi}+\sqrt{a}}{x^{\frac{n}{2}}}.$$

$$\int \frac{dx}{x\sqrt{bx^n-a}} = \frac{2}{n\sqrt{a}}\,\text{arc tg}\sqrt{\frac{bx^n-a}{a}} = \frac{1}{n\sqrt{a}}\,\text{arc sin}\,\frac{bx^n-2a}{bx^n}.$$

$$\int \frac{dx}{x(a+bx^n)^{\frac{3}{2}}} = \frac{2x}{an\sqrt{\Phi}} + \frac{n-2}{an}\int \frac{dx}{\sqrt{\Phi}}.$$

$$\int \frac{x}{(a+bx^n)^2}\,dx = \frac{x^2}{an\Phi} + \frac{n-2}{an}\int \frac{x}{\Phi}\,dx.$$

$$\int \frac{x^2}{(a+bx^n)^2}\,dx = \frac{x^3}{an\Phi} + \frac{n-3}{an}\int \frac{x^2}{\Phi}\,dx.$$

$$\int \frac{x}{\sqrt{a+bx^n}}\,dx = -\frac{2x^2\sqrt{\Phi}}{an} + \frac{n+4}{an}\int x\sqrt{\Phi}\,dx.$$

$$\int \frac{x^2}{\sqrt{a+bx^n}}\,dx = -\frac{2x^3\sqrt{\Phi}}{an} + \frac{n+6}{an}\int x^2\sqrt{\Phi}\,dx.$$

$$\int \frac{x}{(a+bx^n)^{\frac{3}{2}}}\,dx = \frac{2x^2}{an\sqrt{\Phi}} + \frac{n-4}{an}\int \frac{x}{\sqrt{\Phi}}\,dx.$$

$$\int \frac{x^2}{(a+bx^n)^{\frac{3}{2}}}\,dx = \frac{2x^2}{an\sqrt{\Phi}} + \frac{n-6}{an}\int \frac{x^2}{\sqrt{\Phi}}\,dx.$$

$$\int \frac{\sqrt{x^n-1}}{x}\,dx = \frac{2}{n}\sqrt{x^n-1} + \frac{2}{n}\,\text{arc sin}\,\frac{1}{\sqrt{x^n}}.$$

$$\int \frac{\sqrt{a+bx^n}}{x}\,dx = \frac{2}{n}\sqrt{\Phi} + \frac{\sqrt{a}}{n}\log\frac{\sqrt{\Phi}-\sqrt{a}}{\sqrt{\Phi}+\sqrt{a}}.$$

$$\int \frac{\sqrt{bx^n-a}}{x}\,dx = \frac{2}{n}\sqrt{\Phi} - \frac{2\sqrt{a}}{n}\,\text{arc tg}\sqrt{\frac{\Phi}{a}}.$$

$$\int \frac{\sqrt{bx^n+a}}{x^2}\,dx = \frac{\Phi^{\frac{3}{2}}}{ax} - \frac{b(3n-2)}{2a}\int x^{n-2}\sqrt{\Phi}\,dx.$$

$$\int \frac{(a+bx^n)^{\frac{3}{2}}}{x}\,dx = \frac{2}{3n}\Phi^{\frac{3}{2}} + \frac{2a}{n}\sqrt{\Phi} + \frac{a\sqrt{a}}{n}\log\frac{\sqrt{\Phi}-\sqrt{a}}{\sqrt{\Phi}+\sqrt{a}}.$$

$$\int\frac{(a+bx^n)^{\frac{3}{2}}}{x^2}\,dx=-\frac{\Phi^{\frac{5}{2}}}{ax}+\frac{b\,(5n-2)}{2a}\int x^{n-2}\,\Phi^{\frac{3}{2}}dx.$$

$$\int x^m\left(a+bx\right)^{C}dx$$

$$\int x^m\sqrt{a+bx}\,dx=\left\{\frac{\Phi^m}{2m+3}-\frac{m}{1}\frac{a\Phi^{m-1}}{2m+1}+\frac{m\,(m-1)}{1.2}\frac{a^2\Phi^{m-2}}{2m-1}-\frac{m(m-1)(m-2)}{1.2.3}\frac{a^3\Phi^{m-3}}{2m-3}+\cdots\pm \mathrm{N}a^m\right\}\frac{2\Phi^{\frac{3}{2}}}{b^{m+1}}.$$

$$\int x^m\,(a+bx)^{\frac{3}{2}}\,dx=\frac{2x^m\,\Phi^{\frac{5}{2}}}{b\,(2m+5)}-\frac{2am}{b\,(2m+5)}\int x^{m-1}\,\Phi^{\frac{3}{2}}\,dx.$$

$$\int x^{m+\frac{1}{2}}\sqrt{b+cx}\,dx=\frac{x^{m+1}}{m+1}\sqrt{bx+cx^2}-\frac{b}{2\,(m+1)}\int\frac{x^{m+1}}{\sqrt{bx+cx^2}}\,dx-\frac{c}{m+1}\int\frac{x^{m+2}}{\sqrt{bx+cx^2}}\,dx.$$

$$=\left\{\frac{x^{m+1}}{(m+2)c}-\frac{(2m+1)\,bx^{m-2}}{(m+2)(m+1)\,2c^2}+\frac{(2m+1)\,(2m-1)b^2x^{m-3}}{(m+2)\,(m+1)\,m4c^3}-\frac{(2m+1)\,(2m-1)\,(2m-3)\,b^3x^{m-4}}{(m+2)\,(m+1)\,m\,(m-1)\,8c^4}\right.$$

$$\left.+\cdots\pm \mathrm{N}\right\}(bx+cx^2)^{\frac{3}{2}}\mp\frac{3b\mathrm{N}}{2}\int\sqrt{bx+cx^2}\,dx.$$

$$\int\frac{dx}{x^m\,(a+bx)}=-\frac{1}{a\,(m-1)\,x^{m-1}}-\frac{b}{a}\int\frac{dx}{x^{m-1}\,(a+bx)}=-\int\frac{z^{m-1}}{b+az}\,dz.\qquad xz=1.$$

$$=-\frac{1}{(m-1)\,ax^{m-1}}+\frac{b}{(m-2)\,a^2x^{m-2}}-\frac{b^2}{(m-3)\,a^3x^{m-3}}+\cdots\pm\frac{b^{m-1}}{a^m}\log\frac{a+bx}{x}.$$

$$\int\frac{dx}{x^m\,(a+bx)^2}=\frac{1}{ax^{m-1}\Phi}+\frac{m}{a}\int\frac{dx}{x^m\Phi}.$$

$$\int\frac{dx}{x\,(a+bx^n)}=\frac{1}{an}\log\frac{x^n}{a+bx^u}.$$

$$\int\frac{dx}{x^m\sqrt{a+bx}}=2b^{m-1}\int\frac{dz}{(z^2-a)^m}.\qquad a+bx=z^2.$$

$$=-\frac{\sqrt{\Phi}}{a\,(m-1)\,x^{m-1}}-\frac{b\,(2m-3)}{2a\,(m-1)}\int\frac{dx}{x^{m-1}\sqrt{\Phi}}.$$

$$=\left\{-\frac{1}{(m-1)\,ax^{m-1}}+\frac{1}{(m-1)\,a^2}\cdot\frac{2m-3}{2m-4}\,\frac{b}{x^{m-2}}-\frac{1}{(m-1)\,a^3}\,\frac{2m-3}{2m-4}\,\frac{2m-5}{2m-6}\,\frac{b^2}{x^{m-3}}\right.$$

$$\left.+\frac{1}{(m-1)\,a^4}\,\frac{2m-3}{2m-4}\,\frac{2m-5}{2m-6}\,\frac{2m-7}{2m-8}\,\frac{b^3}{x^{m-4}}-\cdots\pm\frac{\mathrm{N}}{x}\right\}\sqrt{\Phi}\pm\frac{\mathrm{N}b}{2}\int\frac{dx}{x\sqrt{\Phi}}.$$

$$\int\frac{dx}{x^m\,(a+bx)^{\frac{3}{2}}}=-\frac{1}{(m-1)\,ax^{m-1}\sqrt{\Phi}}-\frac{b\,(2m-1)}{2a\,(m-1)}\int\frac{dx}{x^{m-1}\Phi^{\frac{3}{2}}}.$$

$$=\left\{-\frac{1}{(m-1)ax^{m-1}}+\frac{1}{(m-1)a^2}\,\frac{2m-1}{2m-4}\,\frac{b}{x^{m-2}}-\frac{1}{(m-1)\,a^3}\,\frac{2m-1}{2m-4}\,\frac{2m-3}{2m-6}\,\frac{b^2}{x^{m-3}}+\cdots\pm\frac{\mathrm{N}}{x}\right\}\frac{1}{\sqrt{\Phi}}\pm\frac{3b\mathrm{N}}{2}\int\frac{dx}{x\Phi^{\frac{3}{2}}}.$$

$$\int\frac{x^m}{x-1}\,dx=\frac{x^m}{m}+\frac{x^{m-1}}{m-1}+\cdots\frac{x^2}{2}+x+\log\,(x-1).$$

$$\int\frac{x^m}{a+bx}\,dx=\frac{x^m}{bm}-\frac{a}{b}\int\frac{x^{m-1}}{a+bx}\,dx=\frac{x^m}{bm}-\frac{ax^{m-1}}{(m-1)\,b^2}+\frac{a^2x^{m-2}}{(m-2)\,b^3}-\cdots\pm\frac{a^m}{b^{m+1}}\log\,(a+bx).$$

$$\int\frac{x^m}{(a+bx)^2}\,dx=\frac{x^m}{b\,(m-1)\,(a+bx)}-\frac{am}{b\,(m-1)}\int\frac{x^{m-1}}{(a+bx)^2}\,dx.$$

$$\int \frac{x^m}{(x-a)^3}\,dx = \int \frac{(z+a)^m}{z^3}\,dz. \qquad x - a = z.$$

$$\int \frac{x^m}{a+bx^4}\,dx = \frac{x^{m-3}}{b\,(m-3)} - \frac{a}{b}\int \frac{x^{m-4}}{a+bx^4}\,dx.$$

$$\int \frac{x^m}{\sqrt{a+bx}}\,dx = \frac{2x^m\sqrt{\Phi}}{(2m+1)\,b} - \frac{2am}{(2m+1)\,b}\int \frac{x^{m-1}}{\sqrt{\Phi}}\,dx.$$

$$= \left\{\frac{\Phi^m}{2m+1} - \frac{m}{1}\,\frac{a\Phi^{m-1}}{2m-1} + \frac{m(m-1)}{1.2}\,\frac{a^2\Phi^{m-2}}{2m-3} - \frac{m\,(m-1)\,(m-2)}{1.2.3}\,\frac{a^3\Phi^{m-3}}{2m-5} + \ldots \pm a^m\right\}\frac{2\sqrt{\Phi}}{b^{m+1}}.$$

$$\int \frac{x^m}{(a+bx)^{\frac{3}{2}}}\,dx = \frac{2x^{m+1}}{a\sqrt{\Phi}} - \frac{2m+1}{a}\int \frac{x^m}{\sqrt{\Phi}}\,dx = \left\{\frac{\Phi^m}{2m-1} - \frac{m}{1}\,\frac{a\Phi^{m-1}}{2m-3} + \frac{m\,(m-1)}{1.2}\,\frac{a^2\Phi^{m-2}}{2m-5} - \ldots \pm a^m\right\}\frac{2}{b^{m+1}\sqrt{\Phi}}.$$

$$\int \frac{x^m}{(a+bx)^{\frac{1}{3}}}\,dx = \frac{3x^m\Phi^{\frac{2}{3}}}{(3m+2)\,b} - \frac{3am}{(3m+2)\,b}\int \frac{x^{m-1}}{\Phi^{\frac{1}{3}}}\,dx.$$

$$\int \frac{a+bx}{x^m}\,dx = -\frac{\Phi^2}{a\,(m-1)\,x^{m-1}} + \frac{b\,(3-m)}{a\,(m-1)}\int \frac{\Phi}{x^{m-1}}\,dx.$$

$$\int \frac{(a+bx)^2}{x^m}\,dx = -\frac{\Phi^3}{a\,(m-1)\,x^{m-1}} + \frac{b\,(4-m)}{a\,(m-1)}\int \frac{\Phi^2}{x^{m-1}}\,dx.$$

$$\int \frac{\sqrt{a+bx}}{x^m}\,dx = -\frac{\Phi^{\frac{3}{2}}}{a\,(m-1)\,x^{m-1}} + \frac{b\,(5-2m)}{2a\,(m-1)}\int \frac{\sqrt{\Phi}}{x^{m-1}}\,dx.$$

$$= \left\{-\frac{1}{(m-1)\,ax^{m-1}} + \frac{1}{m-1}\,\frac{2m-5}{2m-4}\,\frac{b}{a^2x^{m-2}} - \frac{1}{m-1}\,\frac{2m-5}{2m-4}\,\frac{2m-7}{2m-6}\,\frac{b^2}{a^3x^{m-3}} + \ldots \pm \frac{\mathrm{N}}{x}\right\}\Phi^{\frac{3}{2}} + \frac{b\mathrm{N}}{2}\int \frac{\sqrt{\Phi}}{x}\,dx.$$

$$\int \frac{(a+bx)^{\frac{3}{2}}}{x^m}\,dx = -\frac{\Phi^{\frac{5}{2}}}{a\,(m-1)x^{m-1}} + \frac{b\,(7-2m)}{2a\,(m-1)}\int \frac{\Phi^{\frac{3}{2}}}{x^{m-1}}\,dx.$$

$$\int \frac{(a+bx)^{\frac{1}{3}}}{x^m}\,dx = -\frac{\Phi^{\frac{4}{3}}}{a\,(m-1)\,x^{m-1}} + \frac{b\,(7-3m)}{3a\,(m-1)}\int \frac{\Phi^{\frac{1}{3}}}{x^{m-1}}\,dx.$$

$$\int \frac{(a+bx)^{\frac{2}{3}}}{x^m}\,dx = -\frac{\Phi^{\frac{5}{3}}}{a\,(m-1)\,x^{m-1}} + \frac{b\,(8-3m)}{3a\,(m-1)}\int \frac{\Phi^{\frac{2}{3}}}{x^{m-1}}\,dx.$$

$\int x^m \left(a + bx^2\right)^{\mathrm{C}} dx$

$$\int x^{2m}\sqrt{a+bx^2}\,dx = -\,a^{m+1}\int \frac{z^2}{(z^2-b)^{m+2}}\,dz. \qquad a + bx^2 = z^2x^2.$$

$$\int x^{2m+1}\sqrt{a+bx^2}\,dx = \frac{1}{b^{m+1}}\int z^2\,(z^2-a)^m\,dz. \qquad a + bx^2 = z^2.$$

$$\int x^m\sqrt{a+bx^2}\,dx = \frac{x^{m+1}}{m+1}\sqrt{\Phi} - \frac{b}{m+1}\int \frac{x^{m+2}}{\sqrt{\Phi}}\,dx.$$

$$\int x^m\,(a+bx^2)^{\frac{3}{2}}\,dx = \frac{x^{m+1}}{m+1}\,\Phi^{\frac{3}{2}} - \frac{3b}{m+1}\int x^{m+2}\sqrt{\Phi}\,dx.$$

$$\int \frac{dx}{x^{2m}\,(a+bx^2)} = -\frac{1}{(2m-1)\,ax^{2m-1}} + \frac{b}{(2m-3)\,a^2x^{2m-3}} - \frac{b^2}{(2m-5)\,a^3x^{2m-5}} + \ldots \pm \frac{b^m}{a^m}\int \frac{dx}{a+bx^2}.$$

$$\int \frac{dx}{x^{2m+1}\,(a+bx^2)} = -\frac{1}{2max^{2m}} + \frac{b}{(2m-2)\,a^2x^{2m-2}} - \frac{b^2}{(2m-4)\,a^3x^{2m-4}} + \ldots \pm \frac{b^m}{a^m}\log\frac{\sqrt{a+bx^2}}{x}.$$

$$\int \frac{dx}{x^m (a + bx^2)} = - \frac{1}{a(m-1)x^{m-1}} - \frac{b}{a} \int \frac{dx}{x^{m-2}(a + bx^2)} .$$

$$\int \frac{dx}{x^m (a + bx^2)^2} = - \frac{1}{a(m-1)x^{m-1}\Phi} - \frac{b(m+1)}{a(m-1)} \int \frac{dx}{x^{m-2}\Phi^2} .$$

$$\int \frac{dx}{x^m \sqrt{1+x^2}} = - \frac{\sqrt{1+x^2}}{(m-1)x^{m-1}} - \frac{m-2}{m-1} \int \frac{dx}{x^{m-2}\sqrt{1+x^2}} .$$

$$\int \frac{dx}{x^m \sqrt{x^2-1}} = \frac{\sqrt{\Phi}}{(m-1)x^{m-1}} + \frac{m-2}{m-1} \int \frac{dx}{x^{m-2}\sqrt{\Phi}} .$$

$$\int \frac{dx}{x^{2m+1}\sqrt{1-x^2}} = - \left\{ 1 + \frac{2m-1}{2m-2}x^2 + \frac{(2m-1)(2m-3)}{(2m-2)(2m-4)}x^4 + \cdots \frac{(2m-1)\cdots 5.3}{(2m-2)\cdots 4.2} \right\} \frac{\sqrt{1-x^2}}{2mx^{2m}} + \frac{(2m-1)\cdots 3.1}{2m\cdots 4.2} \log \frac{\sqrt{1-x^2}-1}{x} .$$

$$\int \frac{dx}{x^{2m}\sqrt{1-x^2}} = - \left\{ 1 + \frac{2m-2}{2m-3}x^2 + \frac{(2m-2)(2m-4)}{(2m-3)(2m-5)}x^4 + \cdots \frac{(2m-2)\cdots 4.2}{(2m-3)\cdots 3.1}x^{2m-2} \right\} \frac{\sqrt{1-x^2}}{(2m-1)x^{2m-1}} .$$

$$\int \frac{dx}{x^m \sqrt{1-x^2}} = - \frac{\sqrt{1-x^2}}{(m-1)x^{m-1}} + \frac{m-2}{m-1} \int \frac{dx}{x^{m-2}\sqrt{1-x^2}} = \frac{1}{2^{m-1}} \int \frac{(z^2+1)^{m-1}}{z^m} dz. \qquad x(1+z^2) = 2z.$$

$$\int \frac{dx}{x^m \sqrt{a+bx^2}} = - \frac{\sqrt{\Phi}}{(m-1)ax^{m-1}} - \frac{b(m-2)}{a(m-1)} \int \frac{dx}{x^{m-2}\sqrt{\Phi}} .$$

$$\int \frac{x^{2m+1}}{x^2-1} dx = \frac{x^{2m}}{2m} + \frac{x^{2m-2}}{2m-2} + \cdots \frac{x^2}{2} + \frac{1}{2} \log(x^2-1).$$

$$\int \frac{x^m}{a+bx^2} dx = \frac{x^{m-1}}{(m-1)b} - \frac{a}{b} \int \frac{x^{m-2}}{a+bx^2} dx.$$

$$\int \frac{x^{2m}}{a+bx^2} dx = \frac{x^{2m-1}}{(2m-1)b} - \frac{ax^{2m-3}}{(2m-3)b^2} + \frac{a^2x^{2m-5}}{(2m-5)b^3} - \cdots \pm \frac{a^m}{b^m} \int \frac{dx}{a+bx^2} .$$

$$\int \frac{x^{2m+1}}{a+bx^2} dx = \frac{x^{2m}}{2mb} - \frac{ax^{2m-2}}{(2m-2)b^2} + \frac{a^2x^{2m-4}}{(2m-4)b^3} - \cdots - \frac{a^m}{2b^m} \log(a+bx^2).$$

$$\int \frac{x^m}{(a+bx^2)^2} dx = \frac{x^{m-1}}{b(m-3)\Phi} - \frac{a(m-1)}{b(m-3)} \int \frac{x^{m-2}}{\Phi^2} dx = - \frac{x^{m-1}}{2b\Phi} + \frac{m-1}{2b} \int \frac{x^{m-2}}{\Phi} dx.$$

$$\int \frac{x^m}{\sqrt{x^2+1}} dx = \frac{x^{m-1}\sqrt{\Phi}}{m} - \frac{m-1}{m} \int \frac{x^{m-2}}{\sqrt{\Phi}} dx.$$

$$\int \frac{x^m}{\sqrt{x^2-1}} dx = \frac{x^{m-1}\sqrt{\Phi}}{m} + \frac{m-1}{m} \int \frac{x^{m-2}}{\sqrt{\Phi}} dx.$$

$$\int \frac{x^{2m}}{\sqrt{a^2-x^2}} dx = -\sqrt{a^2-x^2} \left\{ \frac{x^{2m-1}}{2m} + \frac{2m-1}{2m(2m-2)}a^2x^{2m-3} + \frac{(2m-1)(2m-3)}{2m(2m-2)(2m-4)}a^4x^{2m-5} + \cdots \frac{(2m-1)(2m-3)\cdots 3}{2m(2m-2)\cdots 2}a^{2m-2}x \right\}$$
$$+ \frac{(2m-1)(2m-3)\cdots 3}{2m(2m-2)\cdots 2} a^{2m} \arcsin \frac{x}{a} .$$

$$\int \frac{x^{2m+1}}{\sqrt{a^2-x^2}} dx = -\sqrt{a^2-x^2} \left\{ \frac{x^{2m}}{2m+1} + \frac{2m}{(2m+1)(2m-1)}a^2x^{2m-2} + \frac{2m(2m-2)}{(2m+1)(2m-1)(2m-3)}a^4x^{2m-4} + \cdots \right.$$
$$\left. + \frac{2m(2m-2)\cdots 2}{(2m+1)(2m-1)\cdots 1} a^{2m} \right\} .$$

$$\int \frac{x^m}{\sqrt{a^2-x^2}} dx = - \frac{x^{m-1}\sqrt{\Phi}}{m} + \frac{(m-1)a^2}{m} \int \frac{x^{m-2}}{\sqrt{\Phi}} dx.$$

$$\int \frac{x^m}{\sqrt{a+bx^2}}\,dx = \frac{x^{m-1}\sqrt{\Phi}}{bm} - \frac{a\,(m-1)}{bm}\int \frac{x^{m-2}}{\sqrt{\Phi}}\,dx.$$

$$\int \frac{x^{2m}}{(a+bx^2)^{\frac{3}{2}}}\,dx = -\,a^{m-1}\int \frac{dz}{z^2\,(z^2-b)^m}\cdot \qquad a+bx^2 = z^2x^2.$$

$$\int \frac{x^{2m+1}}{(a+bx^2)^{\frac{3}{2}}}\,dx = \frac{1}{b^{m+1}}\int \frac{(z^2-a)^m}{z^2}\,dz. \qquad a+bx^2 = z^2.$$

$$\int \frac{x^m}{(a+bx^2)^{\frac{3}{2}}}\,dx = \frac{x^{m+1}}{a\sqrt{\Phi}} - \frac{m}{a}\int \frac{x^m}{\sqrt{\Phi}}\,dx.$$

$$\int \frac{\sqrt{a+bx^2}}{x^m}\,dx = -\frac{\Phi^{\frac{3}{2}}}{(m-1)\,ax^{m-1}} - \frac{b\,(m-4)}{a\,(m-1)}\int \frac{\sqrt{\Phi}}{x^{m-2}}\,dx = -\frac{\sqrt{\Phi}}{(m-1)\,x^{m-1}} + \frac{b}{m-1}\int \frac{dx}{x^{m-2}\sqrt{\Phi}}\cdot$$

$$\int \frac{(a+bx^2)^{\frac{3}{2}}}{x^m}\,dx = -\frac{\Phi^{\frac{3}{2}}}{(m-1)\,x^{m-1}} + \frac{3b}{m-1}\int \frac{\sqrt{\Phi}}{x^{m-2}}\,dx.$$

—— $x^{A}\left(a+bx\right)^{p}\,dx$ ——

$$\int \left(a+bx\right)^{\frac{p}{q}}dx = \frac{q}{b\,(p+q)}\,\Phi^{\frac{p}{q}+1}.$$

$$\int \left(a+bx\right)^{p-\frac{1}{2}}dx = \frac{2}{b\,(2p+1)}\,\Phi^{p+\frac{1}{2}}.$$

$$\int \left(a+bx\right)^{p+\frac{1}{3}}dx = \frac{3}{b\,(3p+4)}\,\Phi^{p+\frac{4}{3}}.$$

$$\int \left(a+bx\right)^{p-\frac{1}{3}}dx = \frac{3}{b\,(3p+2)}\,\Phi^{p+\frac{2}{3}}.$$

$$\int \left(a+bx\right)^{p-\frac{2}{3}}dx = \frac{3}{b\,(3p+1)}\,\Phi^{p+\frac{1}{3}}.$$

$$\int x\left(a+bx\right)^{\frac{p}{q}}dx = \frac{q\Phi^{\frac{p}{q}+1}}{b^2}\left\{\frac{\Phi}{p+2q} - \frac{a}{p+q}\right\}.$$

$$\int x\left(a+bx\right)^{p-\frac{1}{2}}dx = \frac{2\Phi^{p+\frac{1}{2}}}{b\,(2p+3)}\left\{x - \frac{2a}{b\,(2p+1)}\right\}.$$

$$\int x\left(a+bx\right)^{p-\frac{1}{3}}dx = \frac{3\Phi^{p+\frac{2}{3}}}{b\,(3p+5)}\left\{x - \frac{3a}{b\,(3p+2)}\right\}.$$

$$\int x\left(a+bx\right)^{p-\frac{2}{3}}dx = \frac{3\Phi^{p+\frac{1}{3}}}{b\,(3p+4)}\left\{x - \frac{3a}{b\,(3p+1)}\right\}.$$

$$\int x^2\left(a+bx\right)^{p}dx = \frac{\Phi^{p+1}}{b\,(p+3)}\left\{x^2 - \frac{2ax}{b\,(p+2)} + \frac{2a^2}{b^2\,(p+1)\,(p+2)}\right\}.$$

$$\int x^2\left(a+bx\right)^{\frac{p}{q}}dx = \frac{q}{b^3}\int z^{p+q-1}\,(z^q-a)^2\,dz. \qquad a+bx = z^q.$$

$$\int x^2\left(a+bx\right)^{p-\frac{1}{2}}dx = \frac{2\Phi^{p+\frac{1}{2}}}{b\,(2p+5)}\left\{x^2 - \frac{4ax}{b\,(2p+3)} + \frac{8a^2}{b^2\,(2p+1)\,(2p+3)}\right\}.$$

$$\int x^2\left(a+bx\right)^{p-\frac{1}{3}}dx = \frac{3\Phi^{p+\frac{2}{3}}}{b\,(3p+8)}\left\{x^2 - \frac{6ax}{b\,(3p+5)} + \frac{18a^2}{b^2\,(3p+2)\,(3p+5)}\right\}.$$

$$\int x^2 \left(a+bx\right)^{p-\frac{2}{3}} dx = \frac{3\Phi^{p+\frac{1}{3}}}{b(3p+7)}\left\{x^2 - \frac{6ax}{b(3p+4)} + \frac{18a^2}{b^2(3p+1)(3p+4)}\right\}.$$

$$\int x^3 (a+bx)^p \, dx = \frac{\Phi^{p+1}}{b(p+4)}\left\{x^2 - \frac{3ax^2}{b(p+3)} + \frac{6a^2x}{b^2(p+2)(p+3)} - \frac{6a^3}{b^3(p+1)(p+2)(p+3)}\right\}.$$

$$\int \frac{dx}{(a+bx)^{\frac{p}{q}}} = -\frac{q}{b(p-q)\,\Phi^{\frac{p}{q}-1}}.$$

$$\int \frac{dx}{(a+bx)^{p+\frac{1}{2}}} = -\frac{2}{b(2p-1)\,\Phi^{p-\frac{1}{2}}}.$$

$$\int \frac{dx}{x(a+bx)^p} = \frac{1}{a(p-1)\,\Phi^{p-1}} + \frac{1}{a}\int \frac{dx}{x\Phi^{p-1}}.$$

$$= \sum_{k=o}^{k=p-2} \frac{1}{(p-k-1)\,a^{k+1}\,\Phi^{p-k-1}} - \frac{1}{a^p}\log\frac{\Phi}{x}.$$

$$= \frac{1}{a^p}\log x - \frac{b}{a^{p+1}}\int \left\{\left(\frac{a}{\Phi}\right)^p + \left(\frac{a}{\Phi}\right)^{p-1} + \cdots + \frac{a}{\Phi}\right\} dx.$$

$$\int \frac{dx}{x(a+bx)^{\frac{p}{q}}} = q\int \frac{z^{q-p-1}}{z^q - a}\, dz. \qquad a+bx = z^q.$$

$$= -q\int \frac{z^{p-1}}{1-az^q}\, dz. \qquad a+bx = \frac{1}{z^q}.$$

$$\int \frac{dx}{x(a+bx)^{p+\frac{1}{2}}} = \frac{2}{a(2p-1)\,\Phi^{p-\frac{1}{2}}} + \frac{1}{a}\int \frac{dx}{x\Phi^{p-\frac{1}{2}}}$$

$$\int \frac{dx}{x(a+bx)^{p+\frac{3}{2}}} = \frac{2}{a(2p+1)\,\Phi^{p+\frac{1}{2}}} + \frac{1}{a}\int \frac{dx}{x\Phi^{p+\frac{1}{2}}}.$$

$$\int \frac{dx}{x^2(a+bx)^p} = -b\sum_{k=o}^{k=p-2} \frac{k+1}{(p-k-1)\,a^{k+2}\,\Phi^{p-k-1}} - \frac{1}{a^p x} + \frac{bp}{a^{p+1}}\log\frac{\Phi}{x}.$$

$$\int \frac{dx}{x^2(a+bx)^{\frac{p}{q}}} = -bq\int \frac{z^{p+q-1}}{(1-az^q)^2}\, dz. \qquad a+bx = \frac{1}{z^q}.$$

$$\int \frac{dx}{x^2(a+bx)^{p+\frac{1}{2}}} = -\frac{1}{ax\Phi^{p-\frac{1}{2}}} - \frac{b(2p+1)}{2a}\int \frac{dx}{x\Phi^{p+\frac{1}{2}}}.$$

$$\int \frac{dx}{x^3(a+bx)^{\frac{p}{q}}} = -b^2 q\int \frac{z^{p+2q-1}}{(1-az^q)^3}\, dz. \qquad a+bx = \frac{1}{z^q}.$$

$$\int \frac{dx}{x^3(a+bx)^{p+\frac{1}{2}}} = -\frac{2a - b(2p+3)x}{4a^2x^2\Phi^{p-\frac{1}{2}}} + \frac{(2p+1)(2p+3)b^2}{8a^2}\int \frac{dx}{x\Phi^{p+\frac{1}{2}}}.$$

$$\int \frac{x}{(x+1)^p}\, dx = \int \frac{dx}{\Phi^{p-1}} - \int \frac{dx}{\Phi^p}.$$

$$\int \frac{x}{(a+bx)^p}\, dx = \frac{a}{b^2(p-1)\,\Phi^{p-1}} - \frac{1}{b^2(p-2)\,\Phi^{p-2}}.$$

$$\int \frac{x}{(a+bx)^{\frac{p}{q}}}\, dx = \left\{\frac{\Phi}{2q-p} - \frac{a}{q-p}\right\}\frac{q}{b^2\Phi^{\frac{p}{q}-1}}.$$

$$\int \frac{x}{(a+bx)^{p+\frac{1}{2}}}\, dx = -\left\{x + \frac{2a}{b(2p-1)}\right\}\frac{2}{b(2p-3)\,\Phi^{p-\frac{1}{2}}}.$$

$$\int \frac{x^2}{(a+bx)^p}\,dx = \frac{1}{b^3}\left\{-\frac{1}{(p-3)\,\Phi^{p-3}} + \frac{2a}{(p-2)\,\Phi^{p-2}} - \frac{a^2}{(p-1)\,\Phi^{p-1}}\right\}.$$

$$\int \frac{x^2}{(a+bx)^{\frac{p}{q}}}\,dx = \left\{\frac{\Phi^2}{3q-p} - \frac{2a\Phi}{2q-p} + \frac{a^2}{q-p}\right\}\frac{q}{b^3\Phi^{\frac{p}{q}-1}}.$$

$$\int \frac{x^2}{(a+bx)^{p+\frac{1}{2}}}\,dx = \left\{x^2 + \frac{4ax}{(2p-3)\,b} + \frac{4a^2}{(2p-1)(2p-3)\,b^2}\right\}\frac{2}{b\,(2p-5)\,\Phi^{p-\frac{1}{2}}}.$$

$$\int \frac{x^3}{(a+bx)^p}\,dx = \frac{1}{b^4}\left\{-\frac{1}{(p-4)\,\Phi^{p-4}} + \frac{3a}{(p-3)\,\Phi^{p-3}} - \frac{3a^2}{(p-2)\,\Phi^{p-2}} + \frac{a^3}{(p-1)\,\Phi^{p-1}}\right\}.$$

$$\int \frac{(a+bx)^p}{x}\,dx = \frac{\Phi^p}{p} + a\int \frac{\Phi^{p-1}}{x}\,dx.$$

$$\int \frac{(a+bx)^{p+q}}{x}\,dx = \sum_{k=o}^{k=q-1} \frac{a^k\,\Phi^{p+q-k}}{p+q-k} + a^q\int \frac{\Phi^p}{x}\,dx.$$

$$\int \frac{(a+bx)^{\frac{p}{q}}}{x}\,dx = q\int \frac{z^{p+q-1}}{z^q-a}\,dz. \qquad a+bx = z^q.$$

$$\int \frac{(a+bx)^{p-\frac{1}{2}}}{x}\,dx = \frac{2\Phi^{p-\frac{1}{2}}}{2p-1} + a\int \frac{\Phi^{p-\frac{3}{2}}}{x}\,dx = \frac{2}{\sqrt{\Phi}}\sum_{k=o}^{k=p-1}\frac{a^k\,\Phi^{p-k}}{2p-2k-1} + a^p\int \frac{dx}{x\sqrt{\Phi}}.$$

$$\int \frac{(a+bx)^{p+\frac{1}{3}}}{x}\,dx = \frac{3\Phi^{p+\frac{1}{3}}}{3p+1} + a\int \frac{\Phi^{p-\frac{2}{3}}}{x}\,dx.$$

$$\int \frac{(a+bx)^{p-\frac{1}{3}}}{x}\,dx = \frac{3\Phi^{p-\frac{1}{3}}}{3p-1} + a\int \frac{\Phi^{p-\frac{4}{3}}}{x}\,dx = \frac{3}{\Phi^{\frac{1}{3}}}\sum_{k=o}^{k=p-1}\frac{a^k\,\Phi^{p-k}}{3p-3k-1} + a^p\int \frac{dx}{x\Phi^{\frac{1}{3}}}.$$

$$\int \frac{(a+bx)^{p-\frac{2}{3}}}{x}\,dx = \frac{3\Phi^{p-\frac{2}{3}}}{3p-2} + a\int \frac{\Phi^{p-\frac{5}{3}}}{x}\,dx = \frac{3}{\Phi^{\frac{2}{3}}}\sum_{k=o}^{k=p-1}\frac{a^k\,\Phi^{p-k}}{3p-3k-2} + a^p\int \frac{dx}{x\Phi^{\frac{3}{2}}}.$$

$$\int \frac{(a+bx)^p}{x^2}\,dx = -\frac{\Phi^{p+1}}{ax} + \frac{bp}{a}\int \frac{\Phi^p}{x}\,dx.$$

$$\int \frac{(a+bx)^{\frac{p}{q}}}{x^2}\,dx = -\frac{\Phi^{\frac{p}{q}+1}}{ax} + \frac{bp}{aq}\int \frac{\Phi^{\frac{p}{q}}}{x}\,dx = bq\int \frac{z^{p+q-1}}{(z^q-a)^2}\,dz. \qquad a+bx = z^q.$$

$$\int \frac{(a+bx)^{p-\frac{1}{2}}}{x^2}\,dx = -\frac{\Phi^{p+\frac{1}{2}}}{ax} + \frac{b\,(2p-1)}{2a}\int \frac{\Phi^{p-\frac{1}{2}}}{x}\,dx.$$

$$\int \frac{(a+bx)^{p-\frac{1}{3}}}{x^2}\,dx = -\frac{\Phi^{p+\frac{2}{3}}}{ax} + \frac{b\,(3p-1)}{3a}\int \frac{\Phi^{p-\frac{1}{3}}}{x}\,dx.$$

$$\int \frac{(a+bx)^{p-\frac{2}{3}}}{x^2}\,dx = -\frac{\Phi^{p+\frac{1}{3}}}{ax} + \frac{b\,(3p-2)}{3a}\int \frac{\Phi^{p-\frac{2}{3}}}{x}\,dx.$$

$$\int \frac{(a+bx)^p}{x^3}\,dx = -\frac{\Phi^{p+1}}{2a}\left\{\frac{1}{x^2} + \frac{b\,(p-1)}{ax}\right\} + \frac{b^2p\,(p-1)}{2a^2}\int \frac{\Phi^p}{x}\,dx.$$

$$\int \frac{(a+bx)^{p-\frac{1}{2}}}{x^3}\,dx = -\frac{\Phi^{p+\frac{1}{2}}}{2ax}\left\{\frac{1}{x} + \frac{b\,(2p-3)}{2a}\right\} + \frac{b^2\,(2p-1)\,(2p-3)}{8a^2}\int \frac{\Phi^{p-\frac{1}{2}}}{x}\,dx.$$

$\int x^A \left(a+bx^2\right)^p dx$

$$\int (a+bx^2)^p\,dx = \frac{x\Phi^p}{2p+1} + \frac{2ap}{2p+1}\int \Phi^{p-1}\,dx.$$

$$\int (a^2 + x^2)^{\frac{p}{2}}\, dx = x\Phi^{\frac{p}{2}} - p\int x^2\Phi^{\frac{p}{2}-1}\, dx = \frac{x\Phi^{\frac{p}{2}}}{p+1} + \frac{pa^2}{p+1}\int \Phi^{\frac{p}{2}-1}\, dx.$$

$$\int (a^2 - x^2)^{\frac{p}{2}}\, dx = \frac{x\Phi^{\frac{p}{2}}}{p+1} + \frac{pa^2}{p+1}\int \Phi^{\frac{p}{2}-1}\, dx.$$

$$\int (a + bx^2)^{\frac{p}{2}}\, dx = \frac{x\Phi^{\frac{p}{2}}}{p+1} + \frac{ap}{p+1}\int \Phi^{\frac{p}{2}-1}\, dx.$$

$$\int (a^2 - x^2)^{p+\frac{1}{2}}\, dx = \frac{x\Phi^{p+\frac{1}{2}}}{2p+2} + \frac{(2p+1)\,a^2}{2p+2}\int \Phi^{p-\frac{1}{2}}\, dx.$$

$$\int (a + bx^2)^{p+\frac{1}{2}}\, dx = \left\{ \frac{\Phi^p}{2p+2} + \frac{(2p+1)\,a\Phi^{p-1}}{(2p+2)\,2p} + \frac{(2p+1)\,(2p-1)\,a^2\Phi^{p-2}}{(2p+2)\,2p\,(2p-2)} \right.$$

$$\left. + \frac{(2p+1)\,(2p-1)\,(2p-3)\,a^3\Phi^{p-3}}{(2p+2)\,2p\,(2p-2)\,(2p-4)} + \cdots + \mathrm{N}\Phi^{o} \right\} x\sqrt{\Phi} + \mathrm{N}a\int \frac{dx}{\sqrt{\Phi}}$$

$$\int x\,(a + bx^2)^{\frac{p}{q}}\, dx = \frac{q}{2b\,(p+q)}\,\Phi^{\frac{p}{q}+1}.$$

$$\int x^2\,(a + bx^2)^p\, dx = \frac{x\Phi^{p+1}}{b\,(3+2p)} - \frac{a}{b\,(3+2p)}\int \Phi^p\, dx.$$

$$\int x^3\,(a + bx^2)^{\frac{p}{q}}\, dx = \frac{q}{2b^2}\int z^{p+q-1}\,(z^q - a)\, dz. \qquad a + bx^2 = z^q.$$

$$\int \frac{dx}{(a^2 + x^2)^p} = \frac{x}{2a^2\,(p-1)\,\Phi^{p-1}} + \frac{2p-3}{2a^2\,(p-1)}\int \frac{dx}{\Phi^{p-1}}.$$

$$= \frac{1}{2p-1}\sum_{k=o}^{k=p-2} \frac{(2p-1)\,(2p-3)\cdots(2p-1-2k)}{(p-1)\,(p-2)\cdots(p-1-k)}\cdot\frac{x}{(2a^2)^{k+1}\Phi^{p-k-1}}$$

$$+ \frac{(2p-3)\,(2p-5)\cdots 5.3.1}{(p-1)\,(p-2)\cdots 2.1\,(2a^2)^{p-1}}\,\frac{1}{a}\,\mathrm{arc\ tg}\,\frac{x}{a}$$

$$= \frac{1}{(p-1)\,2a^2}\cdot\frac{x}{\Phi^{p-1}} + \frac{2p-3}{(p-1)\,(p-2)\,(2a^2)^2}\cdot\frac{x}{\Phi^{p-2}} + \frac{(2p-3)\,(2p-5)}{(p-1)\,(p-2)\,(p-3)\,(2a^2)^3}\cdot\frac{x}{\Phi^{p-3}} + \cdots$$

$$+ \frac{(2p-3)\,(2p-5)\cdots 5.3}{(p-1)\,(p-2)\cdots 2.1\,(2a^2)^{p-1}}\cdot\frac{x}{\Phi} + \frac{(2p-3)\,(2p-5)\cdots 5.3.1}{(p-1)\,(p-2)\cdots 2.1\,(2a^2)^{p-1}}\cdot\frac{1}{a}\,\mathrm{arc\ tg}\,\frac{x}{a}.$$

$$\int \frac{dx}{(a^2 - x^2)^p} = \frac{x}{2a^2\,(p-1)\,\Phi^{p-1}} + \frac{2p-3}{2a^2\,(p-1)}\int \frac{dx}{\Phi^{p-1}}.$$

$$= \frac{1}{2p-1}\sum_{k=o}^{k=p-2} \frac{(2p-1)\,(2p-3)\cdots(2p-1-2k)}{(p-1)(p-2)\cdots(p-1-k)}\,\frac{x}{(2a^2)^{k+1}\,\Phi^{p-1-k}} + \frac{(2p-3)\,(2p-5)\cdots 3.1}{(p-1)\,(p-2)\cdots 2.1}\,\frac{a}{(2a^2)^p}\log\frac{a+x}{a-x}.$$

$$\int \frac{dx}{(x^2 - a^2)^p} = \frac{1}{(2a)^{2p-1}}\int \frac{(z-1)^{2p-2}}{z^p}\, dz. \qquad x - a = z\,(x+a).$$

$$= -\frac{x}{2a^2\,(p-1)\,\Phi^{p-1}} - \frac{2p-3}{2a^2\,(p-1)}\int \frac{dx}{\Phi^{p-1}}.$$

$$\int \frac{dx}{(a + bx^2)^p} = \frac{x}{2a\,(p-1)\,\Phi^{p-1}} + \frac{2p-3}{2a\,(p-1)}\int \frac{dx}{\Phi^{p-1}}.$$

$$\int \frac{dx}{(a + bx^2)^{\frac{p}{2}}} = \frac{x}{a\,(p-2)\,\Phi^{\frac{p}{2}-1}} + \frac{p-3}{a\,(p-2)}\int \frac{dx}{\Phi^{\frac{p}{2}-1}}.$$

$$\int \frac{dx}{(a^2+x^2)^{p+\frac{1}{2}}} = \frac{x}{a^2(2p-1)\,\Phi^{p-\frac{1}{2}}} + \frac{2p-2}{a^2(2p-1)} \int \frac{dx}{\Phi^{p-\frac{1}{2}}}.$$

$$\int \frac{dx}{(a+bx^2)^{p+\frac{1}{2}}} = \left\{ \frac{1}{a(2p-1)\,\Phi^{p-1}} + \frac{2p-2}{(2p-1)(2p-3)\,a^2\Phi^{p-2}} + \frac{(2p-2)(2p-4)}{(2p-1)(2p-3)(2p-5)\,a^3\,\Phi^{p-3}} + \cdots + \frac{N}{\Phi^0} \right\} \frac{x}{\sqrt{\Phi}}.$$

$$\int \frac{dx}{x\,(a+bx^2)^{\frac{p}{2}}} = \frac{1}{a\,(p-2)\,\Phi^{\frac{p}{2}-1}} + \frac{1}{a} \int \frac{dx}{x\,\Phi^{\frac{p}{2}-1}}$$

$$\int \frac{dx}{x\,(a+bx^2)^{\frac{p}{q}}} = -\frac{q}{2} \int \frac{z^{p-1}}{1-az^q}\,dz. \qquad a+bx^2 = \frac{1}{z^q}.$$

$$\int \frac{dx}{x\,(a+bx^2)^{p+\frac{1}{2}}} = \left\{ \frac{1}{(2p-1)\,a\Phi^{p-1}} + \frac{1}{(2p-3)\,a^2\Phi^{p-2}} + \frac{1}{(2p-5)\,a^3\Phi^{p-3}} + \cdots \frac{1}{a^p} \right\} \frac{1}{\sqrt{\Phi}} + \frac{1}{a^p} \int \frac{dx}{x\sqrt{\Phi}}.$$

$$\int \frac{dx}{x^2\,(a+bx^2)^p} = \frac{1}{a} \int \frac{dx}{x^2\,\Phi^{p-1}} - \frac{b}{a} \int \frac{dx}{\Phi^p}.$$

$$\int \frac{dx}{x^3\,(a+bx^2)^{\frac{p}{q}}} = \frac{bq}{2} \int \frac{z^{q-p-1}}{(z^q-a)^2}\,dz. \qquad a+bx^2 = z^q.$$

$$\int \frac{dx}{x^4\,(a+bx^2)^p} = \frac{1}{a} \int \frac{dx}{x^4\,\Phi^{p-1}} - \frac{b}{a} \int \frac{dx}{x^2\Phi^p}.$$

$$\int \frac{x}{(a+bx^2)^{\frac{p}{q}}}\,dx = \frac{q}{2b\,(q-p)\,\Phi^{\frac{p}{q}-1}}.$$

$$\int \frac{x^2}{(a+bx^2)^p}\,dx = -\frac{x}{2b\,(p-1)\,\Phi^{p-1}} + \frac{1}{2b\,(p-1)} \int \frac{dx}{\Phi^{p-1}}.$$

$$\int \frac{x^3}{(a+bx^2)^{\frac{p}{q}}}\,dx = \frac{q}{2b^2} \int z^{q-p-1}\,(z^q-a)\,dz. \qquad a+bx^2 = z^q.$$

$$\int \frac{(a+bx^2)^{\frac{p}{2}}}{x}\,dx = \frac{1}{p}\,\Phi^{\frac{p}{2}} + a \int \frac{\Phi^{\frac{p}{2}-1}}{x}\,dx.$$

$$\int \frac{(a+bx^2)^{\frac{p}{q}}}{x}\,dx = \frac{q}{2} \int \frac{z^{p+q-1}}{z^q-a}\,dz. \qquad a+bx^2 = z^q.$$

$$\int \frac{(a+bx^2)^{p+\frac{1}{2}}}{x}\,dx = \left\{ \frac{\Phi^p}{2p+1} + \frac{a\Phi^{p-1}}{2p-1} + \frac{a^2\Phi^{p-2}}{2p-3} + \cdots \frac{a^p}{1} \right\} \sqrt{\Phi} + a^{p+1} \int \frac{dx}{x\sqrt{\Phi}}.$$

$$\int \frac{(a+bx^2)^{\frac{p}{q}}}{x^2}\,dx = -\frac{\Phi^{\frac{p}{q}+1}}{ax} + \frac{b\,(2p+q)}{aq} \int \Phi^{\frac{p}{q}}\,dx.$$

$\int x^A \left(a+bx^B\right)^p dx.$

$$\int x^2\,(a+bx^3)^{\frac{p}{q}}\,dx = \frac{q}{3b\,(p+q)}\,\Phi^{\frac{p}{q}+1}.$$

$$\int x^5\,(a+bx^3)^p\,dx = \frac{1}{3b^2} \left\{ \frac{\Phi^{p+2}}{p+2} - \frac{a\Phi^{p+1}}{p+1} \right\}.$$

$$\int \frac{dx}{(a+bx^3)^p} = \frac{x}{3a\,(p-1)\,\Phi^{p-1}} + \frac{3p-4}{3a\,(p-1)} \int \frac{dx}{\Phi^{p-1}}.$$

$$\int \frac{dx}{x^2(a+bx^3)^p} = \frac{1}{a}\int \frac{dx}{x^2\phi^{p-1}} - \frac{b}{a}\int \frac{x}{\phi^p}\,dx.$$

$$\int \frac{dx}{x^3(a+bx^3)^p} = \frac{1}{a}\int \frac{dx}{x^3\phi^{p-1}} - \frac{b}{a}\int \frac{dx}{\phi^p}.$$

$$\int \frac{dx}{(a+bx^4)^p} = \frac{x}{4a(p-1)\phi^{p-1}} + \frac{4p-5}{4a(p-1)}\int \frac{dx}{\phi^{p-1}}.$$

$$\int \frac{x}{(a+bx^3)^p}\,dx = \frac{x^2}{3a(p-1)\phi^{p-1}} + \frac{3p-5}{3a(p-1)}\int \frac{x}{\phi^{p-1}}\,dx.$$

$$\int \frac{x^2}{(a+bx^3)^{\frac{p}{q}}}\,dx = -\frac{q}{3b(p-q)\phi^{\frac{p}{q}-1}}.$$

$$\int \frac{x^2}{(a+bx^4)^p}\,dx = \frac{x^3}{4a(p-1)\phi^{p-1}} + \frac{4p-7}{4a(p-1)}\int \frac{x^2}{\phi^{p-1}}\,dx.$$

$$\int \frac{x^3}{(a+bx^3)^p}\,dx = -\frac{x}{3b(p-1)\phi^{p-1}} + \frac{1}{3b(p-1)}\int \frac{dx}{\phi^{p-1}}.$$

$$\text{———} \quad \int x^m\left(a+bx^n\right)^C dx \quad \text{———}$$

$$\int x^{n-1}\sqrt{a+bx^n}\,dx = \frac{2}{3bn}\phi^{\frac{3}{2}}.$$

$$\int x^{2n-1}\sqrt{a+bx^n}\,dx = \frac{1}{n}\int z\sqrt{a+bz}\,dz. \qquad x^n = z.$$

$$\int x^{n-1}(a+bx^n)^{\frac{3}{2}}\,dx = \frac{2}{5bn}\phi^{\frac{5}{2}}.$$

$$\int x^{2n-1}(a+bx^n)^{\frac{3}{2}}\,dx = \frac{1}{n}\int z(a+bz)^{\frac{3}{2}}\,dz. \qquad x^n = z.$$

$$\int \frac{dx}{x^n(x^n-a)} = \frac{1}{a(n-1)x^{n-1}} + \frac{1}{a}\int \frac{dx}{x^n-a}$$

$$\int \frac{dx}{x^{2n+1}(x^n-a)} = \frac{1}{n}\int \frac{dz}{z^3(z-a)}. \qquad x^n = z.$$

$$\int \frac{dx}{x^{2n}(1-x^{2n})} = \frac{1}{2n-1}\left\{\frac{1}{\phi^{2n-1}} - \frac{1}{x^{2n-1}}\right\} + \frac{2n}{2n-2}\left\{\frac{1}{\phi^{2n-2}} - \frac{1}{x^{2n-2}}\right\} + \frac{2n(2n+1)}{1.2(2n-3)}\left\{\frac{1}{\phi^{2n-3}} - \frac{1}{x^{2n-3}}\right\}$$
$$+ \frac{2n(2n+1)(2n+2)}{1.2.3(2n-4)}\left\{\frac{1}{\phi^{2n-4}} - \frac{1}{x^{2n-4}}\right\} + \cdots \frac{2n(2n+1)\cdots(4n-3)}{1.2\cdots(2n-2)}\left\{\frac{1}{\phi} - \frac{1}{x}\right\}$$
$$+ \frac{2n(2n+1)\cdots(4n-2)}{1.2\cdots(2n-1)}\log\frac{x}{\phi}.$$

$$\int \frac{dx}{x^{n+1}(a+bx^n)} = -\frac{1}{anx^n} + \frac{b}{a^2 n}\log\frac{\phi}{x^n}.$$

$$\int \frac{dx}{x^{n-m}(a+bx^n)} = -\frac{1}{a(n-m-1)x^{n-m-1}} - \frac{b}{a}\int \frac{x^m}{\phi}\,dx.$$

$$\int \frac{dx}{x^m(a+bx^n)} = -\frac{1}{a(m-1)x^{m-1}} - \frac{b}{a}\int \frac{x^{n-m}}{\phi}\,dx.$$

$$\int \frac{dx}{x^{n-1}(a+bx^n)^2} = -\frac{1}{a(n-2)\phi x^{n-2}} - \frac{2b(n-1)}{a(n-2)}\int \frac{x}{\phi^2}\,dx.$$

$$\int \frac{dx}{x^m (a + bx^n)^2} = -\frac{1}{a(m-1)\Phi x^{m-1}} - \frac{b(m+n-1)}{a(m-1)} \int \frac{dx}{x^{m-n}\Phi^2}.$$

$$\int \frac{dx}{x^m (a + bx^n)^3} = -\frac{1}{a(m-1)\Phi^2 x^{m-1}} - \frac{b(m+2n-1)}{a(m-1)} \int \frac{dx}{x^{m-n}\Phi^3}.$$

$$\int \frac{dx}{x^{\frac{n}{2}+1}\sqrt{a+bx^n}} = -\frac{2\sqrt{\Phi}}{nax^{\frac{n}{2}}}.$$

$$\int \frac{dx}{x^{n+1}\sqrt{a+bx^n}} = -\frac{\sqrt{\Phi}}{nax^n} - \frac{b}{2a}\int \frac{dx}{x\sqrt{\Phi}} = -\frac{\sqrt{\Phi}}{nax^n} + \frac{b}{na\sqrt{a}} \log \frac{\sqrt{\Phi}+\sqrt{a}}{x^{\frac{n}{2}}}.$$

$$\int \frac{dx}{x^m\sqrt{a+bx^n}} = -\frac{\sqrt{\Phi}}{a(m-1)x^{m-1}} - \frac{b(2m-n-2)}{2a(m-1)}\int \frac{dx}{x^{m-n}\sqrt{\Phi}}.$$

$$\int \frac{dx}{x^m(a+bx^n)^{\frac{3}{2}}} = \frac{2}{anx^{m-1}\sqrt{\Phi}} + \frac{2m+n-2}{an}\int \frac{dx}{x^m\sqrt{\Phi}}.$$

$$\int \frac{x^n}{a+bx^n}dx = \frac{x}{b} - \frac{a}{b}\int \frac{dx}{a+bx^n}.$$

$$\int \frac{x^{n-1}}{a+bx^n}dx = \frac{1}{bn}\log(a+bx^n).$$

$$\int \frac{x^{2n}}{1-ax^n}dx = -\frac{x^{n+1}}{a(n+1)} + \frac{1}{a}\int \frac{x^n}{1-ax^n}dx.$$

$$\int \frac{x^{2n-1}}{a+bx^n}dx = \frac{x^n}{bn} - \frac{a}{b^2n}\log(a+bx^n).$$

$$\int \frac{x^{\frac{n}{2}}}{a^2-x^n}dx = \frac{1}{2}\int \frac{dx}{a-x^{\frac{n}{2}}} - \frac{1}{2}\int \frac{dx}{a+x^{\frac{n}{2}}}.$$

$$\int \frac{x^{\frac{n}{2}-1}}{x^n+1}dx = \frac{2}{n}\operatorname{arc\,tg} x^{\frac{n}{2}} = -\frac{1}{n}\operatorname{arc\,cos}\frac{x^n-1}{x^n+1}.$$

$$\int \frac{x^{\frac{n}{2}-1}}{a+bx^n}dx = \frac{2}{n\sqrt{ab}}\operatorname{arc\,tg} x^{\frac{n}{2}}\sqrt{\frac{b}{a}}.$$

$$\int \frac{x^{\frac{n}{2}-1}}{a-bx^n}dx = \frac{2}{n\sqrt{ab}}\log\frac{\sqrt{a}+x^{\frac{n}{2}}\sqrt{b}}{\sqrt{a}-x^{\frac{n}{2}}\sqrt{b}}.$$

$$\int \frac{x^m}{1+x^n}dx = -\frac{1}{n}\cos\frac{(m+1)\pi}{n}\log\left(1-2x\cos\frac{\pi}{n}+x^2\right) - \frac{1}{n}\cos\frac{3(m+1)\pi}{n}\log\left(1-2x\cos\frac{3\pi}{n}+x^2\right)$$
$$-\frac{1}{n}\cos\frac{5(m+1)\pi}{n}\log\left(1-2x\cos\frac{5\pi}{n}+x^2\right) - \cdots + 2\sin\frac{(m+1)\pi}{n}\operatorname{arc\,tg}\frac{x\sin\frac{\pi}{n}}{1-x\cos\frac{\pi}{n}}$$
$$+2\sin\frac{3(m+1)\pi}{n}\operatorname{arc\,tg}\frac{x\sin\frac{3\pi}{n}}{1-x\cos\frac{3\pi}{n}} + 2\sin\frac{5(m+1)\pi}{n}\operatorname{arc\,tg}\frac{x\sin\frac{5\pi}{n}}{1-x\cos\frac{5\pi}{n}} + \cdots$$

$$\int \frac{x^m}{1-x^n}dx = -\frac{1}{n}\cos\frac{2(m+1)\pi}{n}\log\left(1-2x\cos\frac{2\pi}{n}+x^2\right) - \frac{1}{n}\cos\frac{4(m+1)\pi}{n}\log\left(1-2x\cos\frac{4\pi}{n}+x^2\right)$$
$$-\frac{1}{n}\cos\frac{6(m+1)\pi}{n}\log\left(1-2x\cos\frac{6\pi}{n}+x^2\right) - \cdots + \frac{2}{n}\sin\frac{2(m+1)\pi}{n}\operatorname{arc\,tg}\frac{x\sin\frac{2\pi}{n}}{1-x\cos\frac{2\pi}{n}}$$
$$+\frac{2}{n}\sin\frac{4(m+1)\pi}{n}\operatorname{arc\,tg}\frac{x\sin\frac{4\pi}{n}}{1-x\cos\frac{4\pi}{n}} + \frac{2}{n}\sin\frac{6(m+1)\pi}{n}\operatorname{arc\,tg}\frac{x\sin\frac{6\pi}{n}}{1-x\cos\frac{6\pi}{n}} + \cdots - \frac{1}{n}\log(1-x).$$

$$\int \frac{x^m}{a+bx^n}\,dx = \frac{x^{m-n+1}}{b\,(m-n+1)} - \frac{a}{b}\int \frac{x^{m-n}}{a+bx^n}\,dx = \frac{1}{a}\left(\frac{a}{b}\right)^{\frac{m+1}{n}} \int \frac{z^m}{1+z^n}\,dz. \qquad x = \left(\frac{a}{b}\right)^{\frac{1}{n}} z.$$

$$\int \frac{x^m}{a-bx^n}\,dx = \frac{1}{a}\left(\frac{a}{b}\right)^{\frac{m+1}{n}} \int \frac{z^m}{1-z^n}\,dz. \qquad x = \left(\frac{a}{b}\right)^{\frac{1}{n}} z.$$

$$\int \frac{x^n}{(a+bx^n)^2}\,dx = -\frac{x}{bn\Phi} + \frac{1}{bn}\int \frac{dx}{\Phi}.$$

$$\int \frac{x^{n-1}}{(2+x^n)^2}\,dx = \frac{x^n+1}{n\Phi}.$$

$$\int \frac{x^{n-1}}{(x^n-a)^2}\,dx = -\frac{1}{2an}\cdot\frac{x^n+a}{x^n-a}.$$

$$\int \frac{x^{n-1}}{(a+bx^n)^2}\,dx = -\frac{1}{bn\Phi} = \frac{x^n}{an\Phi} = \frac{1}{n\,(ap-bq)}\cdot\frac{q+px^n}{a+bx^n}.$$

$$\int \frac{x^{2n-1}}{(x^n+a)^2}\,dx = \frac{a}{n\Phi} + \frac{1}{n}\log\Phi.$$

$$\int \frac{x^{2n-1}}{(a+bx^n)^2}\,dx = \frac{a}{nb^2\Phi} + \frac{1}{nb^2}\log\Phi = -\int \frac{dz}{z^{n+1}(z^n-b)}. \qquad a+bx^n = z^n x^n.$$

$$\int \frac{x^m}{(a+bx^n)^2}\,dx = -\frac{x^{m-n+1}}{nb\Phi} + \frac{m-n+1}{nb}\int \frac{x^{m-n}}{\Phi}\,dx = \frac{x^{m-n+1}}{b\,(m-2n+1)\,\Phi} - \frac{a\,(m-n+1)}{b\,(m-2n+1)}\int \frac{x^{m-n}}{\Phi^2}\,dx.$$

$$= \frac{x^{m+1}}{na\Phi} + \frac{n-m-1}{na}\int \frac{x^m}{\Phi}\,dx.$$

$$\int \frac{x^{n-1}}{(a+bx^n)^3}\,dx = -\frac{1}{a^2}\int \frac{z^n-b}{z^{2n+1}}\,dz. \qquad a+bx^n = z^n x^n.$$

$$\int \frac{x^{2n-1}}{(a+bx^n)^3}\,dx = \frac{x^{2n}}{2an\Phi^2}.$$

$$\int \frac{x^{3n-1}}{(x^n-b)^3}\,dx = -\int \frac{(1+bz^n)^2}{z}\,dz. \qquad x^n - b - \frac{1}{z^n}.$$

$$\int \frac{x^m}{(a+bx^n)^3}\,dx = \frac{x^{m-n+1}}{b\,(m-3n+1)\,\Phi^2} - \frac{a\,(m-n+1)}{b\,(m-3n+1)}\int \frac{x^{m-n}}{\Phi^3}\,dx.$$

$$\int \frac{x^{n-1}}{\sqrt{a+bx^n}}\,dx = \frac{2}{bn}\sqrt{\Phi}.$$

$$\int \frac{x^{2n-1}}{\sqrt{a+bx^n}}\,dx = \frac{2x^n}{bn}\sqrt{\Phi} - \frac{4}{3b^2n}\Phi^{\frac{3}{2}}.$$

$$\int \frac{x^{n-1}}{\sqrt{a+x^{2n}}}\,dx = \frac{1}{n}\log\left(x^n + \sqrt{\Phi}\right).$$

$$\int \frac{x^{n-1}}{\sqrt{a^2-b^2x^{2n}}}\,dx = \frac{1}{bn}\arcsin\frac{b}{a}x^n = \frac{1}{bn}\arccos\frac{\sqrt{a^2-b^2x^{2n}}}{a}.$$

$$\int \frac{x^{n-1}}{(a+bx^n)^{\frac{3}{2}}}\,dx = -\frac{2}{bn\sqrt{\Phi}} = \frac{x^n}{an\sqrt{\Phi}}.$$

$$\int \frac{x^{n-1}}{(a+bx^{2n})^{\frac{3}{2}}}\,dx = \frac{x^n}{an\sqrt{\Phi}}.$$

$$\int \frac{x^m}{(a+bx^n)^{\frac{3}{2}}}\,dx = \frac{2x^{m+1}}{an\sqrt{\Phi}} - \frac{2m-n+2}{an}\int \frac{x^m}{\sqrt{\Phi}}\,dx.$$

$$\int \frac{x^{m+n}}{(a+bx^n)^{\frac{5}{2}}}\,dx = -\frac{2(m+1)}{3bn}\left\{\frac{x^{m+1}}{(m+1)\Phi^{\frac{3}{2}}} - \frac{2x^{m+1}}{an\Phi^{\frac{1}{2}}} + \frac{2m-n+2}{an}\int \frac{x^m}{\sqrt{\Phi}}\,dx\right\}$$

$$\int \frac{\sqrt{a+bx^n}}{x^n}\,dx = -\frac{\sqrt{\Phi}}{(n-1)x^{n-1}} + \frac{nb}{2(n-1)}\int \frac{dx}{\sqrt{\Phi}}.$$

$$\int \frac{\sqrt{a+bx^n}}{x^{n+1}}\,dx = -\frac{\sqrt{\Phi}}{nx^n} + \frac{b}{2}\int \frac{dx}{x\sqrt{\Phi}} = \frac{2b}{n}\int \frac{z^2}{(z^2-a)^2}\,dz. \qquad a+bx^n = z^2.$$

$$\int \frac{\sqrt{a+bx^{2n}}}{x^{3n+1}}\,dx = -\frac{\Phi^{\frac{3}{2}}}{3anx^{3n}}.$$

$$\int \frac{\sqrt{ax^n+b}}{x^m}\,dx = a\int \frac{x^{n-m}}{\sqrt{\Phi}}\,dx + b\int \frac{dx}{\sqrt{\Phi}}.$$

$$\int \frac{(a+bx^n)^{\frac{3}{2}}}{x^m}\,dx = -\frac{\Phi^{\frac{5}{2}}}{a(m-1)x^{m-1}} - \frac{b(2m-5n-2)}{2a(m-1)}\int \frac{\Phi^{\frac{3}{2}}}{x^{m-n}}\,dx = b^3\int \frac{x^{3n-m}}{\Phi^{\frac{3}{2}}}\,dx + 3ab^2\int \frac{x^{2n-m}}{\Phi^{\frac{3}{2}}}\,dx$$

$$+\,3a^2b\int \frac{x^{n-m}}{\Phi^{\frac{3}{2}}}\,dx + a^3\int \frac{dx}{x^m\Phi^{\frac{3}{2}}}.$$

$x^A(a+bx^n)^p\,dx$

$$\int (a+bx^n)^{\frac{n-1}{n}}\,dx = -a\int \frac{z^{2n-2}}{(z^n-b)^2}\,dz. \qquad a+bx^n = z^nx^n.$$

$$\int (a+bx^n)^{\frac{2n-1}{n}}\,dx = -a^2\int \frac{z^{3n-2}}{(z^n-b)^3}\,dz. \qquad a+bx^n = z^nx^n.$$

$$\int (a+bx^n)^p\,dx = a\int \Phi^{p-1}\,dx + b\int x^n\Phi^{p-1}\,dx = x\,\Phi^p - bnp\int x^n\Phi^{p-1}\,dx.$$

$$\int x\,(a+bx^n)^{\frac{n-2}{n}}\,dx = -a\int \frac{z^{2n-3}}{(z^n-b)^2}\,dz. \qquad a+bx^n = z^nx^n.$$

$$\int x\,(a+bx^n)^p\,dx = \frac{x^2\Phi^p}{np+2} + \frac{anp}{np+2}\int x\Phi^{p-1}\,dx.$$

$$\int x\,(a+bx^n)^{\frac{p}{q}}\,dx = \frac{qx^{2-n}\,\Phi^{\frac{p}{q}+1}}{b(np+2q)} + \frac{aq(n-2)}{b(np+2q)}\int \frac{\Phi^{\frac{p}{q}}}{x^{n-1}}\,dx = \frac{qx^2\Phi^{\frac{p}{q}}}{np+2q} + \frac{anp}{np+2q}\int x\,\Phi^{\frac{p}{q}-1}\,dx.$$

$$\int x^2\,(a+bx^n)^p\,dx = \frac{x^3\Phi^p}{np+3} + \frac{anp}{np+3}\int x^2\Phi^{p-1}\,dx.$$

$$\int x^2\,(a+bx^n)^{\frac{p}{q}}\,dx = \frac{q\Phi^{\frac{p}{q}+1}}{b(np+3q)x^{n-3}} + \frac{aq(n-3)}{b(np+3q)}\int \frac{\Phi^{\frac{p}{q}}}{x^{n-2}}\,dx = \frac{qx^3\Phi^{\frac{p}{q}}}{np+3q} + \frac{anp}{np+3q}\int x^2\Phi^{\frac{p}{q}-1}\,dx.$$

$$\int \frac{dx}{(a+bx^n)^{\frac{1}{n}}} = \int \frac{z^{n-2}}{b-z^n}\,dz. \qquad a+bx^n = z^nx^n.$$

$$\int \frac{dx}{(a+bx^n)^{\frac{n+1}{n}}} = \frac{x}{a\Phi^{\frac{1}{n}}}.$$

$$\int \frac{dx}{(a+bx^n)^{\frac{2n+1}{n}}} = -\frac{1}{a^2}\int \frac{z^n-b}{z^{n+2}}\,dz. \qquad a+bx^n = z^nx^n.$$

$$\int \frac{dx}{(a+bx^n)^{\frac{p}{q}}} = \frac{qx}{an(p-q)\,\Phi^{\frac{p}{q}-1}} - \frac{nq+q-np}{an(p-q)} \int \frac{dx}{\Phi^{\frac{p}{q}-1}}.$$

$$\int \frac{dx}{x(a+bx^n)^{\frac{n-1}{n}}} = \int \frac{z^{n-2}}{az^n-1}\,dz. \qquad a+bx^n = \frac{1}{z^n}.$$

$$\int \frac{dx}{x(1+x^n)^p} = -\frac{1}{n}\int \frac{(z-1)^{p-1}}{z^p}\,dz. \qquad 1+x^n = zx^n.$$

$$\int \frac{dx}{x^2(a+bx^n)^{\frac{n-1}{n}}} = -\frac{\Phi^{\frac{1}{n}}}{ax}.$$

$$\int \frac{x}{(a+bx^n)^{\frac{2}{n}}}\,dx = \int \frac{z^{n-3}}{b-z^n}\,dz. \qquad a+bx^n = z^n x^n.$$

$$\int \frac{x}{(a+bx^n)^{\frac{n+2}{n}}}\,dx = \frac{x^2}{2a\Phi^{\frac{2}{n}}}.$$

$$\int \frac{x}{(a+bx^n)^{\frac{p}{q}}}\,dx = \frac{qx^2}{an(p-q)\,\Phi^{\frac{p}{q}-1}} - \frac{2q+nq-np}{an(p-q)} \int \frac{x}{\Phi^{\frac{p}{q}-1}}\,dx.$$

$$\int \frac{x^2}{(a+bx^n)^{\frac{n+3}{n}}}\,dx = \frac{x^3}{3a\Phi^{\frac{3}{n}}}.$$

$$\int \frac{x^2}{(a+bx^n)^{\frac{p}{q}}}\,dx = \frac{qx^3}{an(p-q)\,\Phi^{\frac{p}{q}-1}} - \frac{3q+nq-np}{an(p-q)} \int \frac{x^2}{\Phi^{\frac{p}{q}-1}}\,dx.$$

$$\int \frac{(a+bx^n)^{\frac{p}{q}}}{x}\,dx = \frac{q}{n}\int \frac{z^{p+q-1}}{z^q-a}\,dz. \qquad a+bx^n = z^q.$$

$$\int \frac{(a+bx^n)^{\frac{1}{n}}}{x^2}\,dx = -\int \frac{z^n}{z^n-b}\,dz. \qquad a+bx^n = z^n x^n.$$

$$\int \frac{(a+bx^n)^{\frac{2n+1}{n}}}{x^2}\,dx = -a^2\int \frac{z^{3n}}{(z^n-b)^3}\,dz. \qquad a+bx^n = z^n x^n.$$

$$\int \frac{(a+bx^n)^{\frac{p}{q}}}{x^2}\,dx = -\frac{\Phi^{\frac{p}{q}+1}}{ax} - \frac{b(q-np-nq)}{aq}\int x^{n-2}\,\Phi^{\frac{p}{q}}\,dx.$$

$\int x^m (a+bx)^p\, dx$

$$\int x^m (x+a)^p\,dx = \frac{1}{p+1}x^m\Phi^{p+1} - \frac{m}{(p+1)(p+2)}x^{m-1}\Phi^{p+2} + \frac{m(m-1)}{(p+1)(p+2)(p+3)}x^{m-2}\Phi^{p+3}$$
$$\frac{m(m-1)(m-2)}{(p+1)(p+2)(p+3)(p+4)}x^{m-3}\Phi^{p+4} + \ldots \pm \mathrm{N}x^o\Phi^{p+m+1}.$$

$$\int x^m (a+bx)^p\,dx = \frac{x^m\Phi^{p+1}}{b(m+p+1)} - \frac{am}{b(m+p+1)}\int x^{m-1}\Phi^p\,dx = \frac{x^{m+1}\Phi^p}{m+p+1} + \frac{ap}{m+p+1}\int x^m\Phi^{p-1}\,dx.$$

$$\int x^m (a+bx)^{\frac{p}{q}}\,dx = \frac{qx^m\Phi^{\frac{p}{q}+1}}{b(mq+p+q)} - \frac{amq}{b(mq+p+q)}\int x^{m-1}\Phi^{\frac{p}{q}}\,dx.$$
$$= \left\{\frac{\Phi^m}{qm+p+q} - \frac{m}{1}\frac{a\Phi^{m-1}}{qm+p} + \frac{m(m-1)}{1.2}\frac{a^2\Phi^{m-2}}{qm+p-q} - \frac{m(m-1)(m-2)}{1.2.3.}\frac{a^3\Phi^{m-3}}{qm+p-2q} + \ldots \pm \mathrm{N}a^m\right\}\frac{q\Phi^{\frac{p+q}{q}}}{b^{m+1}}.$$

$$\int x^m (a+bx)^{p-\frac{1}{2}}\,dx = \frac{2x^m\Phi^{p+\frac{1}{2}}}{b(2m+2p+1)} - \frac{2am}{b(2m+2p+1)}\int x^{m-1}\Phi^{p-\frac{1}{2}}\,dx.$$

$$\int x^m (a+bx)^{p+\frac{1}{3}}\,dx = \frac{3x^m\,\Phi^{p+\frac{4}{3}}}{b\,(3m+3p+4)} - \frac{3am}{b\,(3m+3p+4)}\int x^{m-1}\Phi^{p+\frac{1}{3}}\,dx.$$

$$\int x^m (a+bx)^{p-\frac{1}{3}}\,dx = \frac{3x^m\Phi^{p+\frac{2}{3}}}{b\,(3m+3p+2)} - \frac{3am}{b\,(3m+3p+2)}\int x^{m-1}\,\Phi^{p-\frac{1}{3}}\,dx.$$

$$\int x^m (a+bx)^{p-\frac{2}{3}}\,dx = \frac{3x^m\Phi^{p+\frac{1}{3}}}{b\,(3m+3p+1)} - \frac{3am}{b\,(3m+3p+1)}\int x^{m-1}\,\Phi^{p-\frac{2}{3}}\,dx.$$

$$\int\frac{dx}{x^m(1-x)^p} = -\frac{1}{(m-1)\,x^{m-1}\,\Phi^{p-1}} + \frac{m+p-2}{m-1}\int\frac{dx}{x^{m-1}\Phi^p}.$$

$$\int\frac{dx}{x^m(a+bx)^p} = -\int\frac{z^{m+p-2}}{(az+b)^p}\,dz. \qquad xz=1.$$

$$= \frac{(-b)^{m-1}}{a^m}\left\{\frac{1}{(p-1)\,\Phi^{p-1}} + \frac{m}{(p-2)\,a\Phi^{p-2}} + \frac{m\,(m+1)}{1.2(p-3)a^2\Phi^{p-3}} + \frac{m(m+1)(m+2)}{1.2.3(p-4)\,a^3\Phi^{p-4}} + \cdots \frac{m(m+1)\cdots(m+p-3)}{1.2.3\cdots(p-2)\,a^{p-2}\Phi}\right\}$$

$$+\frac{1}{a^p}\left\{-\frac{1}{(m-1)\,x^{m-1}} + \frac{pb}{(m-2)\,ax^{m-2}} - \frac{p\,(p+1)\,b^2}{1.2\,(m-3)\,a^2x^{m-3}} + \frac{p\,(p+1)(p+2)b^3}{1.2.3\,(m-4)\,a^3x^{m-4}}\cdots \pm \frac{p\,(p+1)\cdots(p+m-3)b^{m-2}}{1.2.3\cdots(m-2)\,a^{m-2}x}\right\}$$

$$+(-1)^m\,\frac{m\,(m+1)\cdots(m+p-2)}{1.2.3\cdots(p-1)}\,\frac{b^{m-1}}{a^{m+p-1}}\log\frac{\Phi}{x}.$$

$$= \frac{1}{a\,(p-1)\,x^{m-1}\Phi^{p-1}} + \frac{m+p-2}{a\,(p-1)}\int\frac{dx}{x^m\Phi^{p-1}}.$$

$$= -\frac{1}{(m-1)\,ax^{m-1}\,\Phi^{p-1}} - \frac{b\,(m+p-2)}{a\,(m-1)}\int\frac{dx}{x^{m-1}\Phi^p}.$$

$$\int\frac{dx}{x^m\,(a+bx)^{\frac{p}{2}}} = \left\{-\frac{1}{(m-1)\,ax^{m-1}} + \frac{1}{(m-1)\,a}\,\frac{(2m+p-4)\,b}{(2m-4)\,ax^{m-2}} - \frac{1}{(m-1)\,a}\cdot\frac{(2m+p-4)\,b}{(2m-4)\,a}\cdot\frac{(2m+p-6)\,b}{(2m-6)\,ax^{m-3}} + \cdots\right.$$

$$\left.\pm\frac{\mathrm{N}}{x}\right\}\frac{1}{\Phi^{\frac{p}{2}-1}} \pm \frac{pb\mathrm{N}}{2}\int\frac{dx}{x\Phi^{\frac{p}{2}}}.$$

$$\int\frac{dx}{x^m\,(a+bx)^{\frac{p}{q}}} = \left\{-\frac{1}{(m-1)\,ax^{m-1}} + \frac{1}{(m-1)a}\,\frac{(qm+p-2q)b}{(m-2)\,qax^{m-2}} - \frac{1}{(m-1)\,a}\cdot\frac{(qm+p-2q)\,b}{(m-2)\,qa}\cdot\frac{(qm+p-3q)\,b}{(m-3)\,qax^{m-3}}\right.$$

$$\left.+\frac{1}{(m-1)\,a}\,\frac{(qm+p-2q)\,b}{(m-2)\,qa}\,\frac{(qm+p-3q)\,b}{(m-3)\,qa}\,\frac{(qm+p-4q)\,b}{(m-4)\,qax^{m-4}} - \cdots \pm \frac{\mathrm{N}}{x}\right\}\frac{1}{\Phi^{\frac{p-q}{q}}} \pm \frac{pb\mathrm{N}}{q}\int\frac{dx}{x\Phi^{\frac{p}{q}}}.$$

$$\int\frac{dx}{x^m(a+bx)^{p+\frac{1}{2}}} = -\frac{1}{a\,(m-1)\,x^{m-1}\Phi^{p-\frac{1}{2}}} - \frac{b\,(2m+2p-3)}{2a\,(m-1)}\int\frac{dx}{x^{m-1}\,\Phi^{p+\frac{1}{2}}}.$$

$$\int\frac{x^p}{(a+bx)^p}\,dx = \frac{\Phi}{b^{p+1}}\left\{1 - \frac{p}{1}\,\frac{a}{\Phi}\log\Phi - \frac{p\,(p-1)}{1.2}\left(\frac{a}{\Phi}\right)^2 + \frac{1}{2}\,\frac{p\,(p-1)\,(p-2)}{1.2.3}\left(\frac{a}{\Phi}\right)^3\right.$$

$$\left.-\frac{1}{3}\,\frac{p\,(p-1)(p-2)(p-3)}{1.2.3.4}\left(\frac{a}{\Phi}\right)^4 + \cdots \pm \frac{1}{1-p}\left(\frac{a}{\Phi}\right)^p\right\}.$$

$$\int\frac{x^p}{(a+bx)^{2p}}\,dx = -\frac{1}{b^{p+1}\Phi^{p-1}}\left\{\frac{1}{p-1} - \frac{p}{1}\,\frac{1}{p}\left(\frac{a}{\Phi}\right) + \frac{p\,(p-1)}{1.2}\,\frac{1}{p+1}\left(\frac{a}{\Phi}\right)^2 - \frac{p(p-1)\,(p-2)}{1.2.3}\,\frac{1}{p+2}\left(\frac{a}{\Phi}\right)^3 + \cdots \pm \frac{1}{2p-1}\left(\frac{a}{\Phi}\right)^p\right\}.$$

$$\int\frac{x^{p-1}}{(a+bx)^p}\,dx = \frac{1}{b^p}\left\{\log\Phi + \frac{p-1}{1}\,\frac{a}{\Phi} - \frac{1}{2}\,\frac{(p-1)(p-2)}{1.2}\left(\frac{a}{\Phi}\right)^2 + \frac{1}{3}\,\frac{(p-1)\,(p-2)\,(p-3)}{1.2.3}\left(\frac{a}{\Phi}\right)^3 - \cdots \pm \frac{1}{p-1}\left(\frac{a}{\Phi}\right)^{p-1}\right\}.$$

$$\int\frac{x^{p-2}}{(a+bx)^p}\,dx = \frac{x^{p-1}}{a\,(p-1)\,\Phi^{p-1}}.$$

$$\int \frac{x^m}{(x-a)^p}\,dx = -\frac{x^m}{(p-1)\,\Phi^{p-1}} + \frac{m}{p-1}\int \frac{x^{m-1}}{\Phi^{p-1}}\,dx.$$

$$\int \frac{x^m}{(a+bx)^p}\,dx = \frac{1}{b^{m+1}}\int \frac{(z-a)^m}{z^p}\,dz. \qquad a+bx=z.$$

$$= \frac{x^{m+1}}{a\,(p-1)\,\Phi^{p-1}} + \frac{p-m-2}{a\,(p-1)}\int \frac{x^m}{\Phi^{p-1}}\,dx.$$

$$= -\frac{1}{b^{m+1}\,\Phi^{p-m-1}}\left\{\frac{1}{p-m-1} - \frac{m}{p-m}\cdot\frac{a}{\Phi} + \frac{m(m-1)}{1.2\,(p-m+1)}\left(\frac{a}{\Phi}\right)^2 - \frac{m(m-1)(m-2)}{1.2.3\,(p-m+2)}\left(\frac{a}{\Phi}\right)^3 + \cdots\right.$$

$$\left. \pm \frac{1}{p-1}\left(\frac{a}{\Phi}\right)^m\right\}. \qquad p > m+1.$$

$$\int \frac{x^m}{(a+bx)^{\frac{p}{2}}}\,dx = \left\{\frac{\Phi^m}{2m-p+2} - \frac{m}{1}\,\frac{a\Phi^{m-1}}{2m-p} + \frac{m\,(m-1)}{1.2}\,\frac{a^2\Phi^{m-2}}{2m-p-2} - \cdots \pm \mathrm{N}a^m\right\}\frac{2}{b^{m+1}\Phi^{\frac{p}{2}-1}}.$$

$$\int \frac{x^m}{(a+bx)^{\frac{p}{q}}}\,dx = \frac{qx^{m+1}}{a\,(p-q)\,\Phi^{\frac{p}{q}-1}} + \frac{p-mq-2q}{a\,(p-q)}\int \frac{x^m}{\Phi^{\frac{p}{q}-1}}\,dx.$$

$$= \left\{\frac{\Phi^m}{qm-p+q} - \frac{m}{1}\,\frac{a\Phi^{m-1}}{qm-p} + \frac{m\,(m-1)}{1.2}\,\frac{a^2\Phi^{m-2}}{qm-p-q} - \cdots \pm \mathrm{N}a^m\right\}\frac{q}{b^{m+1}\Phi^{\frac{p}{q}-1}}.$$

$$\int \frac{x^m}{(a+bx)^{p+\frac{1}{2}}}\,dx = \frac{2x^m}{b\,(2m-2p+1)\,\Phi^{p-\frac{1}{2}}} - \frac{2am}{b\,(2m-2p+1)}\int \frac{x^{m-1}}{\Phi^{p+\frac{1}{2}}}\,dx.$$

$$\int \frac{(a-x)^p}{x^{p+2}}\,dx = -\frac{\Phi^{p+1}}{a\,(p+1)\,x^{p+1}}.$$

$$\int \frac{(a+bx)^p}{x^{p+1}}\,dx = b^p\log x - \frac{p}{1}\,a\,\frac{b^{p-1}}{x} - \frac{p\,(p-1)}{1.2}\,a^2\,\frac{b^{p-2}}{2x^2} - \frac{p\,(p-1)\,(p-2)}{1.2.3}\,a^3\,\frac{b^{p-3}}{3x^3} - \cdots$$

$$\int \frac{(a+bx)^p}{x^{p+q}}\,dx = -\frac{b^p}{(q-1)\,x^{q-1}} - \frac{p}{1}\,\frac{ab^{p-1}}{qx^q} - \frac{p\,(p-1)}{1.2}\,\frac{a^2b^{p-2}}{(q+1)\,x^{q+1}} - \cdots$$

$$\int \frac{(a+bx)^p}{x^m}\,dx = -\frac{\Phi^{p+1}}{a\,(m-1)\,x^{m-1}} + \frac{b\,(p-m+2)}{a\,(m-1)}\int \frac{\Phi^p}{x^{m-1}}\,dx.$$

$$\int \frac{(a+bx)^{\frac{p}{2}}}{x^m}\,dx = \left\{-\frac{1}{(m-1)\,ax^{m-1}} + \frac{1}{(m-1)\,a}\,\frac{(2m-p-4)\,b}{(2m-4)\,ax^{m-2}} - \frac{1}{(m-1)\,a}\,\frac{(2m-p-4)\,b}{(2m-4)\,a}\cdot\frac{(2m-p-6)\,b}{(2m-6)\,ax^{m-3}} + \cdots\right.$$

$$\left. \pm \frac{\mathrm{N}}{x}\right\}\Phi^{\frac{p}{2}+1} \pm \frac{pb\mathrm{N}}{2}\int \frac{\Phi^{\frac{p}{2}}}{x}\,dx.$$

$$\int \frac{(a+bx)^{\frac{p}{q}}}{x^m}\,dx = -\frac{\Phi^{\frac{p}{q}+1}}{a\,(m-1)\,x^{m-1}} + \frac{b\,(p-mq+2q)}{aq\,(m-1)}\int \frac{\Phi^{\frac{p}{q}}}{x^{m-1}}\,dx.$$

$$= \left\{-\frac{1}{(m-1)\,ax^{m-1}} + \frac{1}{(m-1)\,a}\,\frac{(qm-p-2q)\,b}{(m-2)\,qax^{m}} - \frac{1}{(m-1)\,a}\,\frac{(qm-b-2q)\,b}{(m-2)\,qa}\cdot\frac{(qm-b-3q)\,b}{(m-3)\,qax^{m-3}}\right.$$

$$\left. + \frac{1}{(m-1)\,a}\cdot\frac{(qm-p-2q)\,b}{(m-2)\,qa}\cdot\frac{(qm-p-3q)\,b}{(m-3)\,qa},\frac{(qm-p-4q)\,b}{(m-4)\,qax^{m-4}} - \cdots \pm \frac{\mathrm{N}}{x}\right\}\Phi^{\frac{p}{q}+1} \pm \frac{pb\mathrm{N}}{q}\int \frac{\Phi^{\frac{p}{q}}}{x}\,dx.$$

$$\int \frac{(a+bx)^{p-\frac{1}{2}}}{x^m}\,dx = -\frac{\Phi^{p+\frac{1}{2}}}{a\,(m-1)\,x^{m-1}} + \frac{b\,(2p-2m+3)}{2a\,(m-1)}\int \frac{\Phi^{p-\frac{1}{2}}}{x^{m-1}}\,dx.$$

$$\int \frac{(a+bx)^{p+\frac{1}{3}}}{x^m}\,dx = \frac{\Phi^{p+\frac{4}{3}}}{a\,(m-1)\,x^{m-1}} + \frac{b\,(3p-3m+7)}{3a\,(m-1)}\int \frac{\Phi^{p+\frac{1}{3}}}{x^{m-1}}\,dx.$$

$$\int \frac{(a+bx)^{p-\frac{1}{3}}}{x^m} dx = -\frac{\Phi^{p+\frac{2}{3}}}{a(m-1)x^{m-1}} + \frac{b(3p-3m+5)}{3a(m-1)} \int \frac{\Phi^{p-\frac{1}{3}}}{x^{m-1}} dx.$$

$$\int \frac{(a+bx)^{p-\frac{2}{3}}}{x^m} dx = -\frac{\Phi^{p+\frac{1}{3}}}{a(m-1)x^{m-1}} + \frac{b(3p-3m+4)}{3a(m-1)} \int \frac{\Phi^{p-\frac{2}{3}}}{x^{m-1}} dx.$$

$$\int x^m \left(a+bx^2\right)^p dx$$

$$\int x^m (a+bx^2)^{\frac{p}{2}} dx = \frac{x^{m+1}}{m+1} \Phi^{\frac{p}{2}} - \frac{bp}{m+1} \int x^{m+2} \Phi^{\frac{p}{2}-1} dx = \frac{x^{m-1}\Phi^{\frac{p}{2}+1}}{b(m+p+1)} - \frac{a(m-1)}{b(m+p+1)} \int x^{m-2} \Phi^{\frac{p}{2}} dx.$$

$$= \frac{x^{m+1}\Phi^{\frac{p}{2}}}{m+p+1} + \frac{ap}{m+p+1} \int x^m \Phi^{\frac{p}{2}-1} dx.$$

$$\int x^{2m} (a+bx^2)^{p+\frac{1}{2}} dx = -a^{m+p+1} \int \frac{z^{2p+2}}{(z^2-b)^{m+p+2}} dz. \qquad a+bx^2 = z^2x^2.$$

$$\int x^{2m+1} (a+bx^2)^{p+\frac{1}{2}} dx = \frac{1}{b^{m+1}} \int z^{2p+2} (z^2-a)^m dz. \qquad a+bx^2 = z^2.$$

$$\int x^m (a+bx^2)^{p+\frac{1}{2}} dx = \frac{x^{m-1}\Phi^{p+\frac{3}{2}}}{b(m+2p+2)} - \frac{a(m-1)}{b(m+2p+2)} \int x^{m-2} \Phi^{p+\frac{1}{2}} dx.$$

$$\int \frac{dx}{x^m (a+bx^2)^p} = \frac{1}{a} \int \frac{dx}{x^m \Phi^{p-1}} - \frac{b}{a} \int \frac{dx}{x^{m-2}\Phi^p} = -\frac{1}{a(m-1)x^{m-1}\Phi^{p-1}} - \frac{b(m+2p-3)}{a(m-1)} \int \frac{dx}{x^{m-2}\Phi^p}.$$

$$\int \frac{dx}{x^m (a+bx^2)^{\frac{p}{2}}} = \frac{1}{a(p-2)x^{m-1}\Phi^{\frac{p}{2}-1}} + \frac{m+p-3}{a(p-2)} \int \frac{dx}{x^m \Phi^{\frac{p}{2}-1}} = -\frac{1}{a(m-1)x^{m-1}\Phi^{\frac{p}{2}-1}} - \frac{b(m+p-3)}{a(m-1)} \int \frac{dx}{x^{m-2}\Phi^{\frac{p}{2}}}.$$

$$\int \frac{dx}{x^{2m}(a+bx^2)^{p-\frac{1}{2}}} = -\frac{1}{a^{m+p-1}} \int \frac{(z^2-b)^{m+p-2}}{z^{2p-2}} dz. \qquad a+bx^2 = z^2x^2.$$

$$\int \frac{dx}{x^{2m-1}(a+bx^2)^{p-\frac{1}{2}}} = b^{m-1} \int \frac{dz}{z^{2p-2}(z^2-a)^m}. \qquad a+bx^2 = z^2.$$

$$\int \frac{x^m}{(a^2+x^2)^p} dx = -\frac{x^{m-1}}{2(p-1)\Phi^{p-1}} + \frac{m-1}{2(p-1)} \int \frac{x^{m-2}}{\Phi^{p-1}} dx.$$

$$\int \frac{x^m}{(a+bx^2)^p} dx = \frac{1}{b} \int \frac{x^{m-2}}{\Phi^{p-1}} dx - \frac{a}{b} \int \frac{x^{m-2}}{\Phi^p} dx = -\frac{x^{m-1}}{2b(p-1)\Phi^{p-1}} + \frac{m-1}{2b(p-1)} \int \frac{x^{m-2}}{\Phi^{p-1}} dx.$$

$$= -\frac{x^{m-1}}{b(2p-m-1)\Phi^{p-1}} + \frac{a(m-1)}{b(2p-m-1)} \int \frac{x^{m-2}}{\Phi^p} dx.$$

$$\int \frac{x^m}{(a+bx^2)^{\frac{p}{2}}} dx = -\frac{x^{m-1}}{b(p-2)\Phi^{\frac{p}{2}-1}} + \frac{m-1}{b(p-2)} \int \frac{x^{m-2}}{\Phi^{\frac{p}{2}-1}} dx = \frac{x^{m-1}}{b(m-p+1)\Phi^{\frac{p}{2}-1}} - \frac{a(m-1)}{b(m-p+1)} \int \frac{x^{m-2}}{\Phi^{\frac{p}{2}}} dx.$$

$$= \frac{x^{m+1}}{a(p-2)\Phi^{\frac{p}{2}-1}} - \frac{m-p+3}{a(p-2)} \int \frac{x^m}{\Phi^{\frac{p}{2}-1}} dx.$$

$$\int \frac{x^{2p}}{(1-x^2)^{p+\frac{1}{2}}} dx = \frac{x^{2p-1}}{(2p-1)\Phi^{p-\frac{1}{2}}} - \frac{x^{2p-3}}{(2p-3)\Phi^{p-\frac{3}{2}}} + \frac{x^{2p-5}}{(2p-5)\Phi^{p-\frac{5}{2}}} - \cdots \pm \frac{x}{\sqrt{\Phi}} \mp \operatorname{arc\,sin} x$$

$$\int \frac{x^{2p}}{(a+bx^2)^{p-\frac{1}{2}}} dx = -a \int \frac{dz}{z^{2p-2}(z^2-b)^2}. \qquad a+bx^2 = z^2x^2.$$

$$\int \frac{x^{2m+1}}{(a+bx^2)^{p-\frac{1}{2}}} dx = \frac{1}{b^{m+1}} \int \frac{(z^2-a)^m}{z^{2p-2}} dz. \qquad a+bx^2 = z^2.$$

$$\int \frac{(a+bx^2)^{p-1}}{x^p} dx = -\frac{\Phi^{p-1}}{(p-1)x^{p-1}} + 2b \int \frac{\Phi^{p-2}}{x^{p-2}} dx.$$

$$\int \frac{(a+bx^2)^{\frac{p}{2}}}{x^m} dx = -\frac{\Phi^{\frac{m}{2}+1}}{a(m-1)x^{m-1}} - \frac{b(m-p-3)}{a(m-1)} \int \frac{\Phi^{\frac{p}{2}}}{x^{m-2}} dx = -\frac{\Phi^{\frac{p}{2}}}{(m-p-1)x^{m-1}} - \frac{ap}{m-p-1} \int \frac{\Phi^{p-2}}{x^m} dx.$$

$$\int \frac{(a+bx^2)^{p+\frac{1}{2}}}{x^m} dx = -\frac{\Phi^{p+\frac{1}{2}}}{(m-1)x^{m-1}} + \frac{b(2p+1)}{m-1} \int \frac{\Phi^{p-\frac{1}{2}}}{x^{m-2}} dx.$$

$\int x^m \left(a + bx^{\mathrm{B}}\right)^p dx$

$$\int x^m (a+bx^4)^p dx = \frac{x^{m+1}\Phi^p}{m+1} - \frac{4bp}{m+1} \int x^{m+4} \Phi^{p-1} dx.$$

$$\int \frac{dx}{x^m (a+bx^3)^p} = \frac{1}{3a(p-1)x^{m-1}\Phi^{p-1}} + \frac{m+3p-4}{3a(p-1)} \int \frac{dx}{x^m \Phi^{p-1}}$$

$$\int \frac{dx}{x^m (a+bx^4)^p} = \frac{1}{a} \int \frac{dx}{x^m \Phi^{p-1}} - \frac{b}{a} \int \frac{dx}{x^{m-4}\Phi^p}.$$

$$\int \frac{x^m}{(a+bx^3)^p} dx = \frac{x^{m+1}}{3a(p-1)\Phi^{p-1}} - \frac{m-3p+4}{3a(p-1)} \int \frac{x^m}{\Phi^{p-1}} dx.$$

$$\int \frac{x^m}{(a+bx^4)^p} dx = \frac{x^{m+1}}{4a(p-1)\Phi^{p-1}} + \frac{4p-m-5}{4a(p-1)} \int \frac{x^m}{\Phi^{p-1}} dx = -\frac{x^{m-3}}{4b(p-1)\Phi^{p-1}} + \frac{m-3}{4b(p-1)} \int \frac{x^{m-4}}{\Phi^{p-1}} dx$$

$$\int \frac{x^m}{(a+bx^5)^p} dx = -\frac{x^{m-4}}{5b(p-1)\Phi^{p-1}} + \frac{m-4}{5b(p-1)} \int \frac{x^{m-5}}{\Phi^{p-1}} dx.$$

$\int x^m \left(a + bx^n\right)^p dx.$

$$\int x^n (a+bx^n)^{\frac{p}{q}} dx = \frac{qx\Phi^{\frac{p}{q}+1}}{bn(p+q)} - \frac{q}{bn(p+q)} \int \Phi^{\frac{p}{q}+1} dx = \frac{qx\Phi^{\frac{p}{q}+1}}{b(np+nq+q)} - \frac{aq}{b(np+nq+q)} \int \Phi^{\frac{p}{q}} dx.$$

$$\int x^{n-1} (a+bx^n)^{\frac{p}{q}} dx = \frac{q\Phi^{\frac{p}{q}+1}}{bn(p+q)}.$$

$$\int x^{2n-1} (a+bx^n)^{\frac{p}{q}} dx = \frac{1}{n} \int z (a+bz)^{\frac{p}{q}} dz. \qquad z = x^n.$$

$$= \frac{q}{b^2 n} \int z^{p+q-1} (z^q - a) dz. \qquad a + bx^n = z^q.$$

$$= \frac{qx^n \Phi^{\frac{p}{q}+1}}{bn(p+q)} - \frac{q^2 \Phi^{\frac{p}{q}+2}}{b^2 n (p+q)(p+2q)}.$$

$$\int x^{kn-1} (a+bx^n)^{\frac{p}{q}} dx = \frac{q}{nb^k} \int z^{p+q-1} (z^q - a)^{k-1} dz. \qquad a + bx^n = z^q.$$

$$\int x^m (a+bx^n)^p dx = \frac{x^{m+1}\Phi^p}{m+np+1} + \frac{anp}{m+np+1} \int x^m \Phi^{p-1} dx.$$

$$= \frac{x^{m-n+1}\Phi^{p+1}}{b(np+m+1)} - \frac{a(m-n+1)}{b(np+m+1)} \int x^{m-n} \Phi^p dx.$$

$$= \frac{x^{m-n+1}\Phi^{p+1}}{bn(p+1)} - \frac{m-n+1}{bn(p+1)} \int x^{m-n} \Phi^{p+1} dx.$$

$$= \frac{x^{m+1}\Phi^p}{m+1} - \frac{bnp}{m+1}\int x^{m+n}\Phi^{p-1}\,dx.$$

$$= \frac{x^{m+1}\Phi^{p+1}}{a(m+1)} - \frac{b(np+m+n+1)}{a(m+1)}\int x^{m+n}\Phi^{p}\,dx.$$

$$= -\frac{x^{m+1}\Phi^{p+1}}{an(p+1)} + \frac{np+m+n+1}{an(p+1)}\int x^{m}\Phi^{p+1}\,dx.$$

$$\int x^m(a+bx^n)^{\frac{p}{q}}\,dx = \frac{q}{nb^{\frac{m+1}{n}}}\int z^{p+q-1}(z^q-a)^{\frac{m+1}{n}-1}\,dz. \qquad a+bx^n = z^q$$

$$= -\frac{q}{nb^{\frac{m+1}{n}}}\int\frac{(1-az^q)^{\frac{m+1}{n}-1}}{z^{\frac{q(m+1)}{n}+p+1}}\,dz. \qquad a+bx^n = \frac{1}{z^q}.$$

$$= -\frac{qa^{\frac{m+1}{n}+\frac{p}{q}}}{n}\int\frac{z^{p+q-1}}{(z^q-b)^{\frac{m+1}{n}+\frac{p}{q}+1}}\,dz. \qquad a+bx^n = z^qx^n.$$

$$= -a^{\frac{m+1}{n}+\frac{p}{q}}\int\frac{z^{\frac{np}{q}+n-1}}{(z^n-b)^{\frac{m+1}{n}+\frac{p}{q}+1}}\,dz. \qquad a+bx^n = z^nx^n.$$

$$= \frac{qx^{m+1}\Phi^{\frac{p}{q}}}{np+mq+q} + \frac{anp}{np+mq+q}\int x^m\Phi^{\frac{p}{q}-1}\,dx.$$

$$= \frac{qx^{m-n+1}\Phi^{\frac{p}{q}+1}}{b(np+mq+q)} - \frac{aq(m-n+1)}{b(np+mq+q)}\int x^{m-n}\Phi^{\frac{p}{q}}\,dx.$$

$$= \frac{qx^{m-n+1}\Phi^{\frac{p}{q}+1}}{bn(p+q)} - \frac{q(m-n+1)}{bn(p+q)}\int x^{m-n}\Phi^{\frac{p}{q}+1}\,dx.$$

$$= \frac{x^{m+1}\Phi^{\frac{p}{q}}}{m+1} - \frac{bnp}{q(m+1)}\int x^{m+n}\Phi^{\frac{p}{q}-1}\,dx.$$

$$= \frac{x^{m+1}\Phi^{\frac{p}{q}+1}}{a(m+1)} - \frac{b(np+mq+nq+q)}{aq(m+1)}\int x^{m+n}\Phi^{\frac{p}{q}}\,dx.$$

$$= -\frac{qx^{m+1}\Phi^{\frac{p}{q}+1}}{an(p+q)} + \frac{np+mq+nq+q}{an(p+q)}\int x^m\Phi^{\frac{p}{q}+1}\,dx.$$

$$\int\frac{dx}{x^n(a+bx^n)^{\frac{1}{n}}} = -\frac{\Phi^{\frac{n-1}{n}}}{a(n-1)x^{n-1}}.$$

$$\int\frac{dx}{x^{kn}(a+bx^n)^{\frac{1}{n}}} = -\frac{1}{a^k}\int z^{n-2}(z^n-b)^{k-1}\,dz. \qquad a+bx^n = z^nx^n.$$

$$\int\frac{dx}{x^n(a+bx^n)^{\frac{n+1}{n}}} = -\frac{a+nbx^n}{a^2(n-1)x^{n-1}\Phi^{\frac{1}{n}}}.$$

$$\int\frac{dx}{x^{kn}(a+bx^n)^{\frac{n+1}{n}}} = -\frac{1}{a^{k+1}}\int\frac{(z^n-b)^k}{z^2}\,dz. \qquad a+bx^n = z^nx^n.$$

$$\int\frac{dx}{x^p(a+bx^n)^{\frac{p}{n}}} = \frac{1}{a}\int\frac{\Phi^{\frac{n-p}{p}}}{x^p}\,dx - \frac{b}{a}\int\frac{x^{n-p}}{\Phi^{\frac{p}{n}}}\,dx.$$

$$\int\frac{dx}{x^{n-p}(a+bx^n)^{\frac{n-p}{n}}} = \frac{1}{a}\int\frac{\Phi^{\frac{p}{n}}}{x^{n-p}}\,dx - \frac{b}{a}\int\frac{x^p}{\Phi^{\frac{n-p}{n}}}\,dx.$$

$$\int\frac{dx}{x^n(a+bx^n)^{\frac{p}{q}}} = -\frac{1}{(n-1)x^{n-1}\Phi^{\frac{p}{q}}} - \frac{bnp}{q(n-1)}\int\frac{dx}{\Phi^{\frac{p}{q}+1}} = -\frac{1}{a(n-1)x^{n-1}\Phi^{\frac{p}{q}-1}} - \frac{b(np-q)}{aq(n-1)}\int\frac{dx}{\Phi^{\frac{p}{q}}}$$

$$\int \frac{dx}{x^{n+1-np}(a+bx^n)^p} = \frac{x^{n(p-1)}}{an(p-1)\Phi^{p-1}}.$$

$$\int \frac{dx}{x^m(a+bx^n)^{\frac{2n-m+1}{n}}} = \frac{1}{a^2}\int \frac{z^n-b}{z^{n\,m+2}}\,dz. \qquad a+bx^n = z^n x^n.$$

$$\int \frac{dx}{x^m(a+bx^n)^p} = \frac{1}{a}\int \frac{dx}{x^m\Phi^{p-1}} - \frac{b}{a}\int \frac{dx}{x^{m-n}\Phi^p}.$$

$$\int \frac{dx}{x^m(a+bx^n)^{\frac{p}{q}}} = \frac{qb^{\frac{m-1}{n}}}{n}\int \frac{dz}{z^{p-q+1}(z^q-a)^{\frac{m-1}{n}+1}} \qquad a+bx^n = z^q.$$

$$= -\frac{q}{na^{\frac{m+1}{n}+\frac{p}{q}}}\int \frac{(z^q-b)^{\frac{m-1}{n}+\frac{p}{q}-1}}{z^{p-q+1}}\,dz. \quad a+bx^n = z^q x^n.$$

$$= -\frac{q}{(np+mq-q)\,x^{m-1}\Phi^{\frac{p}{q}}} + \frac{anp}{np+mq-q}\int \frac{dx}{x^m\Phi^{\frac{p}{q}+1}}$$

$$= -\frac{q}{b\,(np+mq-q)\,x^{m+n-1}\Phi^{\frac{p}{q}-1}} - \frac{aq\,(m+n-1)}{b\,(np+mq-q)}\int \frac{dx}{x^{m+n}\Phi^{\frac{p}{q}}}.$$

$$= -\frac{q}{bn\,(p-q)\,x^{m+n-1}\Phi^{\frac{p}{q}-1}} - \frac{q\,(m+n-1)}{bn\,(p-q)}\int \frac{dx}{x^{m+n}\Phi^{\frac{p}{q}-1}}.$$

$$= -\frac{1}{(m-1)\,x^{m-1}\Phi^{\frac{p}{q}}} - \frac{bnp}{q\,(m-1)}\int \frac{dx}{x^{m-n}\Phi^{\frac{p}{q}+1}}.$$

$$= -\frac{1}{a\,(m-1)\,x^{m-1}\Phi^{\frac{p}{q}-1}} - \frac{b\,(np+mq-nq-q)}{aq\,(m-1)}\int \frac{dx}{x^{m-n}\Phi^{\frac{p}{q}}}.$$

$$= \frac{q}{an\,(p-q)\,x^{m-1}\Phi^{\frac{p}{q}-1}} + \frac{np+mq-nq-q}{an\,(p-q)}\int \frac{dx}{x^m\Phi^{\frac{p}{q}-1}}.$$

$$\int \frac{x^n}{(a+bx^n)^{\frac{2n+1}{n}}}\,dx = \frac{x^{n+1}}{a\,(n+1)\,\Phi^{\frac{n+1}{n}}}.$$

$$\int \frac{x^{n-2}}{(a+bx^n)^{\frac{n-1}{n}}}\,dx = -\int \frac{dz}{z^n-b}. \qquad a+bx^n = z^n x^n.$$

$$\int \frac{x^{kn}}{(a+bx^n)^{\frac{1}{n}}}\,dx = -a^k\int \frac{z^{n-2}}{(z^n-b)^{k+1}}\,dz. \qquad a+bx^n = z^n x^n.$$

$$\int \frac{x^{kn}}{(a+bx^n)^{\frac{n+1}{n}}}\,dx = -a^{k-1}\int \frac{dz}{z^2(z^n-b)^k}. \qquad a+bx^n = z^n x^n.$$

$$\int \frac{x^{pn-1}}{(a+bx^n)^p}\,dx = -\int \frac{z^{n-pn-1}}{z^n-b}\,dz. \qquad a+bx^n = z^n x^n.$$

$$\int \frac{x^{pn-1}}{(a+bx^n)^{p+2}}\,dx = -\frac{1}{a^2}\int \frac{z^n-b}{z^{pn+n+1}}\,dz. \qquad a+bx^n = z^n x^n.$$

$$\int \frac{x^{pn-n-1}}{(a+bx^n)^p}\,dx = \frac{x^{n(p-1)}}{an\,(p-1)\,\Phi^{p-1}}.$$

$$\int \frac{x^{p-1}}{(a+bx^n)^{\frac{p}{n}}}\,dx = \int \frac{z^{n-p-1}}{b-z^n}\,dz. \qquad a+bx^n = z^n x^n.$$

$$\int \frac{x^{p-n-1}}{(a+bx^n)^{\frac{p}{n}}}\,dx = \frac{x^{p-n}}{a\,(p-n)\,\Phi^{\frac{p}{n}-1}}.$$

$$\int \frac{x^m}{(a+bx^n)^{\frac{2n+m+1}{n}}}\,dx = -\frac{1}{a^2}\int \frac{z^n-b}{z^{m+n+2}}\,dz. \qquad a+bx^n = z^n x^n.$$

$$\int \frac{x^n}{(a+bx^n)^{\frac{p}{q}}}\,dx = -\frac{qx}{bn(p-q)\,\Phi^{\frac{p}{q}-1}} - \frac{q}{bn(p-q)}\int \frac{dx}{\Phi^{\frac{p}{q}-1}} = \frac{qx}{b(qn-np+q)\,\Phi^{\frac{p}{q}-1}} - \frac{aq}{b(qn-np+q)}\int \frac{dx}{\Phi^{\frac{p}{q}}}.$$

$$\int \frac{x^{n-1}}{(a+bx^n)^{\frac{p}{q}}}\,dx = -\frac{q}{bn(p-q)\,\Phi^{\frac{p}{q}-1}}.$$

$$\int \frac{x^{2n-1}}{(a+bx^n)^{1-\frac{p}{q}}}\,dx = \frac{q(bpx^n-aq)}{b^2np\,(p+q)}\,\Phi^{\frac{p}{q}}.$$

$$\int \frac{x^{2n-1}}{(a+bx^n)^{\frac{p}{q}}}\,dx = \frac{1}{n}\int \frac{z}{(a+bz)^{\frac{p}{q}}}\,dz. \qquad x^n = z.$$

$$= -\frac{q}{b^2n}\int z^{p-2q-1}(1-az^q)\,dz. \qquad a+bx^n = \frac{1}{z^q}.$$

$$= -\frac{qx^n}{bn(p-q)\,\Phi^{\frac{p}{q}-1}} - \frac{q^2}{b^2n(p-q)(p-2q)\,\Phi^{\frac{p}{q}-2}}.$$

$$\int \frac{x^{3n-1}}{(a+bx^n)^{\frac{p}{q}}}\,dx = \frac{1}{n}\int \frac{z^2}{(a+bz)^{\frac{p}{q}}}\,dz. \qquad x^n = z.$$

$$\int \frac{x^{kn-1}}{(a+bx^n)^{\frac{p}{q}}}\,dx = -\frac{q}{nb^k}\int z^{p-kq-1}(1-az^q)^{k-1}\,dz. \qquad a+bx^n = \frac{1}{z^q}.$$

$$= \frac{1}{nb^k}\int \frac{(z-a)^{k-1}}{z^{\frac{p}{q}}}\,dz. \qquad a+bx^n = z.$$

$$\int \frac{x^m}{(a+bx^n)^{\frac{p}{q}}}\,dx = \frac{q}{nb^{\frac{m+1}{n}}}\int \frac{(z^q-a)^{\frac{m+1}{n}-1}}{z^{p-q+1}}\,dz. \qquad a+bx^n = z^q.$$

$$= -\frac{qa^{\frac{m+1}{n}-\frac{p}{q}}}{n}\int \frac{dz}{z^{p-q+1}(z_q-b)^{\frac{m+1}{n}-\frac{p}{q}+1}} \qquad a+bx^n = z^q x^n.$$

$$\int \frac{x^m}{(a+bx^n)^p}\,dx = \frac{1}{b}\int \frac{x^{m-n}}{\Phi^{p-1}}\,dx - \frac{a}{b}\int \frac{x^{m-n}}{\Phi^p}\,dx.$$

$$\int \frac{x^m}{(a+bx^n)^{\frac{p}{q}}}\,dx = -\frac{qx^{m+1}}{(np-mq-q)\,\Phi^{\frac{p}{q}}} + \frac{anp}{np-mq-q}\int \frac{x^m}{\Phi^{\frac{p}{q}+1}}\,dx.$$

$$= -\frac{qx^{m-n+1}}{b(np-mq-q)\,\Phi^{\frac{p}{q}-1}} + \frac{aq(m-n+1)}{b(np-mq-q)}\int \frac{x^{m-n}}{\Phi^{\frac{p}{q}}}\,dx.$$

$$= -\frac{qx^{m-n+1}}{bn(p-q)\,\Phi^{\frac{p}{q}-1}} + \frac{q(m-n+1)}{bn(p-q)}\int \frac{x^{m-n}}{\Phi^{\frac{p}{q}-1}}\,dx.$$

$$= \frac{x^{m+1}}{(m+1)\,\Phi^{\frac{p}{q}}} + \frac{bnp}{q(m+1)}\int \frac{x^{m+n}}{\Phi^{\frac{p}{q}+1}}\,dx.$$

$$= \frac{x^{m+1}}{a(m+1)\,\Phi^{\frac{p}{q}-1}} + \frac{b(np-mq-nq-q)}{aq(m+1)}\int \frac{x^{m+n}}{\Phi^{\frac{p}{q}}}\,dx.$$

$$= \frac{qx^{m+1}}{an(p-q)\,\Phi^{\frac{p}{q}-1}} + \frac{np-mq-nq-q}{an(p-q)}\int \frac{x^m}{\Phi^{\frac{p}{q}-1}}\,dx.$$

$$\int \frac{(a+bx^n)^{\frac{n-1}{n}}}{x^n}\,dx = -\int \frac{z^{2n-2}}{z^n-b}\,dz. \qquad a+bx^n = z^n x^n.$$

$$\int \frac{(a+bx^n)^p}{x^{2n+1}}\,dx = \frac{b^2}{n}\int \frac{z^p}{(z-a)^3}\,dz. \qquad a+bx^n = z.$$

$$= -\frac{\Phi^p}{2nx^{2n}} + \frac{bp}{2}\int \frac{\Phi^{p-1}}{x^{n+1}}\,dx.$$

$$\int \frac{(a+bx^n)^p}{x^{pn+1}}\,dx = -\int \frac{z^{pn+n-1}}{z^n-b}\,dz. \qquad a+bx^n = z^n x^n.$$

$$\int \frac{(a+bx^n)^{p-2}}{x^{pn+1}}\,dx = -\frac{1}{a^2}\int z^{pn-n-1}\,(z^n-b)\,dz. \qquad a+bx^n = z^n x^n.$$

$$\int \frac{(a+bx^n)^p}{x^{np+n+1}}\,dx = -\frac{\Phi^{p+1}}{an\,(p+1)\,x^{n(p+1)}}.$$

$$\int \frac{(a+bx^n)^{\frac{p}{q}}}{x^n}\,dx = -\frac{\Phi^{\frac{p}{q}}}{(n-1)\,x^{n-1}} + \frac{bnp}{q\,(n-1)}\int \Phi^{\frac{p}{q}-1}\,dx = -\frac{\Phi^{\frac{p}{q}+1}}{a\,(n-1)\,x^{n-1}} + \frac{b\,(np+q)}{aq\,(n-1)}\int \Phi^{\frac{p}{q}}\,dx.$$

$$\int \frac{(a+bx^n)^{\frac{p}{q}}}{x^{n+1}}\,dx = \frac{bq}{n}\int \frac{z^{p+q-1}}{(z^q-a)^2}\,dz. \qquad a+bx^n = z^q.$$

$$\int \frac{(a+bx^n)^{\frac{p}{n}}}{x^{p+n+1}}\,dx = -\frac{\Phi^{\frac{p}{n}+1}}{a\,(p+n)\,x^{p+n}}.$$

$$\int \frac{(a+bx^n)^{\frac{p-1}{n}}}{x^p}\,dx = -\int \frac{z^{p+n-2}}{z^n-b}\,dz. \qquad a+bx^n = z^n x^n.$$

$$\int \frac{(a+bx^n)^{\frac{p}{q}}}{x^m}\,dx = \frac{9b^{\frac{m-1}{n}}}{n}\int \frac{z^{p+q-1}}{(z^q-a)^{\frac{m-1}{n}+1}}\,dz. \qquad a+bx^n = z^q.$$

$$= -\frac{q}{na^{\frac{m-1}{n}-\frac{p}{q}}}\int z^{p+q-1}\,(z^q-b)^{\frac{m-1}{n}-\frac{p}{q}-1}\,dz. \qquad a+bx^n = z^q x^n.$$

$$= \frac{q\Phi^{\frac{p}{q}}}{(np-mq+q)\,x^{m-1}} + \frac{anp}{np-mq+q}\int \frac{\Phi^{\frac{p}{q}-1}}{x^m}\,dx.$$

$$= \frac{q\Phi^{\frac{p}{q}+1}}{b\,(np-mq+q)\,x^{m+n-1}} + \frac{aq\,(m+n-1)}{b\,(np-mq+q)}\int \frac{\Phi^{\frac{p}{q}}}{x^{m+n}}\,dx.$$

$$= \frac{q\Phi^{\frac{p}{q}+1}}{bn\,(p+q)\,x^{m+n-1}} + \frac{q\,(m+n-1)}{bn\,(p+q)}\int \frac{\Phi^{\frac{p}{q}+1}}{x^{m+n}}\,dx.$$

$$= -\frac{\Phi^{\frac{p}{q}}}{(m-1)\,x^{m-1}} + \frac{bnp}{q\,(m-1)}\int \frac{\Phi^{\frac{p}{q}-1}}{x^{m-n}}\,dx$$

$$= -\frac{\Phi^{\frac{p}{q}+1}}{a\,(m-1)\,x^{m-1}} + \frac{b\,(np-mq+nq+q)}{aq\,(m-1)}\int \frac{\Phi^{\frac{p}{q}}}{x^{m-n}}\,dx.$$

$$= -\frac{q\Phi^{\frac{p}{q}+1}}{an\,(p+q)\,x^{m-1}} + \frac{np-mq+nq+q}{an\,(p+q)}\int \frac{\Phi^{\frac{p}{q}+1}}{x^m}\,dx.$$

$\int x^m \left(ax^r + bx^{r+n}\right)^p dx$

$$\int (bx+cx^2)^p\,dx = \int \left(cz^2 - \frac{b^2}{4c}\right)^p dz. \qquad x = z - \frac{b}{2c}.$$

$$\int (bx+cx^2)^{\frac{p}{2}}\,dx = \frac{2cx+b}{2c\,(p+1)}\Phi^{\frac{p}{2}} - \frac{pb^2}{4c\,(p+1)}\int \Phi^{\frac{p}{2}-1}\,dx.$$

$$\int (ax^r + bx^{r+n})^p\, dx = \frac{x\Phi^p}{pr + np + 1} + \frac{anp}{pr + np + 1} \int x^r \Phi^{p-1}\, dx.$$

$$\int x^m \sqrt{bx + cx^2}\, dx = \frac{x^{m+1}}{m+1} \sqrt{\Phi} - \frac{b}{2(m+1)} \int \frac{x^{m+1}}{\sqrt{\Phi}}\, dx - \frac{c}{m+1} \int \frac{x^{m+2}}{\sqrt{\Phi}}\, dx.$$

$$= \left\{ \frac{x^{m-1}}{(m+2)c} - \frac{(2m+1)bx^{m-2}}{(m+2)(m+1)2c^2} + \frac{(2m+1)(2m-1)b^2x^{m-3}}{(m+2)(m+1)m4c^3} - \frac{(2m+1)(2m-1)(2m-3)b^3x^{m-4}}{(m+2)(m+1)m(m-1)8c^4} + \dots \pm \mathrm{N} \right\} \Phi^{\frac{3}{2}}$$

$$\mp \frac{3b\mathrm{N}}{2} \int \sqrt{\Phi}\, dx.$$

$$\int x^m (bx + cx^2)^{\frac{p}{2}}\, dx = \frac{x^{m+1}\Phi^{\frac{p}{2}}}{m + p + 1} + \frac{pb}{2(m + p + 1)} \int x^{m+1} \Phi^{\frac{p}{2}-1}\, dx.$$

$$= \frac{x^{m-1}\Phi^{\frac{p}{2}+1}}{(m + p + 1)c} - \frac{(2m + p)b}{(m + p + 1)2c} \int x^{m-1} \Phi^{\frac{p}{2}}\, dx.$$

$$= \frac{2x^{m+1}\Phi^{\frac{p}{2}}}{2m + p + 2} - \frac{pc}{2m + p + 2} \int x^{m+2} \Phi^{\frac{p}{2}-1}\, dx.$$

$$\int x^m (ax^r + bx^{r+n})^p\, dx = \frac{x^{m+1}\Phi}{m + pr + 1} - \frac{pnb}{m + pr + 1} \int x^{m+r+n} \Phi^{p-1}\, dx.$$

$$= \frac{x^{m-r-n+1}\Phi^{p+1}}{(m + pr + np + 1)b} - \frac{(m + pr - n + 1)a}{(m + pr + np + 1)b} \int x^{m-n} \Phi^p\, dx.$$

$$= \frac{x^{m+1}\Phi^p}{m + pr + np + 1} + \frac{anp}{m + pr + np + 1} \int x^{m+r} \Phi^{p-1}\, dx.$$

$$\int \frac{dx}{(bx + cx^2)^p} = \int \frac{dz}{\left(cz^2 - \frac{b^2}{4c}\right)^p}. \qquad x = z - \frac{b}{2c}.$$

$$\int \frac{dx}{(bx + cx^2)^{\frac{p}{2}}} = -\frac{2(2cx + b)}{(p-2)b^2\Phi^{\frac{p}{2}-1}} - \frac{(p-3)4c}{(p-2)b^2} \int \frac{dx}{\Phi^{\frac{p}{2}-1}}.$$

$$= \left\{ -\frac{1}{(p-2)b^2\Phi^{\frac{p-3}{2}}} + \frac{(p-3)4c}{(p-2)(p-4)b^4\Phi^{\frac{p-5}{2}}} - \frac{(p-3)(p-5)16c^2}{(p-2)(p-4)(p-6)b^6\Phi^{\frac{p-7}{2}}} + \frac{(p-3)(p-5)(p-7)64c^3}{(p-2)(p-4)(p-6)(p-8)b^8\Phi^{\frac{p-9}{2}}} - \dots \right.$$

$$\left. \pm \frac{\mathrm{N}}{\Phi} \right\} \frac{2(2cx + b)}{\sqrt{\Phi}} \pm 8c\mathrm{N} \int \frac{dx}{\Phi^{\frac{3}{2}}}.$$

$$\int \frac{dx}{(ax^r + bx^{r+n})^p} = \frac{1}{(p-1)nax^{r-1}\Phi^{p-1}} + \frac{pr + np - n - 1}{(p-1)na} \int \frac{dx}{x^r\Phi^{p-1}}.$$

$$\int \frac{dx}{x(bx + cx^2)^{p+\frac{1}{2}}} = -\frac{2}{(2p+1)bx\Phi^{p-\frac{1}{2}}} - \frac{4pc}{(2p+1)b} \int \frac{dx}{\Phi^{p+\frac{1}{2}}}.$$

$$\int \frac{dx}{x^2(bx + cx^2)^{p+\frac{1}{2}}} = \frac{2}{(2p+3)bx\Phi^{p-\frac{1}{2}}} \left\{ -\frac{1}{x} + \frac{2c}{b} \right\} + \frac{8pc^2}{(2p+3)b^2} \int \frac{dx}{\Phi^{p+\frac{1}{2}}}.$$

$$\int \frac{dx}{x^3(bx + cx^2)^{p+\frac{1}{2}}} = \frac{2}{(2p+5)bx\Phi^{p-\frac{1}{2}}} \left\{ -\frac{1}{x^2} + \frac{4(p+1)c}{(2p+3)bx} - \frac{8(p+1)c^2}{(2p+3)b^2} \right\} - \frac{32p(p+1)c^3}{(2p+3)(2p+5)b^3} \int \frac{dx}{\Phi^{p+\frac{1}{2}}}.$$

$$\int \frac{dx}{x^m\sqrt{bx + cx^2}} = -\frac{2\sqrt{\Phi}}{b(2m-1)x^m} - \frac{2c(m-1)}{b(2m-1)} \int \frac{dx}{x^{m-1}\sqrt{\Phi}}.$$

$$= \left\{ -\frac{2}{(2m-1)bx^m} + \frac{(m-1)4c}{(2m-1)(2m-3)b^2x^{m-1}} - \frac{(m-1)(m-2)8c^2}{(2m-1)(2m-3)(2m-5)b^3x^{m-2}} \right.$$

$$+\frac{(m-1)(m-2)(m-3)\,16c^3}{(2m-1)(2m-3)(2m-5)(2m-7)\,b^4x^{m-3}}-\dots\pm\frac{N}{x}\Bigg\}\sqrt{\Phi}.$$

$$\int\frac{dx}{x^m(bx+cx^2)^{\frac{3}{2}}}=-\frac{2\sqrt{\Phi}}{b(2m+1)x^m\Phi}-\frac{2c(m+1)}{b(2m+1)}\int\frac{dx}{x^{m-1}\Phi^{\frac{3}{2}}}.$$

$$\int\frac{dx}{x^m(bx+cx^2)^{\frac{p}{2}}}=\frac{2}{(p-2)\,bx^m\Phi^{\frac{p}{2}-1}}+\frac{2(m+p-2)}{(p-2)\,b}\int\frac{dx}{x^{m+1}\Phi^{\frac{p}{2}-1}}.$$

$$\int\frac{dx}{x^m(bx+cx^2)^{p+\frac{1}{2}}}=-\frac{2}{(2p+2m-1)\,bx^m\Phi^{p-\frac{1}{2}}}-\frac{2(2p+m-1)\,c}{(2p+2m-1)\,b}\int\frac{dx}{x^{m-1}\Phi^{p+\frac{1}{2}}}.$$

$$\int\frac{dx}{x^m(ax^r+bx^{r+n})^p}=\frac{1}{(p-1)\,nax^{m+r-1}\Phi^{p-1}}+\frac{m-n+pr+np-1}{(p-1)\,na}\int\frac{dx}{x^{m+r}\Phi^{p-1}}.$$

$$=-\frac{1}{(m+pr-1)\,ax^{m+r-1}\Phi^{p-1}}-\frac{(m-n+np+pr-1)\,b}{(m+pr-1)\,a}\int\frac{dx}{x^{m-n}\Phi^p}$$

$$\int\frac{x^m}{\sqrt{ax-x^2}}dx=2a^m\int\frac{z^{2m}}{\sqrt{1-z^2}}dz.\qquad x=az^2.$$

$$=-2a^m\int\frac{dz}{(1+z^2)^{m+1}}.\qquad \sqrt{ax-x^2}=zx.$$

$$=-\frac{x^{m-1}}{m}\sqrt{\Phi}+\frac{a(2m-1)}{2m}\int\frac{x^{m-1}}{\sqrt{\Phi}}dx.$$

$$\int\frac{x^m}{\sqrt{bx+cx^2}}dx=\frac{x^{m-1}\sqrt{\Phi}}{cm}-\frac{b(2m-1)}{2cm}\int\frac{x^{m-1}}{\sqrt{\Phi}}dx.$$

$$=\left\{\frac{x^{m-1}}{mc}-\frac{(2m-1)\,bx^{m-2}}{m(m-1)\,2c^2}+\frac{(2m-1)(2m-3)\,b^2x^{m-3}}{m(m-1)(m-2)\,4c^3}-\dots\pm Nx^0\right\}\sqrt{\Phi}\pm\frac{bN}{2}\int\frac{dx}{\sqrt{\Phi}}.$$

$$\int\frac{x^m}{(bx+cx^2)^{\frac{p}{2}}}dx=-\frac{2x^{m-1}}{(p-2)\,c\Phi^{\frac{p}{2}-1}}+\frac{2m-p}{(p-2)\,c}\int\frac{x^{m-2}}{\Phi^{\frac{p}{2}-1}}dx=\frac{x^{m-1}}{(m-p+1)\,c\Phi^{\frac{p}{2}}}-\frac{(2m-p)\,b}{(m-p+1)\,2c}\int\frac{x^{m-1}}{\Phi^{\frac{p}{2}}}dx.$$

$$=\frac{2x^m}{(p-2)\,b\Phi^{\frac{p}{2}-1}}-\frac{2(m-p+2)}{(p-2)\,b}\int\frac{x^{m-1}}{\Phi^{\frac{p}{2}-1}}dx.$$

$$\int\frac{x^m}{(ax^r+bx^{r+n})^p}dx=-\frac{x^{m-r-n+1}}{(p-1)\,nb\Phi^{p-1}}+\frac{m-pr-n+1}{(p-1)\,nb}\int\frac{x^{m-r-n}}{\Phi^{p-1}}dx.$$

$$=\frac{x^{m-r-n+1}}{(m-pr-np+1)\,b\Phi^{p-1}}-\frac{(m-pr-n+1)\,a}{(m-pr-np+1)\,b}\int\frac{x^{m-n}}{\Phi^p}dx.$$

$$=\frac{x^{m-r+1}}{(p-1)\,na\Phi^{p-1}}-\frac{m+n-pr-np+1}{(p-1)\,na}\int\frac{x^{m-r}}{\Phi^{p-1}}dx.$$

$$\int\frac{\sqrt{bx+cx^2}}{x^m}dx=-\frac{\sqrt{\Phi}}{(m-1)\,x^{m-1}}+\frac{b}{2(m-1)}\int\frac{dx}{x^{m-1}\sqrt{\Phi}}+\frac{c}{m-1}\int\frac{dx}{x^{m-2}\sqrt{\Phi}}.$$

$$=\Bigg\{-\frac{2}{(2m-3)\,bx^m}+\frac{(m-3)\,4c}{(2m-3)(2m-5)\,b^2x^{m-1}}-\frac{(m-3)(m-4)\,8c^2}{(2m-3)(2m-5)(2m-7)\,b^3x^{m-2}}$$

$$+\frac{(m-3)(m-4)(m-5)\,16c^3}{(2m-3)(2m-5)(2m-7)(2m-9)\,b^4x^{m-3}}-\dots\pm\frac{N}{x^4}\Bigg\}\Phi^{\frac{3}{2}}\pm cN\int\frac{\sqrt{\Phi}}{x^3}dx.$$

$$\int\frac{(bx+cx^2)^{\frac{p}{2}}}{x^m}dx=-\frac{\Phi^{\frac{p}{2}}}{(m-p-1)\,x^{m-1}}-\frac{pb}{2(m-p-1)}\int\frac{\Phi^{\frac{p}{2}-1}}{x^{m-1}}dx.$$

$$= -\frac{2\Phi^{\frac{p}{2}+1}}{(2m-p-2)\,bx^m} - \frac{(m-p-2)\,2c}{(2m-p-2)\,b}\int\frac{\Phi^{\frac{p}{2}}}{x^{m-1}}\,dx.$$

$$\int\frac{(ax^r+bx^{r+n})^p}{x^m}\,dx = -\frac{\Phi^{p+1}}{(m-pr-1)\,ax^{m+r-1}} - \frac{(m-n-pr-np-1)\,b}{(m-pr-1)\,a}\int\frac{\Phi^p}{x^{m-n}}\,dx.$$

$$= -\frac{\Phi^p}{(m-pr-np-1)\,x^{m-1}} - \frac{pna}{m-pr-np-1}\int\frac{\Phi^{p-1}}{x^{m-r}}\,dx.$$

$$\int \mathrm{U}\left(a+bx^m+cx^n\right)^p\,dx$$

$$\int(a+bx+cx^2)^{\frac{p}{2}}\,dx = \frac{(b+2cx)\,\Phi^{\frac{p}{2}}}{(p+1)\,2c} - \frac{p\,(b^2-4ac)}{(p+1)\,4c}\int\Phi^{\frac{p}{2}-1}dx.$$

$$=\left\{\frac{\Phi^{\frac{p-1}{2}}}{(p+1)2c} + \frac{p(4ac-b^2)\,\Phi^{\frac{p-3}{2}}}{(p+1)(p-1)8c^2} + \frac{p(p-2)(4ac-b^2)^2\,\Phi^{\frac{p-5}{2}}}{(p+1)(p-1)(p-3)32c^3} + \frac{p(p-2)(p-4)(4ac-b^2)^3\,\Phi^{\frac{p-7}{2}}}{(p+1)(p-1)(p-3)(p-5)128c^4} + \cdots\right.$$

$$\left.+\,\mathrm{N}\Phi\right\}(2cx+b)\sqrt{\Phi} + \frac{3}{2}\,(4ac-b^2)\,\mathrm{N}\int\sqrt{\Phi}\,dx.$$

$$\int(a+bx+cx^2)^{p+\frac{1}{2}}\,dx = \frac{(2cx+b)\,\Phi^{p+\frac{1}{2}}}{4c\,(p+1)} + \frac{(2p+1)\,(4ac-b^2)}{8c\,(p+1)}\int\Phi^{p-\frac{1}{2}}\,dx.$$

$$\int(a+bx^n+cx^m)^p\,dx = a\int\Phi^{p-1}dx + b\int x^n\Phi^{p-1}\,dx + c\int x^m\Phi^{p-1}\,dx.$$

$$\int x\,(a+bx+cx^2)^{\frac{p}{2}}\,dx = \frac{\Phi^{\frac{p}{2}+1}}{c\,(p+2)} - \frac{b}{2c}\int\Phi^{\frac{p}{2}}\,dx.$$

$$\int x\,(a+bx+cx^2)^{p-\frac{1}{2}}\,dx = \frac{\Phi^{p+\frac{1}{2}}}{c\,(2p+1)} - \frac{b}{2c}\int\Phi^{p-\frac{1}{2}}\,dx.$$

$$\int x^2\,(a+bx+cx^2)^{p-\frac{1}{2}}\,dx = \frac{x\Phi^{p+\frac{1}{2}}}{2c\,(p+1)} - \frac{b\,(2p+3)}{4c\,(p+1)}\int x\Phi^{p-\frac{1}{2}}\,dx - \frac{a}{2c\,(p+1)}\int\Phi^{p-\frac{1}{2}}\,dx.$$

$$=\left\{x - \frac{b\,(2p+3)}{2c\,(2p+1)}\right\}\frac{\Phi^{p+\frac{1}{2}}}{2c\,(p+1)} + \frac{(2p+3)\,b^2-4ac}{8c^2\,(p+1)}\int\Phi^{p-\frac{1}{2}}\,dx.$$

$$\int x^3\,(a+bx+cx^2)^{p-\frac{1}{2}}\,dx = \frac{x^2\Phi^{p+\frac{1}{2}}}{c\,(2p+3)} - \frac{b\,(2p+5)}{2c\,(2p+3)}\int x^2\Phi^{p-\frac{1}{2}}\,dx - \frac{2a}{c\,(2p+3)}\int x\Phi^{p-\frac{1}{2}}\,dx.$$

$$\int x^m\,(a+bx+cx^2)^p\,dx = \frac{x^{m-1}\Phi^{p+1}}{c\,(m+2p+1)} - \frac{a\,(m-1)}{c\,(m+2p+1)}\int x^{m-2}\,\Phi^p\,dx - \frac{b\,(m+p)}{c\,(m+2p+1)}\int x^{m-1}\Phi^p\,dx.$$

$$\int x^m\sqrt{a+bx+cx^2}\,dx = \frac{x^{m+1}}{m+1}\sqrt{\Phi} - \frac{b}{2\,(m+1)}\int\frac{x^{m+1}}{\sqrt{\Phi}}\,dx - \frac{c}{m+1}\int\frac{x^{m+2}}{\sqrt{\Phi}}\,dx.$$

$$\int x^m\,(a+bx+cx^2)^{\frac{p}{2}}\,dx = \frac{x^{m-1}\,\Phi^{\frac{p}{2}-1}}{c\,(m+p+1)} - \frac{b\,(2m+p)}{2c\,(m+p+1)}\int x^{m-1}\,\Phi^{\frac{p}{2}}\,dx.$$

$$\int x^m\,(a+bx+cx^2)^{p-\frac{1}{2}}\,dx = \frac{x^{m-1}\,\Phi^{p+\frac{1}{2}}}{c\,(2p+m)} - \frac{b\,(2p+2m-1)}{2c\,(2p+m)}\int x^{m-1}\Phi^{p-\frac{1}{2}}\,dx - \frac{a\,(m-1)}{c\,(2p+m)}\int x^{m-2}\Phi^{p-\frac{1}{2}}\,dx.$$

$$\int x^m\,(a+bx^n+cx^{2n})^p\,dx = \frac{1}{m+1}\left\{x^{m+1}\,\Phi^p - bnp\int x^{m+n}\,\Phi^{p-1}\,dx - 2cnp\int x^{m+2n}\,\Phi^{p-1}\,dx\right\}.$$

$$= \frac{1}{bn\,(p+1)}\left\{x^{m-n+1}\,\Phi^{p+1} - (m-n+1)\int x^{m-n}\,\Phi^{p+1}\,dx - 2cn\,(p+1)\int x^{m+n}\,\Phi^p\,dx\right\}.$$

$$= \frac{1}{2cn(p+1)} \left\{ x^{m-2n+1} \Phi^{p+1} - (m-2n+1) \int x^{m-2n} \Phi^{p+1} dx - bn(p+1) \int x^{m-n} \Phi^p dx \right\}.$$

$$= \frac{1}{m+np+1} \left\{ x^{m+1} \Phi^p + anp \int x^m \Phi^{p-1} dx - cnp \int x^{m+2n} \Phi^{p-1} dx \right\}.$$

$$= \frac{1}{m+2np+1} \left\{ x^{m+1} \Phi^p + 2anp \int x^m \Phi^{p-1} dx + bnp \int x^{m+n} \Phi^{p-1} dx \right\}.$$

$$= \frac{1}{b(m+np+1)} \left\{ x^{m-n+1} \Phi^{p+1} - a(m-n+1) \int x^{m-n} \Phi^p dx - c(m+2np+n+1) \int x^{m+n} \Phi^p dx \right\}.$$

$$= \frac{1}{bn(p+1)} \left\{ -x^{m-n+1} \Phi^{p+1} + (m+2np+n+1) \int x^{m-n} \Phi^{p+1} dx - 2an(p+1) \int x^{m-n} \Phi^p dx \right\}.$$

$$= \frac{1}{cn(p+1)} \left\{ x^{m-2n+1} \Phi^{p+1} + an(p+1) \int x^{m-2n} \Phi^p dx - (m+np-n+1) \int x^{m-2n} \Phi^{p+1} dx \right\}.$$

$$= \frac{1}{an(p+1)} \left\{ -x^{m+1} \Phi^{p+1} + (m+np+n+1) \int x^m \Phi^{p+1} dx + cn(p+1) \int x^{m+2n} \Phi^p dx \right\}.$$

$$= \frac{1}{2an(p+1)} \left\{ -x^{m+1} \Phi^{p+1} + (m+2np+2n+1) \int x^m \Phi^{p+1} dx - bn(p+1) \int x^{m+n} \Phi^p dx \right\}.$$

$$= \frac{1}{a(m+1)} \left\{ x^{m+1} \Phi^{p+1} - b(m+np+n+1) \int x^{m+n} \Phi^p dx - c(m+2np+2n+1) \int x^{m+2n} \Phi^p dx \right\}.$$

$$= \frac{1}{c(m+2np+1)} \left\{ x^{m-2n+1} \Phi^{p+1} - b(m+np-n+1) \int x^{m-n} \Phi^p dx - a(m-2n+1) \int x^{m-2n} \Phi^p dx \right\}.$$

$$\int \frac{dx}{(a+bx+cx^2)^p} = 2^{2p-1} c^{p-1} \int \frac{dz}{(z^2+4ac-b^2)^p}. \qquad 2cx+b=z.$$

$$= \frac{2^{2p-1} c^{p-1}}{(4ac-b^2)^{p-\frac{1}{2}}} \int \frac{dz}{(z^2+1)^p}. \qquad x+\frac{b}{2c} = \frac{z}{2c}\sqrt{4ac-b^2}.$$

$$= \frac{2cx+b}{(p-1)(4ac-b^2)\Phi^{p-1}} + \frac{2c(2p-3)}{(p-1)(4ac-b^2)} \int \frac{dx}{\Phi^{p-1}}.$$

$$\int \frac{dx}{(a+bx-cx^2)^p} = 2^{2p-1} c^{p-1} \int \frac{dz}{(4ac+b^2-z^2)^p}. \qquad 2cx-b=z.$$

$$\int \frac{dx}{x(a+bx+cx^2)^p} = \frac{1}{2a(p-1)\Phi^{p-1}} + \frac{1}{a} \int \frac{dx}{x\Phi^{p-1}} - \frac{b}{2a} \int \frac{dx}{\Phi^p}.$$

$$\int \frac{dx}{x^2(a+bx+cx^2)^p} = -\frac{1}{ax\Phi^{p-1}} - \frac{bp}{a} \int \frac{dx}{x\Phi^p} - \frac{c(2p-1)}{a} \int \frac{dx}{\Phi^p}.$$

$$\int \frac{dx}{x^3(a+bx+cx^2)^p} = -\frac{1}{2ax^2\Phi^{p-1}} - \frac{b(p+1)}{2a} \int \frac{dx}{x^2\Phi} - \frac{cp}{a} \int \frac{dx}{x\Phi^p}$$

$$\int \frac{dx}{x^m(a+bx+cx^2)} = -\frac{1}{a(m-1)x^{m-1}} - \frac{b}{a} \int \frac{dx}{x^{m-1}\Phi} - \frac{c}{a} \int \frac{dx}{x^{m-2}\Phi}.$$

$$\int \frac{dx}{x^m(a+bx+cx^2)^2} = -\frac{1}{a(m-1)x^{m-1}\Phi} - \frac{bm}{a(m-1)} \int \frac{dx}{x^{m-1}\Phi^2} - \frac{c(m+1)}{a(m-1)} \int \frac{dx}{x^{m-2}\Phi^2}.$$

$$\int \frac{dx}{x^m(a+bx+cx^2)^p} = \frac{1}{a} \int \frac{dx}{x^m\Phi^{p-1}} - \frac{b}{a} \int \frac{dx}{x^{m-1}\Phi^p} - \frac{c}{a} \int \frac{dx}{x^{m-2}\Phi^p}.$$

$$= -\frac{1}{a(m-1)x^{m-1}\Phi^{p-1}} - \frac{b(p+m-2)}{a(m-1)}\int\frac{dx}{x^{m-1}\Phi^p} - \frac{c(2p+m-3)}{a(m-1)}\int\frac{dx}{x^{m-2}\Phi^p}.$$

$$\int\frac{dx}{(a+bx+cx^2)^{\frac{p}{2}}} = -\frac{2(b+2cx)}{(p-2)(b^2-4ac)\Phi^{\frac{p}{2}-1}} - \frac{4c(p-3)}{(p-2)(b^2-4ac)}\int\frac{dx}{\Phi^{\frac{p}{2}-1}}.$$

$$= \left\{\frac{1}{(p-2)(4ac-b^2)\Phi^{\frac{p-3}{2}}} + \frac{(p-3)4c}{(p-2)(p-4)(4ac-b^2)^2\Phi^{\frac{p-5}{2}}}\right.$$

$$+ \frac{(p-3)(p-5)16c^2}{(p-2)(p-4)(p-6)(4ac-b^2)^3\Phi^{\frac{p-7}{2}}} + \cdots + \frac{\mathrm{N}}{\Phi}\left\{\frac{2(2cx+b)}{\sqrt{\Phi}} + 8c\mathrm{N}\int\frac{dx}{\Phi^{\frac{3}{2}}}.\right.$$

$$\int\frac{dx}{x(a+bx+cx^2)^{\frac{p}{2}}} = \frac{1}{a(p-2)\Phi^{\frac{p}{2}-1}} + \frac{1}{a}\int\frac{dx}{x\Phi^{\frac{p}{2}-1}} - \frac{b}{2a}\int\frac{dx}{\Phi^{\frac{p}{2}}}.$$

$$\int\frac{dx}{x^m\sqrt{a+bx+cx^2}} = 2\sqrt{a}\int\frac{(c-az^2)^{m-1}}{(2az-b)^m}\,dz. \qquad x = \frac{2az-b}{c-az^2}.$$

$$= -\frac{\sqrt{\Phi}}{a(m-1)x^{m-1}} - \frac{b(2m-3)}{2a(m-1)}\int\frac{dx}{x^{m-1}\sqrt{\Phi}} - \frac{c(m-2)}{a(m-1)}\int\frac{dx}{x^{m-2}\sqrt{\Phi}}.$$

$$\int\frac{dx}{x^m(a+bx+cx^2)^{\frac{p}{2}}} = -\frac{1}{a(m-1)x^{m-1}\Phi^{\frac{p}{2}-1}} - \frac{b(p+2m-4)}{2a(m-1)}\int\frac{dx}{x^{m-1}\Phi^{\frac{p}{2}}} - \frac{c(p+m-3)}{a(m-1)}\int\frac{dx}{x^{m-2}\Phi^{\frac{p}{2}}}.$$

$$\int\frac{dx}{(a+bx+cx^2)^{p+\frac{1}{2}}} = \frac{2(2cx+b)}{(2p-1)(4ac-b^2)\Phi^{p-\frac{1}{2}}} + \frac{8c(p-1)}{(2p-1)(4ac-b^2)}\int\frac{dx}{\Phi^{p-\frac{1}{2}}}.$$

$$\int\frac{dx}{x(a+bx+cx^2)^{p+\frac{1}{2}}} = \frac{1}{(2p-1)a\Phi^{p-\frac{1}{2}}} + \frac{1}{a}\int\frac{dx}{x\Phi^{p-\frac{1}{2}}} - \frac{b}{2a}\int\frac{dx}{\Phi^{p+\frac{1}{2}}}.$$

$$\int\frac{dx}{x^2(a+bx+cx^2)^{p+\frac{1}{2}}} = -\frac{1}{ax\Phi^{p-\frac{1}{2}}} - \frac{b(2p+1)}{2a}\int\frac{dx}{x\Phi^{p+\frac{1}{2}}} - \frac{2pc}{a}\int\frac{dx}{\Phi^{p+\frac{1}{2}}}.$$

$$\int\frac{dx}{x^m(a+bx+cx^2)^{p+\frac{1}{2}}} = \frac{1}{a}\int\frac{dx}{x^m\Phi^{p-\frac{1}{2}}} - \frac{b}{a}\int\frac{dx}{x^{m-1}\Phi^{p+\frac{1}{2}}} - \frac{c}{a}\int\frac{dx}{x^{m-2}\Phi^{p+\frac{1}{2}}}.$$

$$= -\frac{1}{(m-1)ax^{m-1}\Phi^{p-\frac{1}{2}}} - \frac{b(2p+2m-3)}{2a(m-1)}\int\frac{dx}{x^{m-1}\Phi^{p+\frac{1}{2}}} - \frac{c(2p+m-2)}{a(m-1)}\int\frac{dx}{x^{m-2}\Phi^{p+\frac{1}{2}}}.$$

$$\int\frac{dx}{(x+h)^m\sqrt{a+bx+cx^2}} = -\int\frac{z^{m-1}}{\sqrt{(a-bh+ch^2)z^2-(b-2ch)z+c}}\,dz. \qquad zx = 1-hz.$$

$$\int\frac{dx}{(x-h)^m\sqrt{a+bx+cx^2}} = -\int\frac{z^{m-1}}{\sqrt{(a+bh+ch^2)z^2+(b+2ch)z+c}}\,dz. \qquad zx = 1+hz.$$

$$\int\frac{dx}{x^p(mx+n)\sqrt{a+bx+cx^2}} = \frac{1}{n}\int\frac{dx}{x^p\sqrt{\Phi}} - \frac{m}{n^2}\int\frac{dx}{x^{p-1}\sqrt{\Phi}} + \frac{m^2}{n^3}\int\frac{dx}{x^{p-2}\sqrt{\Phi}} - \cdots \pm \frac{m^{p-1}}{n^p}\int\frac{dx}{x\sqrt{\Phi}} \mp \frac{m^p}{n^p}\int\frac{dx}{(mx+n)\sqrt{\Phi}}.$$

$$\int\frac{dx}{(a+bx^2+cx^4)^p} = -\frac{bcx^3+(b^2-2ac)x}{2a(p-1)(4ac-b^2)\Phi^{p-1}} - \frac{bc(4p-7)}{2a(p-1)(4ac-b^2)}\int\frac{x^2}{\Phi^{p-1}}\,dx.$$

$$+ \frac{2ac(4p-5)-b^2(2p-3)}{2a(p-1)(4ac-b^2)}\int\frac{dx}{\Phi^{p-1}}.$$

$$\int\frac{dx}{x^2(a+bx^2+cx^4)^p} = \frac{2cx^2+b}{2(p-1)(4ac-b^2)x^3\Phi^{p-1}} + \frac{c(4p-3)}{(p-1)(4ac-b^2)}\int\frac{dx}{x^2\Phi^{p-1}} + \frac{3b}{2(p-1)(4ac-b^2)}\int\frac{dx}{x^4\Phi^{p-1}}.$$

$$\int\frac{dx}{x^m(a+bx^2+cx^4)^p} = \frac{1}{a}\int\frac{dx}{x^m\Phi^{p-1}} - \frac{b}{a}\int\frac{dx}{x^{m-2}\Phi^p} - \frac{c}{a}\int\frac{dx}{x^{m-4}\Phi^p}$$

$$= \frac{bcx^2 + b^2 - 2ac}{2a(p-1)(b^2-4ac)x^{m-1}\Phi^{p-1}} + \frac{bc(4p+m-7)}{2a(p-1)(b^2-4ac)}\int\frac{dx}{x^{m-2}\Phi^{p-1}}$$

$$+ \frac{b^2(2p+m-3) - 2ac(4p+m-5)}{2a(p-1)(b^2-4ac)}\int\frac{dx}{x^m\Phi^{p-1}}$$

$$\int\frac{dx}{x^m(a+bx^n+cx^{2n})^p} = \frac{1}{a}\int\frac{dx}{x^m\Phi^{p-1}} - \frac{b}{a}\int\frac{dx}{x^{m-n}\Phi^p} - \frac{c}{a}\int\frac{dx}{x^{m-2n}\Phi^p}$$

$$= -\frac{bcx^n + b^2 - 2ac}{an(p-1)(4ac-b^2)x^{m-1}\Phi^{p-1}} - \frac{bc(2pn-3n+m-1)}{an(p-1)(4ac-b^2)}\int\frac{dx}{x^{m-n}\Phi^{p-1}}$$

$$+ \frac{2(2pn-2n+m-1)ac - (pn-n+m-1)b^2}{an(p-1)(4ac-b^2)}\int\frac{dx}{x^m\Phi^{p-1}}.$$

$$\int\frac{x}{(a+bx+cx^2)^p}dx = -\frac{1}{2c(p-1)\Phi^{p-1}} - \frac{b}{2c}\int\frac{dx}{\Phi^{p-1}} = -\frac{bx+2a}{(p-1)(4ac-b^2)\Phi^{p-1}} - \frac{b(2p-3)}{(p-1)(4ac-b^2)}\int\frac{dx}{\Phi^{p-1}}.$$

$$\int\frac{x}{(a+bx+cx^2)^{\frac{p}{2}}}dx = -\frac{1}{(p-2)c\Phi^{\frac{p}{2}-1}} - \frac{b}{2c}\int\frac{dx}{\Phi^{\frac{p}{2}}}.$$

$$\int\frac{x}{(a+bx+cx^2)^{p+\frac{1}{2}}}dx = -\frac{1}{(2p-1)c\Phi^{p-\frac{1}{2}}} - \frac{b}{2c}\int\frac{dx}{\Phi^{p+\frac{1}{2}}}.$$

$$\int\frac{2cx+b}{(a+bx+cx^2)^p}dx = -\frac{1}{(p-1)\Phi^{p-1}}.$$

$$\int\frac{mx+n}{(a+bx+cx^2)^p}dx = \frac{m}{2c}\int\frac{2cx+b}{\Phi^p}dx + \frac{2cn-bm}{2c}\int\frac{dx}{\Phi^p}.$$

$$= -\frac{m}{2c(p-1)\Phi^{p-1}} + \frac{2cn-bm}{2c}\int\frac{dx}{\Phi^p}.$$

$$\int\frac{Ax+B}{\{(x-\alpha)^2+\beta^2\}^p}dx = A\int\frac{z}{(z^2+\beta^2)^p}dz + (A\alpha+B)\int\frac{dz}{(z^2+\beta^2)^p}. \qquad x-\alpha = z.$$

$$\int\frac{a+bx}{(x^2+2\alpha x+\beta)^p}dx = -\frac{b}{(2p-2)\Phi^{p-1}} + (a-b\alpha)(x+\alpha)\left\{\frac{1}{(2p-2)(\beta-\alpha^2)\Phi^{p-1}}\right.$$

$$\left.+\frac{2p-3}{(2p-2)(2p-4)(\beta-\alpha^2)^2\Phi^{p-2}} + \frac{(2p-3)(2p-5)}{(2p-2)(2p-4)(2p-6)(\beta-\alpha^2)^3\Phi^{p-3}} + \cdots + \frac{(2p-3)(2p-5)\cdots3}{(2p-2)(2p-4)\cdots2(\beta-\alpha^2)^{p-1}\Phi}\right\}$$

$$+\frac{(2p-3)(2p-5)\cdots3.1}{(2p-2)(2p-4)\cdots4.2}\,\frac{a-b\alpha}{(\beta-\alpha^2)^{p-1}\sqrt{\beta-\alpha^2}}\,\text{arc tg}\,\frac{x+\alpha}{\sqrt{\beta-\alpha^2}}. \qquad \beta > \alpha^2.$$

$$= -\frac{b}{(2p-2)\Phi^{p-1}} - (a-b\alpha)(x+\alpha)\left\{\frac{1}{(2p-2)(\alpha^2-\beta)\Phi^{p-1}}\right.$$

$$\left.-\frac{2p-3}{(2p-2)(2p-4)(\alpha^2-\beta)^2\Phi^{p-2}} + \frac{(2p-3)(2p-5)}{(2p-2)(2p-4)(2p-6)(\alpha^2-\beta)^3\Phi^{p-3}} - \cdots + \frac{(-1)^{p-2}(2p-3)(2p-5)\cdots3}{(2p-2)(2p-4)\cdots2(\alpha^2-\beta)^{p-1}\Phi}\right\}$$

$$+\frac{(-1)^{p-1}(2p-3)(2p-5)\cdots3}{(2p-2)(2p-4)\cdots2}\,\frac{a-b\alpha}{2(\alpha^2-\beta)^{p-1}\sqrt{\alpha^2-\beta}}\log\frac{x+\alpha-\sqrt{\alpha^2-\beta}}{x+\alpha+\sqrt{\alpha^2-\beta}}. \qquad \beta < \alpha^2.$$

$$\int\frac{x^2}{(a+bx+cx^2)^p}dx = \frac{(b^2-2ac)x+ab}{c(p-1)(4ac-b^2)\Phi^{p-1}} + \frac{2ac+b^2(p-2)}{c(p-1)(4ac-b^2)}\int\frac{dx}{\Phi^{p-1}}.$$

$$\int\frac{x^2}{(a+bx+cx^2)^{p+\frac{1}{2}}}dx = \frac{(2b^2-4ac)x+2ab}{c(2p-1)(4ac-b^2)\Phi^{p-\frac{1}{2}}} + \frac{4ac+b^2(2p-3)}{c(4ac-b^2)(2p-1)}\int\frac{dx}{\Phi^{p-\frac{1}{2}}}$$

$$\int\frac{x^2}{(a+bx^2+cx^4)^p}dx=\frac{2cx^3+bx}{2(p-1)(4ac-b^2)\Phi^{p-1}}+\frac{c(4p-7)}{(p-1)(4ac-b^2)}\int\frac{x^2}{\Phi^{p-1}}dx-\frac{b}{2(p-1)(4ac-b^2)}\int\frac{dx}{\Phi^{p-1}}.$$

$$\int\frac{x^3}{(a+bx+cx^2)^p}dx=\frac{1}{c}\int\frac{x}{\Phi^{p-1}}dx-\frac{b}{c}\int\frac{x^2}{\Phi^p}dx-\frac{a}{c}\int\frac{x}{\Phi^p}dx.$$

$$\int\frac{x^m}{a+bx+cx^2}dx=\frac{x^{m-1}}{c(m-1)}-\frac{a}{c}\int\frac{x^{m-2}}{\Phi}dx-\frac{b}{c}\int\frac{x^{m-1}}{\Phi}dx.$$

$$\int\frac{x^m}{(a+bx+cx^2)^2}dx=\frac{x^{m-1}}{c(m-3)\Phi}-\frac{a(m-1)}{c(m-3)}\int\frac{x^{m-2}}{\Phi}dx-\frac{b(m-2)}{c(m-3)}\int\frac{x^{m-1}}{\Phi^2}dx.$$

$$\int\frac{x^m}{(a+bx+cx^2)^p}dx=-\frac{x^{m-1}}{c(2p-m-1)\Phi^{p-1}}-\frac{b(p-m)}{c(2p-m-1)}\int\frac{x^{m-1}}{\Phi^p}dx+\frac{a(m-1)}{c(2p-m-1)}\int\frac{x^{m-2}}{\Phi^p}dx.$$

$$\int\frac{x^m}{\sqrt{a+bx+cx^2}}dx=\frac{x^{m-1}\sqrt{\Phi}}{cm}-\frac{b(2m-1)}{2cm}\int\frac{x^{m-1}}{\sqrt{\Phi}}dx-\frac{a(m-1)}{cm}\int\frac{x^{m-2}}{\sqrt{\Phi}}dx.$$

$$\int\frac{x^m}{(c+dx)\sqrt{a+bx+\gamma x^2}}dx=\frac{1}{d}\int\frac{x^{m-1}}{\sqrt{\Phi}}dx-\frac{c}{d^2}\int\frac{x^{m-2}}{\sqrt{\Phi}}dx+\frac{c^2}{d^3}\int\frac{x^{m-3}}{\sqrt{\Phi}}dx-\dots\pm\frac{c^{m-1}}{d^m}\int\frac{dx}{\sqrt{\Phi}}\mp\frac{c^m}{d^m}\int\frac{dx}{(c+dx)\sqrt{\Phi}}.$$

$$\int\frac{x^m}{(a+bx+cx^2)^{\frac{p}{2}}}dx=-\frac{x^{m-1}}{c(p-m-1)\Phi^{\frac{p}{2}-1}}-\frac{b(p-2m)}{2c(p-m-1)}\int\frac{x^{m-1}}{\Phi^{\frac{p}{2}}}dx+\frac{a(m-1)}{c(p-m-1)}\int\frac{x^{m-2}}{\Phi^{\frac{p}{2}}}dx.$$

$$\int\frac{x^m}{(a+bx+cx^2)^{p+\frac{1}{2}}}dx=-\frac{x^{m-1}}{c(2p-m)\Phi^{p-\frac{1}{2}}}-\frac{b(2p-2m+1)}{2c(2p-m)}\int\frac{x^{m-1}}{\Phi^{p+\frac{1}{2}}}dx+\frac{a(m-1)}{c(2p-m)}\int\frac{x^{m-2}}{\Phi^{p+\frac{1}{2}}}dx.$$

$$\int\frac{x^m}{(a+bx^2+cx^4)^p}dx=\frac{1}{c}\int\frac{x^{m-4}}{\Phi^{p-1}}dx-\frac{b}{c}\int\frac{x^{m-2}}{\Phi^p}dx-\frac{a}{c}\int\frac{x^{m-4}}{\Phi^p}dx.$$

$$=\frac{(2cx^2+b)x^{m-1}}{2(p-1)(4ac-b^2)\Phi^{p-1}}+\frac{c(4p-m-5)}{(p-1)(4ac-b^2)}\int\frac{x^m}{\Phi^{p-1}}dx-\frac{b(m-1)}{2(p-1)(4ac-b^2)}\int\frac{x^{m-2}}{\Phi^{p-1}}dx.$$

$$\int\frac{x^{m-1}}{ax^{2m}+bx^m+c}dx=\frac{1}{m\sqrt{b^2-4ac}}\log\frac{2ax^m+b-\sqrt{b^2-4ac}}{2ax^m+b+\sqrt{b^2-4ac}}.\qquad b^2-4ac>0.$$

$$=\frac{2}{m\sqrt{4ac-b^2}}\operatorname{arc\,tg}\frac{2ax^m+b}{\sqrt{4ac-b^2}}.\qquad b^2-4ac<0.$$

$$\int\frac{x^{n+m-1}}{x^{2n}+2x^n\cos\alpha+1}dx=-\frac{1}{n\sin\alpha}\sum_{k=o}^{k=n-1}\operatorname{arc\,tg}\frac{x\sin\frac{(2k+1)\pi+\alpha}{n}}{1-x\cos\frac{(2k+1)\pi+\alpha}{n}}\cos\frac{(2k+1)m\pi+m\alpha}{n}$$

$$\int\frac{x^m}{x^{2n}+2x^n\cos\alpha+1}dx=\frac{1}{n\sin\alpha}\sum_{k=o}^{k=n-1}\left[\cos\frac{(2k+1)(m+1)\pi-(n-m-1)\alpha}{n}.\operatorname{arc\,tg}\frac{x\sin\frac{(2k+1)\pi+\alpha}{n}}{1-x\cos\frac{(2k+1)\pi+\alpha}{n}}\right.$$

$$\left.+\sin\frac{(2k+1)(m+1)\pi-(n-m-1)\alpha}{n}\cdot\frac{1}{2}\log\left(x^2-2x\cos\frac{(2k+1)\pi+\alpha}{n}+1\right)\right]$$

$$\int\frac{x^m}{x^{2n}-2x^n\cos n\theta+1}dx=\frac{1}{n\sin n\theta}\sum\cos(n-m-1)\omega.\operatorname{arc\,tg}\frac{x-\cos\omega}{\sin\omega}$$

$$-\frac{1}{2}\sum\sin(n-m-1)\omega.\log(x^2-2x\cos\omega+1).$$

where $\omega=\theta+\frac{k\pi}{n}$ and $k=0,2,4,\dots 2(n-1)$ successively.

$$\int \frac{x^m}{a + bx^n + cx^{2n}} dx = \frac{1}{c(\alpha - \beta)} \left[\int \frac{x^m}{x^n - \alpha} dx - \int \frac{x^m}{x^n - \beta} dx \right].$$

α, β being the real roots of $a + bx^n + cx^{2n} = 0$.

$$= \frac{1}{a} \left(\frac{a}{c}\right)^{\frac{m+1}{2n}} \int \frac{z^m}{1 - 2z^n \cos n\theta + z^{2n}} dz. \qquad z = \left(\frac{c}{a}\right)^{\frac{1}{2n}} x.$$

where $\cos n\theta = -\dfrac{b}{2\sqrt{ac}}$; the roots being imaginary.

$$= \frac{2c}{\sqrt{b^2 - 4ac}} \left[\int \frac{x^m}{2cx^n + b - \sqrt{b^2 - 4ac}} dx - \int \frac{x^m}{2cx^n + b + \sqrt{b^2 - 4ac}} dx \right] \quad b^2 - 4ac > 0,\ m < n$$

$$= \frac{1}{ncq^{2n-m-1} \sin \varepsilon} \sum_{k=0}^{k=n-1} \left[-\sin(n - m - 1)\varepsilon_k . \frac{1}{2} \log (x^2 - 2qx \cos \varepsilon_k + q^2) \right.$$

$$\left. + \cos(n - m - 1)\varepsilon_k . \operatorname{arc\,tg} \frac{x \sin \varepsilon_k}{q - x \cos \varepsilon_k} \right].$$

where $q = \left(\dfrac{a}{c}\right)^{\frac{1}{2n}}$, $\cos \varepsilon = -\dfrac{b}{2\sqrt{ac}}$, $\varepsilon_k = \dfrac{2k\pi + \varepsilon}{n}$, $b^2 - 4ac < 0$, $m < 2n$.

$$\int \frac{x^m}{(a + bx^n + cx^{2n})^p} dx = \frac{1}{c} \int \frac{x^{m-2n}}{\Phi^{p-1}} dx - \frac{b}{c} \int \frac{x^{m-n}}{\Phi^p} dx - \frac{a}{c} \int \frac{x^{m-2n}}{\Phi^p} dx.$$

$$= \frac{(2cx^n + b)\, x^{m-n+1}}{n(p-1)(4ac - b^2)\, \Phi^{p-1}} + \frac{2c(2pn - 2n - m - 1)}{n(p-1)(4ac - b^2)} \int \frac{x^m}{\Phi^{p-1}} dx$$

$$- \frac{b(m - n + 1)}{n(p-1)(4ac - b^2)} \int \frac{x^{m-n}}{\Phi^{p-1}} dx.$$

$$\int \frac{x^m(1 + x - 2x^{m+1})}{1 - x^2} dx = \frac{x^m(x^m - 1)}{m} + \frac{x^{m-1}(x^{m-1} - 1)}{m - 1} + \dots + \frac{x(x - 1)}{1} + \log(x + 1).$$

$$\int \frac{(a + bx + cx^2)^p}{x} dx = \frac{\Phi^p}{2p} + \frac{b}{2} \int \Phi^{p-1} dx + a \int \frac{\Phi^{p-1}}{x} dx.$$

$$\int \frac{(a + bx + cx^2)^{\frac{p}{2}}}{x} dx = \frac{\Phi^{\frac{p}{2}}}{p} + a \int \frac{\Phi^{\frac{p}{2}-1}}{x} dx + \frac{b}{2} \int \Phi^{\frac{p}{2}-1} dx.$$

$$\int \frac{(a + bx + cx^2)^{p-\frac{1}{2}}}{x} dx = \frac{\Phi^{p-\frac{1}{2}}}{2p - 1} + a \int \frac{\Phi^{p-\frac{3}{2}}}{x} dx + \frac{b}{2} \int \Phi^{p-\frac{3}{2}} dx.$$

$$\int \frac{(a + bx + cx^2)^{p-\frac{1}{2}}}{x^2} dx = -\frac{\Phi^{p-\frac{1}{2}}}{x} + \frac{b(2p - 1)}{2} \int \frac{\Phi^{p-\frac{3}{2}}}{x} dx + c(2p - 1) \int \Phi^{p-\frac{3}{2}} dx.$$

$$\int \frac{(a + bx + cx^2)^p}{x^m} dx = -\frac{\Phi^{p+1}}{a(m-1)\, x^{m-1}} + \frac{b(p - m + 2)}{a(m - 1)} \int \frac{\Phi^p}{x^{m-1}} dx + \frac{c(2p - m + 3)}{a(m - 1)} \int \frac{\Phi^p}{x^{m-2}} dx.$$

$$\int \frac{\sqrt{a + bx + cx^2}}{x^m} dx = -\frac{\sqrt{\Phi}}{(m-1)\, x^{m-1}} + \frac{b}{2(m-1)} \int \frac{dx}{x^{m-1}\sqrt{\Phi}} + \frac{c}{m - 1} \int \frac{dx}{x^{m-2}\sqrt{\Phi}}.$$

$$\int \frac{(a + bx + cx^2)^{\frac{p}{2}}}{x^m} dx = -\frac{\Phi^{\frac{p}{2}+1}}{a(m-1)\, x^{m-1}} + \frac{b(p - 2m + 4)}{2a(m - 1)} \int \frac{\Phi^{\frac{p}{2}}}{x^{m-1}} dx + \frac{c(p - m + 3)}{a(m - 1)} \int \frac{\Phi^{\frac{p}{2}}}{x^{m-2}} dx.$$

$$\int \frac{(a + bx + cx^2)^{p-\frac{1}{2}}}{x^m} dx = -\frac{\Phi^{p-\frac{1}{2}}}{(m-1)\, x^{m-1}} + \frac{b(2p-1)}{2(m-1)} \int \frac{\Phi^{p-\frac{3}{2}}}{x^{m-1}} dx + \frac{c(2p - 1)}{m - 1} \int \frac{x^{p-\frac{3}{2}}}{x^{m-2}} dx.$$

$$\int \frac{(a+bx+cx^2)^{p-\frac{1}{2}}}{(x+h)^m}\,dx = -\frac{\Phi^{p+\frac{1}{2}}}{(m-1)\,A\,(x+h)^{m-1}} + \frac{B\,(2p-2m+3)}{2A\,(m-1)} \int \frac{\Phi^{p-\frac{1}{2}}}{(x+h)^{m-1}}\,dx + \frac{c(2p-m+2)}{A\,(m-1)} \int \frac{\Phi^{p-\frac{1}{2}}}{(x+h)^{m-2}}\,dx$$

where $A = a - bh + ch^2$, $\qquad B = b - 2ch.$

$$\int \frac{(a+bx^n+cx^{2n})^p}{x^m}\,dx = -\frac{\Phi^{p+1}}{a(m-1)x^{m-1}} - \frac{b\,(m-pn-n-1)}{a\,(m-1)} \int \frac{\Phi^p}{x^{m-n}}\,dx - \frac{c\,(m-2pn-2n-1)}{a\,(m-1)} \int \frac{\Phi^p}{x^{m-2n}}\,dx.$$

$$\int \frac{\sqrt{a+bx^n+cx^m}}{x^p}\,dx = c \int \frac{x^{m-p}}{\sqrt{\Phi}}\,dx + b \int \frac{x^{n-p}}{\sqrt{\Phi}}\,dx + a \int \frac{dx}{x^p\sqrt{\Phi}}.$$

$\int U\left(X^3\right)^p\,dx$

In that which follows, $\Phi = c_0x^3 + c_1x^2 + c_2x + c_3$.

$$\left\{\begin{aligned} \int \frac{dx}{\Phi^{p+1}} &= \left\{\frac{1}{3}U_0x^2 + \left(N_0U_0 + \frac{1}{3}U_1\right)x + N_0U_1 + \frac{N_1U_0}{3c_0}\right\}\frac{1}{p\Phi^p} + \frac{3p-2}{3p}U_0 \int \frac{x}{\Phi^p}\,dx + \frac{(3p-1)\,U_1 - 3N_0U_0}{3p} \int \frac{dx}{\Phi^p}. \\ \int \frac{x}{\Phi^{p+1}}\,dx &= \left\{\frac{1}{3}V_0x^2 + \left(N_0V_0 + \frac{1}{3}V_1\right)x + N_0V_1 + \frac{N_1V_0}{3c_0}\right\}\frac{1}{p\Phi^p} + \frac{3p-2}{3p}V_0 \int \frac{x}{\Phi^p}\,dx + \frac{(3p-1)\,V_1 - 3N_0V_0}{3p} \int \frac{dx}{\Phi^p}. \end{aligned}\right.$$

where $N_0 = \dfrac{c_1}{9c_0}$, $\qquad N_1 = \dfrac{2}{3}c_2 - \dfrac{2c_1^2}{9c_0}$, $\qquad N_2 = 3c_0c_3 - \dfrac{5}{3}c_1c_2 + \dfrac{4c_1^3}{9c_0}$,

$N_3 = -6c_0c_1c_3 - 2c_0c_2^2 + 4c_1^2c_2 - \dfrac{8c_1^4}{9c_0}$, $\qquad D = N_1N_3 - N_2^2$,

$u_0 = \dfrac{9c_0^2N_1}{D}$, $\quad u_1 = -\dfrac{3c_0N_2}{D}$, $\quad V_0 = -\dfrac{3c_0\,(N_2 + 2c_1N_1)}{D}$, $\quad V_1 = \dfrac{N_3 + 2c_1N_2}{D}$.

$$\int \frac{dx}{x\Phi^{p+1}} = \frac{1}{3pc_3\Phi^p} - \frac{c_1}{3c_3} \int \frac{x}{\Phi^{p+1}}\,dx - \frac{2c_2}{3c_3} \int \frac{dx}{\Phi^{p+1}} + \frac{1}{c_3} \int \frac{dx}{x\Phi^p}.$$

$$\int \frac{dx}{x^2\Phi^{p+1}} = -\frac{1}{c_3x\Phi^p} - \frac{(3p+1)\,c_0}{c_3} \int \frac{x}{\Phi^{p+1}}\,dx - \frac{(2p+1)\,c_1}{c_3} \int \frac{dx}{\Phi^{p+1}} - \frac{(p+1)\,c_2}{c_3} \int \frac{dx}{x\Phi^{p+1}}.$$

$$\int \frac{dx}{x^m\Phi^{p+1}} = \frac{1}{c_3} \int \frac{dx}{x^m\Phi^p} - \frac{c_2}{c_3} \int \frac{dx}{x^{m-1}\Phi^{p+1}} - \frac{c_1}{c_3} \int \frac{dx}{x^{m-2}\Phi^{p+1}} - \frac{c_0}{c_3} \int \frac{dx}{x^{m-3}\Phi^{p+1}}.$$

$$= -\frac{1}{(m-1)\,c_3x^{m-1}\Phi^p} - \frac{(3p+m-1)c_0}{(m-1)\,c_3} \int \frac{dx}{x^{m-3}\Phi^{p+1}} - \frac{(2p+m-1)\,c_1}{(m-1)\,c_3} \int \frac{dx}{x^{m-2}\Phi^{p+1}} - \frac{(p+m-1)c_2}{(m-1)c_3} \int \frac{dx}{x^{m-1}\Phi^{p+1}}$$

$\int \dfrac{x}{\Phi^{p+1}}\,dx \qquad$ See above.

$$\int \frac{x^2}{\Phi^{p+1}}\,dx = -\frac{1}{3pc_0\Phi^p} - \frac{2c_1}{3c_0} \int \frac{x}{\Phi^{p+1}}\,dx - \frac{c_2}{3c_0} \int \frac{dx}{\Phi^{p+1}}.$$

$$\int \frac{x^m}{\Phi^{p+1}}\,dx = \frac{1}{c_0} \int \frac{x^{m-3}}{\Phi^p}\,dx - \frac{c_1}{c_0} \int \frac{x^{m-1}}{\Phi^{p+1}}\,dx - \frac{c_2}{c_0} \int \frac{x^{m-2}}{\Phi^{p+1}}\,dx - \frac{c_3}{c_0} \int \frac{x^{m-3}}{\Phi^{p+1}}\,dx.$$

$$= -\frac{x^{m-2}}{(3p-m+2)\,c_0\Phi^p} - \frac{2p-m+2}{3p-m+2}\cdot\frac{c_1}{c_0} \int \frac{x^{m-1}}{\Phi^{p+1}}\,dx - \frac{p-m+2}{3p-m+2}\cdot\frac{c_2}{c_0} \int \frac{x^{m-2}}{\Phi^{p+1}}\,dx$$

$$+ \frac{m-2}{3p-m+2}\cdot\frac{c_3}{c_2} \int \frac{x^{m-3}}{\Phi^{p+1}}\,dx.$$

$\int U\left(X^n\right)^p dx$

In that which follows, $\Phi = c_0x^n + c_1x^{n-1} + \cdots c_n = f(x)$, $\quad f^k(x) = \frac{d^k\Phi}{dx^k}$.

$$\int\frac{dx}{x^m\Phi^{p+1}} = \frac{1}{c_n}\int\frac{dx}{x^m\Phi^p} - \sum_{k=1}^{k=n}\frac{c_{n-k}}{c_n}\int\frac{dx}{x^{m-k}\Phi^{p+1}}.$$

$$= -\frac{1}{(m-1)\,c_nx^{m-1}\Phi^p} - \sum_{k=1}^{k=n}\frac{kp+m-1}{m-1}\cdot\frac{c_{n-k}}{c_n}\int\frac{dx}{x^{m-k}\Phi^{p+1}}.$$

$$\int\frac{dx}{(x-a)^2\,\Phi^{p+1}} = -\frac{1}{f(a)\,(x-a)\,\Phi^p} - (p+1)\frac{f'(a)}{f(a)}\int\frac{dx}{(x-a)\,\Phi^{p+1}} - \sum_{k=2}^{k=n}\frac{kp+1}{k!}\cdot\frac{f^k(a)}{f(a)}\int\frac{(x-a)^{k-2}}{\Phi^{p+1}}\,dx.$$

$$\int\frac{dx}{(x-a)^m\,\Phi^{p+1}} = -\frac{1}{(m-1)\,f(a)\,(x-a)^{m-1}\Phi^p} - \sum_{k=1}^{k=n}\frac{kp+m-1}{k!\,(m-1)}\cdot\frac{f^k(a)}{f(a)}\int\frac{dx}{(x-a)^{m-k}\,\Phi^{p+1}}.$$

$$\int\frac{x^{n-1}}{\Phi}\,dx = \frac{1}{nc_0}\log\Phi - \sum_{k=n}^{k=n-1}\frac{(n-k)\,c_k}{nc_0}\int\frac{x^{n-k-1}}{\Phi}\,dx.$$

$$\int\frac{x^{n-1}}{\Phi^{p+1}}\,dx = \frac{1}{npc_0\,\Phi^p} - \sum_{k=1}^{k=n-1}\frac{(n-k)\,c_k}{nc_0}\int\frac{x^{n-k-1}}{\Phi^{p+1}}\,dx.$$

$$\int\frac{x^m}{\Phi^{p+1}}\,dx = \frac{1}{c_0}\int\frac{x^{m-n}}{\Phi^p}\,dx - \frac{1}{c_0}\sum_{k=1}^{k=n}c_k\int\frac{x^{m-k}}{\Phi^{p+1}}\,dx.$$

$$= -\frac{x^{m-n+1}}{(np-m+n-1)\,c_0\,\Phi^p} - \sum_{k=1}^{k=n}\frac{(n-k)\,p-m+n-1}{np-m+n-1}\cdot\frac{c_k}{c_0}\int\frac{x^{m-k}}{\Phi^{p+1}}\,dx.$$

Irrational Monomials

$$\int\frac{dx}{x^{\frac{3}{2}}-x} = 2\int\frac{dz}{z^2-z}. \qquad x = z^2.$$

$$\int\frac{dx}{\sqrt{x}+x^{\frac{1}{3}}} = 6\int\frac{z^3}{z+1}\,dz. \qquad x = z^6.$$

$$\int\frac{dx}{\sqrt{x}+x^{\frac{1}{6}}} = 2\sqrt{x} - 6x^{\frac{1}{6}} + 6\,\text{arc tg}\,x^{\frac{1}{6}}.$$

$$\int\frac{\sqrt{x}}{1-\sqrt{x}}\,dx = -x - 2\sqrt{x} - 2\log\left(1-\sqrt{x}\right).$$

$$\int\frac{x^{\frac{1}{4}}}{1+\sqrt{x}}\,dx = \frac{4}{3}x^{\frac{3}{4}} - 4x^{\frac{1}{4}} + 4\,\text{arc tg}\,x^{\frac{1}{4}}.$$

$$\int\frac{1+\sqrt{x}}{1-\sqrt{x}}\,dx = -x - 4\sqrt{x} - 4\log\left(\sqrt{x}-1\right).$$

$$\int\frac{1-\sqrt{x}}{1-x^{\frac{1}{3}}}\,dx = 6\int\frac{z^5-z^8}{1-z^2}\,dz = 6\int\left\{z^6+z^4-z^3+z^2-z+1-\frac{1}{1+z}\right\}dz. \qquad x = z^6.$$

$$\int\frac{\sqrt{x}-1}{1+x^{\frac{1}{3}}}\,dx = 6\int\frac{z^8-z^5}{z^2+1}\,dz = 6\int\left\{z^6-z^4-z^3+z^2+z-1+\frac{1-z}{z^2+1}\right\}dz. \qquad x = z^6.$$

$$\int\frac{1+\sqrt{x}}{1+x^{\frac{1}{3}}}\,dx = 6\int\frac{z^8+z^5}{z^2+1}\,dz = 6\int\left\{z^6-z^4+z^3+z^2-z-1+\frac{z+1}{z^2+1}\right\}dz. \qquad x = z^6$$

$$\int \frac{\sqrt{x} - a}{x^{\frac{1}{3}} - \sqrt{x}}\, dx = 6 \int \frac{z^6 - az^3}{1 - z}\, dz. \qquad x = z^6.$$

$$\int \frac{1 + x^{\frac{1}{3}}}{1 + x^{\frac{1}{4}}}\, dx = 12 \int \frac{z^{15} + z^{11}}{z^3 + 1}\, dz. \qquad x = z^{12}.$$

$$\int \frac{1 + x^{\frac{1}{4}}}{1 + x^{\frac{1}{3}}}\, dx = \frac{1}{2^{\frac{5}{2}}} \log \frac{x^{\frac{1}{6}} - x^{\frac{1}{12}}\sqrt{2} + 1}{x^{\frac{1}{6}} + x^{\frac{1}{12}}\sqrt{2} + 1} + 2 \text{ arc tg } \frac{x^{\frac{1}{12}}\sqrt{2}}{1 - x^{\frac{1}{6}}}$$

$$\int \frac{x - \sqrt{x} + x^{\frac{2}{3}}}{1 - \sqrt{x}}\, dx = -6 \int z^8\, dz - 6 \int \frac{z^9 - 1}{z^3 - 1}\, dz - 6 \int \frac{dz}{z^3 - 1}. \qquad x = z^6.$$

See also $\int U\sqrt{x}\, dx$, p. 39; $\int U x^{\frac{3}{2}}\, dx$, p. 66; $\int x^{A} (a + bx^{B})^{C}\, dx$, p. 74.

Sums and Differences of Binomials

$$\int \frac{dx}{\sqrt{x + a} + \sqrt{x + b}} = \frac{2}{3(a - b)} \left\{ (x + a)^{\frac{3}{2}} - (x + b)^{\frac{3}{2}} \right\}.$$

$$\int \frac{dx}{\sqrt{x + a} - \sqrt{x - a}} = \frac{1}{3a} \left\{ (x + a)^{\frac{3}{2}} + (x - a)^{\frac{3}{2}} \right\}.$$

$$\int \frac{dx}{\sqrt{1 + x^2} - \sqrt{1 - x^2}} = -\frac{\sqrt{1 + x^2} + \sqrt{1 - x^2}}{2x} + \frac{1}{2} \log (x + \sqrt{1 + x^2}) - \frac{1}{2} \text{ arc sin } x.$$

$$\int \frac{x}{(1 + x)^{\frac{1}{3}} - \sqrt{1 + x}}\, dx = -\left\{ \frac{2}{3} \Phi^{\frac{5}{6}} + \frac{3}{4} \Phi^{\frac{2}{3}} + \frac{6}{7} \Phi^{\frac{1}{2}} + \Phi^{\frac{1}{3}} + \frac{6}{5} \Phi^{\frac{1}{6}} + \frac{3}{2} \right\} \Phi^{\frac{2}{3}}.$$

$$\int \frac{x}{(a + x)^{\frac{1}{4}} + (a + x)^{\frac{1}{2}}}\, dx = 4 \int \frac{(z - 1)^6 - a^2 (z - 1)^2}{z}\, dz. \qquad (a + x)^{\frac{1}{4}} = z - 1.$$

$$\int \frac{x}{(1 - x^2)^{\frac{1}{3}} + \sqrt{1 - x^2}}\, dx = -3 \int \frac{z^3}{z + 1}\, dz. \qquad 1 - x^2 = z^6.$$

$$\int \frac{x^{m-1}}{(ax^m + b)^2 + (px^m + q)^2}\, dx = \frac{1}{m(aq - bp)} \text{ arc tg } \frac{ax^m + b}{px^m + q}.$$

$$\int \frac{x^{m-1}}{\sqrt{ax^m + b} + \sqrt{ax^m + c}}\, dx = \frac{2}{3am(b - c)} \left\{ (ax^m + b)^{\frac{3}{2}} \mp (ax^m + c)^{\frac{3}{2}} \right\}.$$

Products and Quotients of Binomials

$$\int (ax^m + b)(a'x^n + b')\, dx = \frac{aa'}{m + n + 1} x^{m+n+1} + \frac{ab'}{m + 1} x^{m+1} + \frac{ba'}{n + 1} x^{n+1} + bb'x.$$

$$\int (ax + b)^p (a'x + b')^q\, dx = \frac{(ax + b)^p (a'x + b')^{q+1}}{a'(p + q + 1)} - \frac{p(ab' - a'b)}{a'(p + q + 1)} \int (ax + b)^{p-1} (a'x + b')^q\, dx.$$

$$\int \frac{dx}{\sqrt{ax + b}\,(mx + n)^{\frac{3}{2}}} = \frac{2}{an - bm} \sqrt{\frac{ax + b}{mx + n}}.$$

$$\int \frac{dx}{(x - a)^{\frac{3}{4}} (x - b)^{\frac{1}{4}}} = \log \frac{(x - b)^{\frac{1}{4}} + (x - a)^{\frac{1}{4}}}{(x - b)^{\frac{1}{4}} - (x - a)^{\frac{1}{4}}} + 2 \text{ arc tg } \left(\frac{x - a}{x - b}\right)^{\frac{1}{4}}.$$

$$\int \frac{dx}{(x^m + a)(x^m + b)} = \frac{1}{a - b} \int \frac{dx}{x^m + b} - \frac{1}{a - b} \int \frac{dx}{x^m + a}.$$

$$\int\frac{dx}{(1+nx^2)^p\sqrt{a+bx^2}}=\frac{1}{(2p-2)(an-b)}\left\{\frac{nx\sqrt{\Phi}}{(1+nx^2)^{p-1}}+(2p-3)(an-2b)\int\frac{dx}{(1+nx^2)^{p-1}\sqrt{\Phi}}\right.$$

$$\left.+b(2p-4)\int\frac{dx}{(1+nx^2)^{p-2}\sqrt{\Phi}}\right\}.$$

$$\int\frac{dx}{(1-x^n)(2x^n-1)^{\frac{1}{2n}}}=\int\frac{z^{2n-2}}{1-z^{2n}}dz. \qquad (2x^n-1)^{\frac{1}{2n}}=zx.$$

$$\int\frac{dx}{(1+x^n)(1+2x^n)^{\frac{1}{2n}}}=\int\frac{dz}{1+z^{2n}}. \qquad (1+2x^n)^{\frac{1}{2n}}=\frac{x}{z}.$$

$$\int\frac{dx}{x^p(mx+n)\sqrt{ax+b}}=\frac{1}{n}\int\frac{dx}{x^p\sqrt{\Phi}}-\frac{m}{n^2}\int\frac{dx}{x^{p-1}\sqrt{\Phi}}+\frac{m^2}{n^3}\int\frac{dx}{x^{p-2}\sqrt{\Phi}}-\cdots\pm\frac{m^{p-1}}{n^p}\int\frac{dx}{x\sqrt{\Phi}}\mp\frac{m^p}{n^p}\int\frac{dx}{(mx+n)\sqrt{\Phi}}.$$

$$\int\frac{dx}{x\sqrt{(a+bx^m)(a+cx^m)}}=-\frac{2}{am}\log\left\{\sqrt{\frac{a}{x^m}+b}+\sqrt{\frac{a}{x^m}+c}\right\}.$$

$$\int\frac{x-4}{(x-2)^{\frac{4}{3}}\sqrt{x-1}}dx=6\frac{\sqrt{x-1}}{(x-2)^{\frac{1}{3}}}.$$

$$\int\frac{x(2-x^2)}{(1+x^2)^{\frac{5}{2}}\sqrt{1-x^2}}dx=-\frac{1}{2}\sqrt{\frac{1-x^2}{(1+x^2)^3}}.$$

$$\int\frac{x^m}{(x^m+a)(x^m+b)}dx=\frac{a}{a-b}\int\frac{dx}{x^m+a}-\frac{b}{a-b}\int\frac{dx}{x^m+b}.$$

$$\int\frac{x^{2m-1}}{(x^m-a)(x^m-b)}dx=\frac{1}{m(a-b)}\log\frac{(x^m-a)^a}{(x^m-b)^b}.$$

$$\int\frac{x^{m-1}}{(ax^m+b)(px^m+q)}dx=\frac{1}{m(aq-bp)}\log\frac{ax^m+b}{px^m+q}.$$

$$\int\frac{x^{m-1}}{\sqrt{(ax^m+b)(ax^m+c)}}dx=\frac{2}{am}\log\left\{\sqrt{ax^m+b}+\sqrt{ax^m+c}\right\}.$$

$$\int\frac{x^p}{(x-a_1)(x-a_2)\cdots(x-a_n)}dx=\frac{a_1^p\log(x-a_1)}{(a_1-a_2)(a_1-a_3)\cdots(a_1-a_n)}+\frac{a_2^p\log(x-a_2)}{(a_2-a_1)(a_2-a_3)\cdots(a_2-a_n)}+\cdots$$

$$+\frac{a_n^p\log(x-a_n)}{(a_n-a_1)(a_n-a_2)\cdots(a_n-a_{n-1})}.$$

$$\int\frac{x^{m-1}}{(ax^m-1)(ax^m+1)^{\frac{3}{2}}}dx=\frac{1}{am}\left\{\frac{1}{\sqrt{ax^m+1}}+\frac{1}{\sqrt{2}}\log\frac{\sqrt{ax^m+1}-\sqrt{2}}{\sqrt{ax^m-1}}\right\}.$$

$$\int\frac{x^{m-1}}{(1+x^m)(1-x^m)^{\frac{1}{m}}}dx=-\int\frac{z^{m-2}}{2-z^m}dz. \qquad 1-x^m=z^m.$$

$$\int\frac{x^{m-2}}{(1-x^m)(2x^m-1)^{\frac{1}{2m}}}dx=2\int\frac{z^{2m-1}}{1-z^{2m}}dz. \qquad 2x^m-1=z^{2m}.$$

$$\int\frac{x^p}{(mx+n)\sqrt{ax^2+b}}dx=\frac{1}{m}\int\frac{x^{p-1}}{\sqrt{\Phi}}dx-\frac{n}{m^2}\int\frac{x^{p-2}}{\sqrt{\Phi}}dx+\frac{n^2}{m^3}\int\frac{x^{p-3}}{\sqrt{\Phi}}dx-\cdots\mp\frac{n^{p-1}}{m^p}\int\frac{dx}{\sqrt{\Phi}}\pm\frac{n^p}{m^p}\int\frac{dx}{(mx+n)\sqrt{\Phi}}.$$

$$\int\frac{x^{m-1}}{(px^m+q)^{\frac{3}{2}}\sqrt{ax^m+b}}dx=\frac{2}{m(aq-bp)}\sqrt{\frac{ax^m+b}{px^m+q}}.$$

$$\int\frac{\sqrt{x-a}}{(x-b)^{\frac{3}{2}}}dx=-2\left(\frac{x-a}{x-b}\right)^{\frac{1}{2}}+\log\frac{\sqrt{x-b}+\sqrt{x-a}}{\sqrt{x-b}-\sqrt{x-a}}.$$

$$\int\frac{\sqrt{1-x}}{(1+x)^{\frac{3}{2}}}\,dx = -2\left(\frac{1-x}{1+x}\right)^{\frac{1}{2}} + \text{arc cos}\, x.$$

$$\int\frac{(ax+b)^m}{px+q}\,dx = \frac{a^m}{p^{m+1}}\left\{\frac{1}{m}(px+q)^m - \frac{m}{m-1}\,\frac{aq-bp}{a}(px+q)^{m-1} + \frac{m(m-1)}{1.2\,(m-2)}\,\frac{(aq-bp)^2}{a^2}(px+q)^{m-2} - \ldots\right.$$

$$\left.\pm\frac{(aq-bp)^m}{a^m}\log(px+q)\right\}.$$

$$\int\frac{mx+n}{(ax+b)^p}\,dx = -\frac{m}{a^2(p-2)\,\Phi^{p-2}} - \frac{an-bm}{a^2(p-1)\,\Phi^{p-1}}.$$

$$\int\frac{bx+c}{(a^2+x^2)^{m+\frac{1}{2}}}\,dx = -\frac{b}{(2m-1)\,\Phi^{m-\frac{1}{2}}} + c\int\frac{dx}{\Phi^{m+\frac{1}{2}}}.$$

$$\int\left(\frac{ax+b}{mx+n}\right)^p dx = \frac{1}{m^{p+1}}\int\frac{(az+bm-an)^p}{z^p}\,dz. \qquad mx+n=z.$$

$$\int\frac{(ax+b)^m}{(a'x+b')^n}\,dx = -\frac{(ax+b)^{m+1}}{(n-1)(a'b-ab')(a'x+b')^{n-1}} - \frac{a(n-m-2)}{(n-1)(a'b-ab')}\int\frac{(ax+b)^m}{(a'x+b')^{n-1}}\,dx.$$

$$\int\frac{(a+bx)^m}{(a-bx)^n}\,dx = \frac{(2a)^m}{b(n-1)(a-bx)^{n-1}} - \frac{m}{1}\,\frac{(2a)^{m-1}}{b(n-2)(a-bx)^{n-2}} + \frac{m(m-1)}{1.2}\,\frac{(2a)^{m-2}}{b(n-3)(a-bx)^{n-3}} - \ldots$$

$$\int\frac{(ax^2-c)^n}{(b-ax)^{n+1}}\,dx = -\frac{1}{a^{n+1}}\int\frac{(z^2-2bz+b^2-ac)^n}{z^{n+1}}\,dz. \qquad b-ax=z.$$

$$\int\frac{(c-ax^2)^{n-1}}{(2ax-b)^n}\,dx = \frac{1}{2\sqrt{a}}\int\frac{dz}{z^n\sqrt{a+bz+cz^2}}. \qquad \frac{c-ax^2}{2ax-b}=\frac{1}{z}.$$

$$\int\frac{(ax+b)^{m+n-2}}{(x-\alpha)^m(x-\beta)^n}\,dx = \frac{1}{(\alpha-\beta)^{m+n-1}}\int\frac{\{a+b\alpha-(a+b\beta)z\}^{m+n-2}}{z^m}\,dz. \qquad z(\beta-x)=\alpha-x.$$

$\int U\left(X+Y^{\frac{m}{n}}\right)dx$

$$\int\left\{x+\sqrt{1+x^2}\right\}^n dx = \frac{(x+\sqrt{1+x^2})^{n+1}}{2n+2} + \frac{(x+\sqrt{1+x^2})^{n-1}}{2n-2}.$$

$$\int\left\{x+\sqrt{a^2+x^2}\right\}^n dx = \frac{(x+\sqrt{a^2+x^2})^n}{n^2-1}\left(n\sqrt{a^2+x^2}-x\right).$$

$$\int x^m\left\{x+\sqrt{1+x^2}\right\}^n dx = \frac{1}{2^{m+1}}\int\frac{(z^2-1)^m}{z^{m-n}}\,dz + \frac{1}{2^{m+1}}\int\frac{(z^2-1)^m}{z^{m-n+2}}\,dz. \qquad x+\sqrt{1+x^2}=z.$$

$$\int\frac{dx}{x+\sqrt{a^2+x^2}} = \frac{1}{2}\int\frac{z^2+a^2}{z^3}\,dz. \qquad x+\sqrt{a^2+x^2}=z.$$

$$\int\frac{dx}{1+x-\sqrt{1-x^2}} = \frac{1}{2}\log(\sqrt{1-x^2}-1) + \frac{1}{2}\text{arc cos}\, x$$

$$\int\frac{dx}{x^2+2x+4-4\sqrt{1+x}} = -\frac{1}{3(\sqrt{\Phi}-1)} + \frac{1}{18}\log\frac{x+2-2\sqrt{\Phi}}{x+4+2\sqrt{\Phi}} - \frac{5}{9\sqrt{2}}\,\text{arc tg}\,\frac{1+\sqrt{\Phi}}{\sqrt{2}}.$$

$$\int\frac{dx}{\sqrt{1+x^2}\,(\sqrt{1+x^2}-x)^2} = \frac{1}{2}\,\frac{\sqrt{\Phi}+x}{\sqrt{\Phi}-x} = \frac{1}{2}(\sqrt{\Phi}+x)^2.$$

$$\int\frac{dx}{\sqrt{ax^2+b}\,\{mx-n\sqrt{ax^2+b}\}^2} = -\frac{1}{2bmn}\cdot\frac{mx+n\sqrt{\Phi}}{mx-n\sqrt{\Phi}} = -\frac{1}{bn}\cdot\frac{x}{mx-n\sqrt{\Phi}}.$$

$$\int \frac{dx}{(x-2)^{\frac{1}{3}}-5} = 3\Phi^{\frac{1}{3}}\left(\frac{1}{2}\Phi^{\frac{1}{3}}+5\right)+75\log\left(\Phi^{\frac{1}{3}}-5\right).$$

$$\int \frac{x}{1+\sqrt{1+x}}\,dx = (1+x)\left\{\frac{2}{3}\sqrt{1+x}-1\right\}.$$

$$\int \frac{x}{2+\sqrt{1+x}}\,dx = -2(1+x)+\frac{2x+20}{3}\sqrt{1+x}-12\log\left\{2+\sqrt{1+x}\right\}.$$

$$\int \frac{x}{1+x^2+a\sqrt{1+x^2}}\,dx = \log\left(a+\sqrt{1+x^2}\right).$$

$$\int \frac{x}{\sqrt{1+x^2}\left\{1-\sqrt{1+x^2}\right\}^2}\,dx = \frac{1}{1-\sqrt{1+x^2}}.$$

$$\int \frac{x}{\sqrt{x^2-a^2}\left(\sqrt{x^2-a^2}-a\right)}\,dx = \log\left\{\sqrt{x^2-a^2}-a\right\} = \frac{1}{2}\log\left\{x^2-2a\sqrt{x^2-a^2}-a^2\right\}.$$

$$\int \frac{x}{\sqrt{1+x^4}\left\{\sqrt{1+x^4}-x^2\right\}^2}\,dx = \frac{1}{4}\,\frac{\sqrt{1+x^4}+x^2}{\sqrt{1+x^4}-x^2}.$$

$$\int \frac{1+\sqrt{1+x}}{1-\sqrt{1+x}}\,dx = 6\int\left\{z^4+z^2+z+\frac{2}{3(z-1)}+\frac{z+2}{z^2+z+1}\right\}dz. \qquad 1+x=z^6.$$

$$\int \frac{x-\sqrt{1+x}}{x+\sqrt{1+x}}\,dx = 2\int\left\{z-2+2\,\frac{z-1}{z^2+z-1}\right\}dz. \qquad 1+x=z^2.$$

$$\int \frac{x+\sqrt{1-x^2}}{1-\sqrt{1-x^2}}\,dx = -\frac{1+x^2}{x}+\frac{x-1}{x}\sqrt{1-x^2}+\log\frac{x^2}{1+\sqrt{1-x^2}}-\arcsin x.$$

$$\int \frac{4x-\sqrt{1-x^2}}{5+\sqrt{1-x^2}}\,dx = -x-4\sqrt{\Phi}+20\log\left(5+\sqrt{\Phi}\right)+5\arcsin x+\frac{25}{\sqrt{24}}\,\mathrm{arc\,tg}\left\{x\sqrt{24}\,\frac{\sqrt{\Phi}-5}{24\sqrt{\Phi}+5x^2}\right\}.$$

$$\int \frac{x-\sqrt{1+x}}{1+(1+x)^{\frac{1}{4}}}\,dx = 4\int\left\{z^6-z^5-z^2+z-1+\frac{1}{1+z}\right\}dz. \qquad 1+x=z^4$$

$$\int \frac{1+(1+x)^{\frac{1}{3}}}{1-\sqrt{1+x}}\,dx = \frac{6}{5}z^5+2z^3+3z^2-4\log(z-1)+\log\sqrt{1+\frac{(2z+1)^2}{3}}+\frac{8-\sqrt{3}}{2\sqrt{3}}\,\mathrm{arc\,tg}\,\frac{2z+1}{\sqrt{3}}. \quad 1+x=z^6.$$

$$\int \frac{x+\sqrt{1+x^2}}{(1+x^2)^{\frac{3}{2}}}\,dx = -\frac{1}{\sqrt{1+x^2}}+\mathrm{arc\,tg}\,x.$$

$$\int \frac{x^m+\sqrt{1+x^{2m}}}{x\sqrt{1+x^{2m}}}\,dx = \frac{1}{m}\log\frac{x^m}{\sqrt{1+x^{2m}}-x^m}.$$

$$\int \frac{1-\sqrt{1-x^2}}{x^2\sqrt{1-x^2}}\,dx = \frac{1-\sqrt{1-x^2}}{x}.$$

$$\int \frac{\left\{x+\sqrt{1+x^2}\right\}^2}{\sqrt{1+x^2}}\,dx = \left(x+\sqrt{1+x^2}\right)\sqrt{1+x^2} = \left(x+\sqrt{1+x^2}\right)x.$$

$$\int \frac{\left\{\sqrt{1+x^2}-x\right\}^2}{\sqrt{1+x^2}}\,dx = \frac{x}{x+\sqrt{1+x^2}}.$$

$$\int \frac{\left\{\sqrt{1+x^2}+x\right\}^{\frac{m}{n}}}{\sqrt{1+x^2}}\,dx = \frac{n}{m}\left\{\sqrt{1+x^2}+x\right\}^{\frac{m}{n}}.$$

$$\int \frac{x^m\left\{\sqrt{1+x^2}+x\right\}^n}{\sqrt{1+x^2}}\,dx = \frac{1}{2^m}\int\frac{(z^2-1)^m}{z^{m-n+1}}\,dz. \qquad \sqrt{1+x^2}+x=z.$$

$$\int\sqrt{c+\sqrt{ax+b}}\,dx=\frac{4}{15a}\left(3\sqrt{ax+b}-2c\right)\left(\sqrt{ax+b}+c\right)^{\frac{3}{2}}.$$

$$\int\frac{dx}{\sqrt{a+b\sqrt{x}}}=\frac{4}{3}\cdot\frac{b\sqrt{x}-2a}{b^2}\sqrt{a+b\sqrt{x}}.$$

$$\int\frac{dx}{\sqrt{c+\sqrt{ax+b}}}=\frac{4}{3a}\left(\sqrt{ax+b}-2c\right)\sqrt{c+\sqrt{ax+b}}.$$

$$\int\frac{dx}{(1+x^{2m})\sqrt{(1+x^{2m})^{\frac{1}{m}}-x^2}}=\text{arc tg}\,\frac{x}{\sqrt{(1+x^{2m})^{\frac{1}{m}}-x^2}}.$$

$$\int\frac{x}{\sqrt{mx^2+n}\sqrt{amx^2+b\sqrt{mx^2+n}}}\,dx=\frac{1}{m}\int\frac{dz}{\sqrt{az^2+bz-an}}.\qquad mx^2+n=z^2.$$

$$\int\frac{x^{\frac{m}{2}-1}}{(1+x^{2m})\sqrt{\sqrt{1+x^{2m}}-x^m}}\,dx=\frac{2}{m}\,\text{arc tg}\,\frac{x^{\frac{m}{2}}}{\sqrt{\sqrt{1+x^{2m}}-x^m}}.$$

$$\int\frac{\sqrt{x+\sqrt{a^2+x^2}}}{x}\,dx=2\sqrt{z}-\sqrt{\frac{a}{2}}\log\frac{a+z+\sqrt{2az}}{a+z-\sqrt{2az}}-\sqrt{2a}\,\text{arc tg}\,\frac{\sqrt{2az}}{a-z}.\qquad x+\sqrt{a^2+x^2}=z.$$

$$\int\frac{\sqrt{x+\sqrt{1+x^2}}}{\sqrt{1+x^2}}\,dx=2\sqrt{x+\sqrt{1+x^2}}.$$

Function x^m

$$\int x^m\,dx=\frac{x^{m+1}}{m+1}.\qquad\int\frac{dx}{x^m}=-\frac{1}{(m-1)\,x^{m-1}}=-\frac{a+bx^{m-1}}{(m-1)\,ax^{m-1}}.$$

$$\int\frac{x^{m-1}+x^{n-m-1}}{1+x^n}\,dx=\frac{4}{n}\sum_{k=o}^{k=\frac{n}{2}-1}\text{arc tg}\,\frac{x\sin\frac{(2k+1)\pi}{n}}{1-x\cos\frac{(2k+1)\pi}{n}}\sin\frac{(2k+1)\,m\pi}{n}.\qquad\text{For } n \text{ even.}$$

$$=\frac{4}{n}\sum_{k=o}^{k=\frac{n-3}{2}}\text{arc tg}\,\frac{x\sin\frac{(2k+1)\pi}{n}}{1-x\cos\frac{(2k+1)\pi}{n}}\sin\frac{(2k+1)\,m\pi}{n}.\qquad\text{For } n \text{ odd.}$$

$$\int\frac{x^{m-1}-x^{n-m-1}}{1+x^n}\,dx=-\frac{4}{n}\sum_{k=o}^{k=\frac{n}{2}-1}\frac{1}{2}\log\left\{x^2-2x\cos\frac{(2k+1)\pi}{n}+1\right\}\cos\frac{(2k+1)\,m\pi}{n}.\qquad\text{For } n \text{ even.}$$

$$=\frac{2(-1)^{m-1}}{n}\log(1+x)-\frac{4}{n}\sum_{k=o}^{k=\frac{n-3}{2}}\frac{1}{2}\log\left\{x^2-2x\cos\frac{(2k+1)\pi}{n}+1\right\}\cos\frac{(2k+1)\,m\pi}{n}$$

For n odd.

$$\int\frac{x^{m-1}+x^{n-m-1}}{1-x^n}\,dx=-\frac{2}{n}\log(1-x)-\frac{2(-1)^m}{n}\log(1+x)-\frac{4}{n}\sum_{k=1}^{k=\frac{n}{2}-1}\frac{1}{2}\log\left\{x^2-2x\cos\frac{2k\pi}{n}+1\right\}\cos\frac{2km\pi}{n}$$

For n even.

$$\int\frac{x^{m-1}-x^{n-m-1}}{1-x^n}\,dx=\frac{4}{n}\sum_{k=1}^{k=\frac{n}{2}-1}\text{arc tg}\,\frac{x\sin\frac{2k\pi}{n}}{1-x\cos\frac{2k\pi}{n}}\cdot\sin\frac{2km\pi}{n}.\qquad\text{For } n \text{ even.}$$

$$\int\frac{x^{m-1}+x^{2n-m-1}}{x^{2n}+2x^n\cos\alpha+1}\,dx=\frac{2}{n\sin\alpha}\sum_{k=o}^{k=n-1}\text{arc tg}\,\frac{x\sin\frac{(2k+1)\pi+\alpha}{n}}{1-x\cos\frac{(2k+1)\pi+\alpha}{n}}\cos\frac{(2k+1)\,m\pi-(n-m)\,\alpha}{n}$$

$$\int \frac{x^{m-1} - x^{2n-m-1}}{x^{2n} + 2x^n \cos\alpha + 1}\,dx = \frac{2}{n \sin\alpha} \sum_{k=o}^{k=n-1} \frac{1}{2} \log\left\{x^2 - 2x\cos\frac{(2k+1)\pi + \alpha}{n} + 1\right\} \sin\frac{(2k+1)m\pi - (n-m)\alpha}{n}.$$

$$\int \frac{(n-m)\,x^n + m}{x\,(1-x^n)}\,dx = \log\frac{x^m}{1-x^n}.$$

$$\int \frac{(m+n)\,x^{2n} + m}{x\,(1+x^{2n})}\,dx = \log x^m \sqrt{x^{2n}+1}.$$

See also p. 104.

Diverse Functions

$$\int \frac{dx}{(1+\sqrt{x})\sqrt{x-x^2}} = -2\sqrt{\frac{1-\sqrt{x}}{1+\sqrt{x}}}.$$

$$\int \frac{dx}{\sqrt{(a-x)(x-b)}\,\left\{\sqrt{a-x} + m\sqrt{x-b}\right\}} = -\frac{4}{\sqrt{a-b}} \int \frac{dz}{1 - z^2 + 2mz}. \qquad x = \frac{a\,(1-z^2)^2 + 4bz^2}{(1+z^2)^2}.$$

$$\int \frac{dx}{\sqrt{(a-x)(x-b)}\,\left\{m\sqrt{a-x} + n\sqrt{x-b}\right\}} = 2\sqrt{\frac{a-b}{m^2+n^2}} \log\frac{\sqrt{(a-b)(m^2+n^2)} + m\sqrt{x-b} - n\sqrt{a-x}}{m\sqrt{a-x} + n\sqrt{x-b}}$$

$$\int \frac{x^{m-1}}{\sqrt{1-(ax^m+b)^2}}\,dx = \frac{1}{ma} \arcsin\,(ax^m + b).$$

$$\int \frac{x^{\frac{m}{p}-1}}{(1+x^m)^{1-\frac{1}{p}} \left\{(1+x^m)^{\frac{1}{p}} - x^{\frac{m}{p}}\right\}^2}\,dx = \frac{p}{2m} \frac{(1+x^m)^{\frac{1}{p}} + x^{\frac{m}{p}}}{(1+x^m)^{\frac{1}{p}} - x^{\frac{m}{p}}}.$$

$$\int \frac{x^{m-1}\,(x^m+1)}{(x^m-1)\sqrt{x^m\,(x^{2m}+1)}}\,dx = -\frac{\sqrt{2}}{m} \log\frac{\sqrt{x^{2m}+1} + \sqrt{2x^m}}{x^m - 1}.$$

$$\int \frac{x^{m-1}\,(a+bx^m)^{p-1}}{\sqrt{k^2 - (a+bx^m)^{2p}}}\,dx = \frac{1}{bmp} \arcsin\frac{(a+bx^m)^p}{k}.$$

$$\int \frac{amx^2 - anx - 2bn}{(ax+b)(mx+n)\sqrt{(ax+b)(mx+n)^2 - c^2x^2}}\,dx = \frac{2}{c} \arccos\frac{cx}{(mx+n)\sqrt{ax+b}}.$$

$$\int \frac{\sqrt{\sqrt{x} + \sqrt{x-a}}\,(\sqrt{x} - \sqrt{a})}{\sqrt{x}\,(x-a)^{\frac{3}{2}}}\,dx = \frac{2}{a^{\frac{1}{4}}} \left\{-\sqrt{\frac{2\sqrt{a}}{\sqrt{x}+\sqrt{a}}} + \arcsin\frac{\sqrt{x}-\sqrt{a}}{\sqrt{x-a}}\right\}.$$

General Formulas

$$\int f(x)\,dx = xf(o) + \frac{x^2}{1.2}f'(o) + \frac{x^3}{1.2.3}f''(o) + \cdots$$

$$\int U dx = xU - \frac{x^2}{1.2}U' + \frac{x^3}{1.2.3}U'' - \cdots$$

$$\int U dx = xU - \int xU'\,dx.$$

$$\int U'V\,dx = UV - \int UV'\,dx.$$

$$\int UV\,dx = U_1V - \int U_1V'\,dx.$$

$$\int UV\,dx = UV_1 - U'V_2 + U''V_3 - \cdots$$

$$\int x^mV\,dx = x^mV_1 - mx^{m-1}V_2 + m(m-1)x^{m-2}V_3 - \cdots$$

$$\int U^mU'\,dx = \frac{U^{m+1}}{m+1}.$$

$$\int (aU+b)^m U'\,dx = \frac{(aU+b)^{m+1}}{a(m+1)}.$$

$$\int (U+V)^m\,dx = \int U(U+V)^{m-1}\,dx + \int V(U+V)^{m-1}\,dx.$$

$$\int (U'V + UV')\,dx = UV.$$

$$\int \frac{dx}{U(x^2-a^2)} = \frac{1}{2a}\int\frac{dx}{U(x-a)} - \frac{1}{2a}\int\frac{dx}{U(x+a)}.$$

$$\int \frac{dx}{(U-x)^2} = \frac{1}{U-x} + \int\frac{U'}{(U-x)^2}\,dx.$$

$$\int \frac{dx}{U(U\pm V)} = \mp\int\frac{dx}{V(U\pm V)} \pm \int\frac{dx}{UV}.$$

$$\int \frac{U}{(U+a)(U+b)}\,dx = \frac{a}{a-b}\int\frac{dx}{U+a} - \frac{b}{a-b}\int\frac{dx}{U+b}.$$

$$\int \frac{U}{(U+V)^m}\,dx = \int\frac{dx}{(U+V)^{m-1}} - \int\frac{V}{(U+V)^m}\,dx.$$

$$\int \frac{U^{2m}}{U^{2m}-a^2}\,dx = \frac{1}{2}\int\frac{U^m}{U^m-a}\,dx + \frac{1}{2}\int\frac{U^m}{U^m+a}\,dx.$$

$$\int \frac{U^{2m}}{1-U^{2m}}\,dx = -x + \int\frac{dx}{1-U^{2m}}.$$

$$\int \frac{U'}{U}\,dx = \log U = -\frac{1}{m}\log\frac{1}{U^m}.$$

$$\int \frac{U'}{U^m}\,dx = -\frac{1}{(m-1)U^{m-1}}.$$

$$\int \frac{U'}{aU+b}\,dx = \frac{1}{a}\log(aU+b).$$

$$\int \frac{U'}{U^2+1}\,dx = \text{arc tg } U = -\text{arc tg}\frac{1}{U}.$$

$$\int \frac{U'}{a^2+U^2}\,dx = \frac{1}{a}\text{arc tg}\frac{U}{a}.$$

$$\int \frac{U'}{b^2+a^2U^2}\,dx = \frac{1}{ab}\text{arc tg}\frac{aU}{b}.$$

$$\int \frac{U'}{a^2U^2-b^2}\,dx = \frac{1}{2ab}\log\frac{aU-b}{aU+b}.$$

$$\int \frac{U'}{aU^2+bU}\,dx = \frac{1}{b}\log\frac{U}{aU+b}.$$

$$\int \frac{U'}{aU^2-bU}\,dx = \frac{1}{b}\log\frac{aU-b}{U}.$$

$$\int \frac{U'}{(aU+b)^2}\,dx = \frac{U}{b(aU+b)}.$$

$$\int \frac{U'}{(a-U)^2}\,dx = \frac{1}{a-U} = \frac{U}{a(a-U)} = \frac{1}{a-b}\,\frac{U-b}{a-U}.$$

$$\int \frac{U'}{(aU+b)(mU+n)}\,dx = \frac{1}{an-bm}\log\frac{aU+b}{mU+n}.$$

$$\int \frac{U'}{U(1+U^2)}\,dx = \log\frac{U}{\sqrt{1+U^2}}$$

$$\int \frac{U'}{(a^2+U^2)(1+U^2)}\,dx = \frac{1}{a(1-a^2)}\left\{\text{arc tg}\frac{U}{a} - a\,\text{arc tg } U\right\}.$$

$$\int \frac{U'}{\sqrt{U}}\,dx = 2\sqrt{U}.$$

$$\int \frac{U'}{\sqrt{aU+b}}\,dx = \frac{2}{a}\sqrt{aU+b}.$$

$$\int \frac{U'}{\sqrt{U^2+a}}\,dx = \log(U+\sqrt{U^2+a}).$$

$$\int \frac{U'}{\sqrt{1-U^2}}\,dx = \text{arc sin } U.$$

$$\int \frac{U'}{\sqrt{a^2-U^2}}\,dx = \text{arc sin}\frac{U}{a}.$$

$$\int \frac{U'}{\sqrt{U^2+aU}}\,dx = 2\log(\sqrt{U}+\sqrt{U+a}).$$

$$\int \frac{U'}{\sqrt{U - aU^2}} dx = -\frac{1}{\sqrt{a}} \text{arc sin } (1 - 2aU).$$

$$\int \frac{U'}{\sqrt{(U+a)(U+b)}} dx = 2 \log (\sqrt{U+a} + \sqrt{U+b}).$$

$$\int \frac{U'}{U\sqrt{U^2 + a^2}} dx = \frac{1}{a} \log \frac{U}{a + \sqrt{U^2 + a^2}}.$$

$$\int \frac{U'}{U\sqrt{a^2 - U^2}} dx = \frac{1}{a} \log \frac{U}{a + \sqrt{a^2 - U^2}}.$$

$$\int \frac{U'}{U\sqrt{2aU - a^2}} dx = \frac{1}{a} \text{arc sin } \frac{U - a}{U}.$$

$$\int \frac{U'}{U\sqrt{U^2 - 1}} dx = \text{arc séc } U.$$

$$\int \frac{U'}{(U - a^2)\sqrt{U}} dx = \frac{1}{a} \log \frac{\sqrt{U} - a}{\sqrt{U} + a}.$$

$$\int \frac{U'}{U\sqrt{U^4 - 1}} dx = -\frac{1}{2} \text{arc sin } \frac{1}{U^2}.$$

$$\int \frac{U'}{(b - U)^{\frac{3}{2}}\sqrt{U - a}} dx = \frac{2}{b - a} \sqrt{\frac{U - a}{b - U}}.$$

$$\int \frac{U'}{aU^{\frac{p}{q}} + U} dx = \frac{q}{q - p} \log \left(a + U^{\frac{q-p}{q}}\right).$$

$$\int \frac{UU'}{U^4 + 1} dx = \frac{1}{2} \text{arc tg } U^2.$$

$$\int \frac{UU'}{U^4 - 1} dx = \frac{1}{4} \log \frac{U^2 - 1}{U^2 + 1}.$$

$$\int \frac{UU'}{(1 - U^2)^2} dx = \frac{U^2}{2(1 - U^2)} = \frac{1 + U^2}{4(1 - U^2)}.$$

$$\int \frac{UU'}{\sqrt{1 - U^4}} dx = \frac{1}{2} \text{arc sin } U^2.$$

$$\int \frac{UU''}{U'^2\sqrt{1 + U'^2}} dx = -\frac{U\sqrt{1 + U'^2}}{U'} + \int \sqrt{1 + U'^2} dx.$$

$$\int \frac{U^2U'}{1 + U^2} dx = U - \text{arc tg } U.$$

$$\int \frac{U^2U'}{(1 + U^2)(a^2 + b^2U^2)} dx = \frac{1}{b^2 - a^2} \left\{ \text{arc tg } U - \frac{a}{b} \text{arc tg } \frac{b}{a} U \right\}.$$

$$\int \frac{U^{m-1}U'}{U^{2m} + 1} dx = \frac{1}{m} \text{arc tg } U^m.$$

$$\int \frac{U^{m-1}U'}{U^{2m} - a^2} dx = \frac{1}{2am} \log \frac{U^m - a}{U^m + a}.$$

$$\int \frac{U^{m-1}U'}{(a - U^m)^2} dx = \frac{1}{2am} \frac{a + U^m}{a - U^m}.$$

$$\int \frac{U'\sqrt{a^2 - U^2}}{U^2} dx = -\frac{\sqrt{a^2 - U^2}}{U} - \text{arc sin } \frac{U}{a}.$$

$$\int \frac{UU'U''}{\sqrt{1 + U'^2}} dx = U\sqrt{1 + U'^2} - \int U'\sqrt{1 + U'^2} dx.$$

$$\int \frac{VU'}{U^2} dx = -\frac{V}{U} + \int \frac{V'}{U} dx.$$

$$\int \frac{xU' + U}{(xU + a)^2} dx = \frac{1}{a - b} \frac{Ux + b}{Ux + a}.$$

$$\int \frac{U - xU'}{(aU + bx)^2} dx = \frac{1}{an - bm} \frac{mU + nx}{aU + bx}.$$

$$\int \frac{U - xU'}{(aU + bx)(mU + nx)} dx = \frac{1}{an - bm} \log \frac{mU + nx}{aU + bx}.$$

$$\int \frac{U'V - UV'}{V^2} dx = \frac{U}{V}$$

$$\int \frac{U'V - UV'}{UV} dx = \log \frac{U}{V}.$$

$$\int \frac{U'V - UV'}{U^2 + V^2} dx = \text{arc tg } \frac{U}{V}.$$

$$\int \frac{U'V - UV'}{V^2 - U^2} dx = \frac{1}{2} \log \frac{U + V}{U - V}.$$

$$\int \frac{U'V - UV'}{(U + V)^2} dx = \frac{U - V}{2(U + V)} = -\frac{V}{U + V}.$$

$$\int \frac{U'V - UV'}{(U - V)^2} dx = \frac{V + U}{2(V - U)} = \frac{V}{V - U}.$$

$$\int \frac{U'V - UV'}{V\sqrt{V^2 - U^2}} dx = \text{arc sin } \frac{U}{V}.$$

$$\int \frac{U'V + UV'}{\sqrt{U^2V^2 + a}} dx = \log \left\{ UV + \sqrt{U^2V^2 + a} \right\}.$$

$$\int \frac{2U - xU'}{(U + a^2x^2)\sqrt{U}} dx = \frac{2}{a} \text{arc tg } \frac{ax}{\sqrt{U}}.$$

$$\int \frac{2U - xU'}{(U - a^2x^2)\sqrt{U}} dx = \frac{1}{a} \log \frac{\sqrt{U} + ax}{\sqrt{U} - ax}.$$

$$\int \frac{U'V^2 + V'U^2}{UV(U + V)} dx = \log \frac{UV}{U + V}.$$

Transcendental Functions

Sin U

$$\int \sin x \, dx = -\cos x = \sin \operatorname{vers} x.$$

$$\int \sin mx \, dx = -\frac{1}{m} \cos mx \cdot$$

$$\int \sin (ax + b) \, dx = -\frac{1}{a} \cos (ax + b).$$

$$\int x \sin x \, dx = \sin x - x \cos x.$$

$$\int x^2 \sin x \, dx = 2x \sin x - (x^2 - 2) \cos x.$$

$$\int x^3 \sin x \, dx = (3x^2 - 6) \sin x - (x^3 - 6x) \cos x.$$

$$\int x^4 \sin x \, dx = (4x^3 - 24x) \sin x - (x^4 - 12x^2 + 24) \cos x.$$

$$\int x^n \sin x \, dx = -x^n \cos x + n \int x^{n-1} \cos x \, dx = -x^n \cos x + n x^{n-1} \sin x - n(n-1) \int x^{n-2} \sin x \, dx.$$

$$\int x^n \sin ax \, dx = \frac{x^{n-1}}{a^2} (n \sin ax - ax \cos ax) - \frac{n(n-1)}{a^2} \int x^{n-2} \sin ax \, dx.$$

$$\int U \sin x \, dx = -U \cos x - U' \cos \left(x + \frac{\pi}{2}\right) - U'' \cos \left(x + 2\frac{\pi}{2}\right) - \cdots - U^{(m)} \cos \left(x + m\frac{\pi}{2}\right) \cdot$$

$$\int U \sin ax \, dx = \frac{\sin ax}{a} \left\{ \frac{U'}{a} - \frac{U'''}{a^3} + \frac{U^{\text{v}}}{a^5} - \cdots \right\} - \frac{\cos ax}{a} \left\{ U - \frac{U''}{a^2} + \frac{U^{\text{IV}}}{a^4} - \cdots \right\} \cdot$$

$$\int \frac{dx}{\sin x} = \log \operatorname{tg} \frac{x}{2} = -\frac{1}{2} \log \frac{1 + \cos x}{1 - \cos x} = \log \sqrt{\frac{1 - \cos x}{1 + \cos x}} \cdot$$

$$\int \frac{dx}{\sin 2x} = \frac{1}{2} \log \operatorname{tg} x.$$

$$\int \frac{dx}{\sin mx} = \frac{1}{2m} \log \frac{1 - \cos mx}{1 + \cos mx}$$

$$\int \frac{dx}{\sin (x - \alpha)} = \log \operatorname{tg} \frac{1}{2} (x - \alpha).$$

$$\int \frac{dx}{\sin (ax + b)} = \frac{1}{a} \log \operatorname{tg} \frac{1}{2} (ax + b).$$

$$\int \frac{dx}{\sin \alpha + \sin x} = \frac{1}{\cos \alpha} \log \frac{\sin \dfrac{x + \alpha}{2}}{\cos \dfrac{x - \alpha}{2}} \cdot$$

$$\int \frac{dx}{1 + \sin \alpha \sin x} = \frac{2}{\cos \alpha} \operatorname{arc} \operatorname{tg} \left\{ \operatorname{tg} \left(\frac{\pi}{4} + \frac{x}{2}\right) \operatorname{tg} \left(\frac{\pi}{4} + \frac{\alpha}{2}\right) \right\} \cdot$$

$$\int \frac{U'}{\sqrt{U - aU^2}}\,dx = -\frac{1}{\sqrt{a}} \text{ arc sin } (1 - 2aU).$$

$$\int \frac{U'}{\sqrt{(U+a)(U+b)}}\,dx = 2\log\left(\sqrt{U+a} + \sqrt{U+b}\right).$$

$$\int \frac{U'}{U\sqrt{U^2 + a^2}}\,dx = \frac{1}{a}\log\frac{U}{a + \sqrt{U^2 + a^2}}.$$

$$\int \frac{U'}{U\sqrt{a^2 - U^2}}\,dx = \frac{1}{a}\log\frac{U}{a + \sqrt{a^2 - U^2}}.$$

$$\int \frac{U'}{U\sqrt{2aU - a^2}}\,dx = \frac{1}{a} \text{ arc sin } \frac{U-a}{U}.$$

$$\int \frac{U'}{U\sqrt{U^2 - 1}}\,dx = \text{arc séc } U.$$

$$\int \frac{U'}{(U - a^2)\sqrt{U}}\,dx = \frac{1}{a}\log\frac{\sqrt{U} - a}{\sqrt{U} + a}.$$

$$\int \frac{U'}{U\sqrt{U^4 - 1}}\,dx = -\frac{1}{2} \text{ arc sin } \frac{1}{U^2}.$$

$$\int \frac{U'}{(b - U)^{\frac{3}{2}}\sqrt{U - a}}\,dx = \frac{2}{b-a}\sqrt{\frac{U-a}{b-U}}.$$

$$\int \frac{U'}{aU^{\frac{p}{q}} + U}\,dx = \frac{q}{q-p}\log\left(a + U^{\frac{q-p}{q}}\right).$$

$$\int \frac{UU'}{U^4 + 1}\,dx = \frac{1}{2} \text{ arc tg } U^2.$$

$$\int \frac{UU'}{U^4 - 1}\,dx = \frac{1}{4}\log\frac{U^2 - 1}{U^2 + 1}.$$

$$\int \frac{UU'}{(1 - U^2)^2}\,dx = \frac{U^2}{2(1 - U^2)} = \frac{1 + U^2}{4(1 - U^2)}.$$

$$\int \frac{UU'}{\sqrt{1 - U^4}}\,dx = \frac{1}{2} \text{ arc sin } U^2.$$

$$\int \frac{UU''}{U'^2\sqrt{1 + U'^2}}\,dx = -\frac{U\sqrt{1 + U'^2}}{U'} + \int \sqrt{1 + U'^2}\,dx.$$

$$\int \frac{U^2U'}{1 + U^2}\,dx = U - \text{arc tg } U.$$

$$\int \frac{U^2U'}{(1 + U^2)(a^2 + b^2U^2)}\,dx = \frac{1}{b^2 - a^2}\left\{\text{arc tg } U - \frac{a}{b} \text{ arc tg } \frac{b}{a} U\right\}.$$

$$\int \frac{U^{m-1}U'}{U^{2m} + 1}\,dx = \frac{1}{m} \text{ arc tg } U^m.$$

$$\int \frac{U^{m-1}U'}{U^{2m} - a^2}\,dx = \frac{1}{2am}\log\frac{U^m - a}{U^m + a}.$$

$$\int \frac{U^{m-1}U'}{(a - U^m)^2}\,dx = \frac{1}{2am}\,\frac{a + U^m}{a - U^m}.$$

$$\int \frac{U'\sqrt{a^2 - U^2}}{U^2}\,dx = -\frac{\sqrt{a^2 - U^2}}{U} - \text{arc sin } \frac{U}{a}.$$

$$\int \frac{UU'U''}{\sqrt{1 + U'^2}}\,dx = U\sqrt{1 + U'^2} - \int U'\sqrt{1 + U'^2}\,dx.$$

$$\int \frac{VU'}{U^2}\,dx = -\frac{V}{U} + \int \frac{V'}{U}\,dx.$$

$$\int \frac{xU' + U}{(xU + a)^2}\,dx = \frac{1}{a-b}\,\frac{Ux + b}{Ux + a}.$$

$$\int \frac{U - xU'}{(aU + bx)^2}\,dx = \frac{1}{an - bm}\,\frac{mU + nx}{aU + bx}.$$

$$\int \frac{U - xU'}{(aU + bx)(mU + nx)}\,dx = \frac{1}{an - bm}\log\frac{mU + nx}{aU + bx}.$$

$$\int \frac{U'V - UV'}{V^2}\,dx = \frac{U}{V}$$

$$\int \frac{U'V - UV'}{UV}\,dx = \log\frac{U}{V}.$$

$$\int \frac{U'V - UV'}{U^2 + V^2}\,dx = \text{arc tg } \frac{U}{V}.$$

$$\int \frac{U'V - UV'}{V^2 - U^2}\,dx = \frac{1}{2}\log\frac{U + V}{U - V}.$$

$$\int \frac{U'V - UV'}{(U + V)^2}\,dx = \frac{U - V}{2(U + V)} = -\frac{V}{U + V}.$$

$$\int \frac{U'V - UV'}{(U - V)^2}\,dx = \frac{V + U}{2(V - U)} = \frac{V}{V - U}.$$

$$\int \frac{U'V - UV'}{V\sqrt{V^2 - U^2}}\,dx = \text{arc sin } \frac{U}{V}.$$

$$\int \frac{U'V + UV'}{\sqrt{U^2V^2 + a}}\,dx = \log\left\{UV + \sqrt{U^2V^2 + a}\right\}.$$

$$\int \frac{2U - xU'}{(U + a^2x^2)\sqrt{U}}\,dx = \frac{2}{a} \text{ arc tg } \frac{ax}{\sqrt{U}}.$$

$$\int \frac{2U - xU'}{(U - a^2x^2)\sqrt{U}}\,dx = \frac{1}{a}\log\frac{\sqrt{U} + ax}{\sqrt{U} - ax}.$$

$$\int \frac{U'V^2 + V'U^2}{UV(U + V)}\,dx = \log\frac{UV}{U + V}.$$

Transcendental Functions

Sin U

$$\int \sin x\, dx = -\cos x = \sin \text{vers}\, x.$$

$$\int \sin mx\, dx = -\frac{1}{m}\cos mx \cdot$$

$$\int \sin (ax+b)\, dx = -\frac{1}{a}\cos (ax+b).$$

$$\int x \sin x\, dx = \sin x - x\cos x.$$

$$\int x^2 \sin x\, dx = 2x \sin x - (x^2-2)\cos x.$$

$$\int x^3 \sin x\, dx = (3x^2-6)\sin x - (x^3-6x)\cos x.$$

$$\int x^4 \sin x\, dx = (4x^3-24x)\sin x - (x^4-12x^2+24)\cos x.$$

$$\int x^n \sin x\, dx = -x^n\cos x + n\int x^{n-1}\cos x\, dx = -x^n\cos x + nx^{n-1}\sin x - n(n-1)\int x^{n-2}\sin x\, dx.$$

$$\int x^n \sin ax\, dx = \frac{x^{n-1}}{a^2}(n\sin ax - ax\cos ax) - \frac{n(n-1)}{a^2}\int x^{n-2}\sin ax\, dx.$$

$$\int \text{U} \sin x\, dx = -\text{U}\cos x - \text{U}'\cos\left(x+\frac{\pi}{2}\right) - \text{U}''\cos\left(x+2\frac{\pi}{2}\right) - \cdots - \text{U}^{(m)}\cos\left(x+m\frac{\pi}{2}\right)\cdot$$

$$\int \text{U} \sin ax\, dx = \frac{\sin ax}{a}\left\{\frac{\text{U}'}{a} - \frac{\text{U}'''}{a^3} + \frac{\text{U}^{\text{v}}}{a^5} - \cdots\right\} - \frac{\cos ax}{a}\left\{\text{U} - \frac{\text{U}''}{a^2} + \frac{\text{U}^{\text{IV}}}{a^4} - \cdots\right\}\cdot$$

$$\int \frac{dx}{\sin x} = \log \text{tg}\,\frac{x}{2} = -\frac{1}{2}\log\frac{1+\cos x}{1-\cos x} = \log\sqrt{\frac{1-\cos x}{1+\cos x}}\cdot$$

$$\int \frac{dx}{\sin 2x} = \frac{1}{2}\log \text{tg}\, x.$$

$$\int \frac{dx}{\sin mx} = \frac{1}{2m}\log\frac{1-\cos mx}{1+\cos mx}$$

$$\int \frac{dx}{\sin (x-\alpha)} = \log \text{tg}\,\frac{1}{2}(x-\alpha).$$

$$\int \frac{dx}{\sin (ax+b)} = \frac{1}{a}\log \text{tg}\,\frac{1}{2}(ax+b).$$

$$\int \frac{dx}{\sin\alpha + \sin x} = \frac{1}{\cos\alpha}\log\frac{\sin\frac{x+\alpha}{2}}{\cos\frac{x-\alpha}{2}}\cdot$$

$$\int \frac{dx}{1+\sin\alpha\sin x} = \frac{2}{\cos\alpha}\text{arc tg}\left\{\text{tg}\left(\frac{\pi}{4}+\frac{x}{2}\right)\text{tg}\left(\frac{\pi}{4}+\frac{\alpha}{2}\right)\right\}\cdot$$

$$\int \frac{dx}{1+\cos\alpha\sin x} = \frac{2}{\sin\alpha}\,\text{arc tg}\left\{\frac{\cos\alpha + \text{tg}\,\frac{x}{2}}{\sin\alpha}\right\}.$$

$$\int \frac{dx}{5+4\sin x} = \frac{2}{3}\,\text{arc tg}\,\frac{5\,\text{tg}\,\frac{x}{2}+4}{3}$$

$$\int \frac{dx}{4+5\sin x} = \frac{1}{3}\log\frac{2\,\text{tg}\,\frac{x}{2}+1}{2\,\text{tg}\,\frac{x}{2}+4}.$$

$$\int \frac{dx}{a+b\sin x} = \frac{2}{\sqrt{a^2-b^2}}\,\text{arc tg}\left\{\frac{b+a\,\text{tg}\,\frac{x}{2}}{\sqrt{a^2-b^2}}\right\} = \frac{1}{\sqrt{a^2-b^2}}\,\text{arc sin}\,\frac{b+a\sin x}{a+b\sin x}. \qquad a > b.$$

$$= \frac{1}{\sqrt{b^2-a^2}}\log\frac{b+a\,\text{tg}\,\frac{x}{2}-\sqrt{b^2-a^2}}{b+a\,\text{tg}\,\frac{x}{2}+\sqrt{b^2-a^2}}. \qquad a < b.$$

$$\int \frac{\sin x}{x^2}\,dx = -\frac{\sin x}{x} + \int \frac{\cos x}{x}\,dx.$$

$$\int \frac{\sin x}{x^3}\,dx = -\frac{\sin x}{2x^2} - \frac{\cos x}{2x} - \frac{1}{2}\int \frac{\sin x}{x}\,dx.$$

$$\int \frac{\sin x}{x^4}\,dx = -\frac{\sin x}{3x^3} - \frac{\cos x}{6x^2} + \frac{\sin x}{6x} - \frac{1}{6}\int \frac{\cos x}{x}\,dx.$$

$$\int \frac{\sin x}{x^m}\,dx = -\frac{\sin x}{(m-1)\,x^{m-1}} + \frac{1}{m-1}\int \frac{\cos x}{x^{m-1}}\,dx.$$

$$= -\frac{\sin x}{(m-1)\,x^{m-1}} - \frac{\cos x}{(m-1)(m-2)\,x^{m-2}} - \frac{1}{(m-1)(m-2)}\int \frac{\sin x}{x^{m-2}}\,dx.$$

Sin^n U

$$\int \text{Sin}^2 x\,dx = -\frac{1}{2}(\sin x\cos x - x) = \frac{x}{2} - \frac{1}{4}\sin 2x.$$

$$\int \text{Sin}^2 mx\,dx = \frac{x}{2} - \frac{\sin 2mx}{4m}.$$

$$\int x\sin^2 x\,dx = \frac{1}{4}(x^2 + x\cos 2x - \sin x\cos x) = \frac{1}{4}\left(x^2 - x\sin 2x - \frac{1}{2}\cos 2x\right).$$

$$\int \frac{dx}{\sin^2 x} = -\text{cotg}\,x.$$

$$\int \frac{dx}{\sin^2 nx} = -\frac{1}{n}\,\text{cotg}\,nx.$$

$$\int \frac{dx}{\sin^2(px+q)} = -\frac{1}{p}\,\text{cotg}\,(px+q).$$

$$\int \frac{dx}{2-\sin^2 x} = \frac{1}{\sqrt{2}}\,\text{arc tg}\left(\frac{\text{tg}\,x}{\sqrt{2}}\right).$$

$$\int \frac{dx}{1-a^2\sin^2 x} = \frac{1}{\sqrt{1-a^2}}\,\text{arc tg}\,(\sqrt{1-a^2}\,\text{tg}\,x).$$

$$\int \frac{dx}{a+b\sin^2 x} = \frac{1}{\sqrt{a^2+ab}}\,\text{arc tg}\,\frac{\sqrt{a+b}}{\text{cotg}\,x\sqrt{a}}.$$

$$\int \frac{dx}{(a+b\sin x)^2} = \frac{b\cos x}{(a^2-b^2)(a+b\sin x)} + \frac{a}{a^2-b^2}\int\frac{dx}{a+b\sin x}.$$

$$= \frac{1}{a^2-b^2}\left\{\frac{b\cos x}{a+b\sin x} + \frac{2a}{\sqrt{a^2-b^2}}\operatorname{arc\,tg}\left(\frac{a\operatorname{tg}\frac{x}{2}+b}{\sqrt{a^2-b^2}}\right)\right\}. \qquad a > b.$$

$$= -\frac{1}{b^2-a^2}\left\{\frac{b\cos x}{a+b\sin x} + \frac{a}{\sqrt{b^2-a^2}}\log\frac{a\operatorname{tg}\frac{x}{2}+b-\sqrt{b^2-a^2}}{a\operatorname{tg}\frac{x}{2}+b+\sqrt{b^2-a^2}}\right\}. \qquad a < b.$$

$$\int\frac{x}{\sin^2 x}\,dx = -x\operatorname{cotg}x + \log\sin x.$$

$$\int\sin^3 x\,dx = \frac{1}{3}\cos^3 x - \cos x = \frac{1}{4}\left(\frac{\cos 3x}{3} - 3\cos x\right) = -\frac{\cos x}{3}(\sin^2 x + 2).$$

$$\int\frac{dx}{\sin^3 x} = -\frac{\cos x}{2\sin^2 x} + \frac{1}{2}\log\operatorname{tg}\frac{x}{2}.$$

$$\int\frac{x}{\sin^3 x}\,dx = -\frac{\sin x + x\cos x}{2\sin^2 x} + \frac{1}{2}\int\frac{x}{\sin x}\,dx.$$

$$\int\sin^4 x\,dx = \frac{1}{32}\sin 4x - \frac{1}{4}\sin 2x + \frac{3}{8}x = -\frac{\cos x}{4}\left(\sin^3 x + \frac{3}{2}\sin x\right) + \frac{3}{8}x.$$

$$\int\frac{dx}{\sin^4 x} = -\frac{\cos x}{3\sin^3 x} - \frac{2}{3}\operatorname{cotg}x = \frac{\cos x}{3\sin^3 x}(1 + 2\sin^2 x).$$

$$\int\frac{x}{\sin^4 x}\,dx = -\frac{\sin x + 2x\cos x}{6\sin^3 x} - \frac{2}{3}(x\operatorname{cotg}x - \log\sin x).$$

$$\int\sin^n x\,dx = -\frac{\sin^{n-1}x\cos x}{n} + \frac{n-1}{n}\int\sin^{n-2}x\,dx.$$

$$\int\sin^{2n}x\,dx = -\frac{\cos x}{2n}\left\{\sin^{2n-1}x + \frac{2n-1}{2n-2}\sin^{2n-3}x + \frac{(2n-1)(2n-3)}{(2n-2)(2n-4)}\sin^{2n-5}x + \cdots \frac{(2n-1)\cdots 3.1}{(2n-2)\cdots 4.2}\sin x\right\}$$

$$+ \frac{(2n-1)\cdots 3.1}{2n\cdots 4.2}x.$$

$$= \frac{(-1)^n}{2^{2n-1}}\left\{\frac{\sin 2nx}{2n} - 2n\frac{\sin(2n-2)x}{2n-2} + \frac{2n(2n-1)}{1.2}\frac{\sin(2n-4)x}{2n-4} - \cdots\right.$$

$$\left. + (-1)^{n-1}\frac{2n(2n-1)\cdots(n+2)}{(n-1)\cdots 2.1}\frac{\sin 2x}{2} + (-1)^n\frac{2n(2n-1)\cdots(n+1)}{n\cdots 2.1}\cdot\frac{x}{2}\right\}.$$

$$\int\sin^{2n+1}x\,dx = -\frac{\cos x}{2n+1}\left\{\sin^{2n}x + \frac{2n}{2n-1}\sin^{2n-2}x + \frac{2n(2n-2)}{(2n-1)(2n-3)}\sin^{2n-4}x + \cdots + \frac{2n\cdots 4.2}{(2n-1)\cdots 3.1}\right\}.$$

$$= \frac{(-1)^{n+1}}{2^{2n}}\left\{\frac{\cos(2n+1)x}{2n+1} - (2n+1)\frac{\cos(2n-1)x}{2n-1} + \cdots + (-1)^n\frac{(2n+1)\cdots(n+2)}{n\cdots 2.1}\cos x\right\}.$$

$$\int\frac{dx}{\sin^n x} = -\frac{\cos x}{(n-1)\sin^{n-1}x} + \frac{n-2}{n-1}\int\frac{dx}{\sin^{n-2}x}.$$

$$\int\frac{dx}{\sin^{2n+2}x} = -\operatorname{cotg}x - \frac{n_1}{3}\operatorname{cotg}^3 x - \frac{n_2}{5}\operatorname{cotg}^5 x - \cdots - \frac{1}{2n+1}\operatorname{cotg}^{2n+1}x.$$

$$\int\frac{dx}{(a+b\sin x)^n} = \frac{b\cos x}{(n-1)(a^2-b^2)\Phi^{n-1}} + \frac{(2n-3)a}{(n-1)(a^2-b^2)}\int\frac{dx}{\Phi^{n-1}} - \frac{n-2}{(n-1)(a^2-b^2)}\int\frac{dx}{\Phi^{n-2}}.$$

$$\int \frac{x}{\sin^n x}\,dx = -\frac{\sin x + (n-2)\,x\cos x}{(n-2)(n-1)\sin^{n-1} x} + \frac{n-2}{n-1}\int \frac{x}{\sin^{n-2} x}\,dx.$$

$\mathrm{Sin}^n\,\mathrm{U}\ \mathrm{Sin}^m\,\mathrm{V}$

$$\int \sin px \sin nx\,dx = \frac{1}{2}\left\{\frac{\sin(p-n)\,x}{p-n} - \frac{\sin(p+n)\,x}{p+n}\right\}.$$

$$\int \sin(ax+\alpha)\sin(ax+\beta)\,dx = \frac{x}{2}\cos(\alpha-\beta) - \frac{\sin(2ax+\alpha+\beta)}{4a}.$$

$$\int \sin(ax+\alpha)\sin(bx+\beta)\,dx = \frac{\sin\{(a-b)\,x+\alpha-\beta\}}{2(a-b)} - \frac{\sin\{(a+b)\,x+\alpha+\beta\}}{2(a+b)}.$$

$$\int \sin^{n-1} x \sin(n+1)\,dx = \frac{1}{n}\sin nx \sin^n x.$$

$$\int \sin^p x \sin mx\,dx = -\frac{\sin^p x \cos mx}{m+p} + \frac{p}{m+p}\int \sin^{p-1} x \cos(m-1)\,x\,dx.$$

$$= \frac{\sin^{p-1} x}{p^2-m^2}(m \sin x \cos mx - p\cos x \sin mx) + \frac{p(p-1)}{p^2-m^2}\int \sin mx \sin^{p-2} x\,dx.$$

$$\int \sin x \sin 2x \sin 3x\,dx = \frac{1}{8}\left\{\cos 2x + \frac{1}{2}\cos 4x - \frac{1}{3}\cos 6x\right\}.$$

$$\int \frac{\sin x}{\sin 2x}\,dx = \frac{1}{2}\log \operatorname{tg}\left(\frac{\pi}{4}+\frac{x}{2}\right)$$

$$\int \frac{\sin x}{\sin 3x}\,dx = \frac{1}{2\sqrt{3}}\log \frac{\sin\left(\frac{\pi}{3}+x\right)}{\sin\left(\frac{\pi}{3}-x\right)}.$$

$$\int \frac{\sin 2x}{\sin x}\,dx = 2\sin x.$$

$$\int \frac{\sin 2x}{\sin^2 x}\,dx = 2\log \sin x.$$

$$\int \frac{\sin 2x}{\sin^3 x}\,dx = -\frac{2}{\sin x}.$$

$$\int \frac{\sin 2x}{\sin^4 x}\,dx = -\frac{1}{\sin^2 x}.$$

$$\int \frac{\sin 2x}{\sin^n x}\,dx = -\frac{2}{(n-2)\sin^{n-2} x}.$$

$$\int \frac{\sin 3x}{\sin x}\,dx = x + 2\sin x \cos x.$$

$$\int \frac{\sin 3x}{\sin 4x}\,dx = \frac{1}{8}\log\frac{1+\sin x}{1-\sin x} + \frac{1}{4\sqrt{2}}\log\frac{1+\sqrt{2}\sin x}{1-\sqrt{2}\sin x}.$$

$$\int \frac{\sin 3x}{\sin^2 x}\,dx = 3\log \operatorname{tg}\frac{x}{2} + 4\cos x.$$

$$\int \frac{\sin 3x}{\sin^3 x}\,dx = -3\operatorname{cotg} x - 4x.$$

$$\int \frac{\sin 3x}{\sin^4 x}\, dx = -\frac{3 \operatorname{cotg} x}{2 \sin x} - \frac{5}{2} \log \operatorname{tg} \frac{x}{2}.$$

$$\int \frac{\sin ax}{x^m}\, dx = -\frac{ax \cos ax + (m-2) \sin ax}{(m-1)(m-2)\, x^{m-1}} - \frac{a^2}{(m-1)(m-2)} \int \frac{\sin ax}{x^{m-2}}\, dx.$$

$$\int \frac{\sin mx}{\sin^n x}\, dx = 2 \int \frac{\cos (m-1)\, x}{\sin^{n-1} x}\, dx + \int \frac{\sin (m-2)\, x}{\sin^n x}\, dx.$$

$$\int \frac{\sin (x+a)}{\sin x}\, dx = x \cos a + \sin a \log \sin x.$$

$$\int \frac{\sin^2 (x+a)}{\sin x}\, dx = -\cos^2 a \cos x + \sin 2a \sin x + \sin^2 a \left\{ \log \operatorname{tg} \frac{x}{2} + \cos x \right\}.$$

Cos U

$$\int \cos x\, dx = \sin x.$$

$$\int \cos mx\, dx = \frac{1}{m} \sin mx.$$

$$\int \cos (px + q)\, dx = \frac{1}{p} \sin (px + q).$$

$$\int x \cos x\, dx = \cos x + x \sin x.$$

$$\int x \cos ax\, dx = \frac{1}{a} x \sin ax + \frac{1}{a^2} \cos ax.$$

$$\int x^2 \cos x\, dx = 2x \cos x + (x^2 - 2) \sin x.$$

$$\int x^2 \cos ax\, dx = \frac{x^2 \sin ax}{a} + \frac{2x \cos ax}{a^2} - \frac{2 \sin ax}{a^3}.$$

$$\int x^3 \cos x\, dx = (3x^2 - 6) \cos x + (x^3 - 6x) \sin x.$$

$$\int x^4 \cos x\, dx = (4x^3 - 24x) \cos x + (x^4 - 12x^2 + 24) \sin x.$$

$$\int x^4 \cos ax\, dx = (4a^2x^2 - 24) \frac{x \cos ax}{a^4} + (a^4x^4 - 12a^2x^2 + 24) \frac{\sin ax}{a^5}.$$

$$\int x^m \cos x\, dx = x^m \sin x - m \int x^{m-1} \sin x\, dx.$$

$$= x^m \sin x + m x^{m-1} \cos x - m (m-1) \int x^{m-2} \cos x\, dx.$$

$$= \sin x \left\{ x^m - m (m-1)\, x^{m-2} + m (m-1)(m-2)(m-3)\, x^{m-4} - \cdots \right\}$$

$$+ \cos x \left\{ m x^{m-1} - m (m-1)(m-2)\, x^{m-2} + \cdots \right\}.$$

$$\int x^m \cos ax\, dx = \frac{x^m \sin ax}{a} - \frac{m}{a}\int x^{m-1} \sin ax\, dx.$$

$$= \frac{x^{m-1}}{a^2}(m \cos ax + ax \sin ax) - \frac{m(m-1)}{a^2}\int x^{m-2} \cos ax\, dx.$$

$$\int U \cos x\, dx = \sin x \left\{ U - U'' + U^{IV} - \dots \right\} + \cos x \left\{ U' - U''' + U^{V} - \dots \right\}.$$

$$\int U \cos ax\, dx = \frac{\sin ax}{a}\left\{ U - \frac{U''}{a^2} + \frac{U^{IV}}{a^4} - \dots \right\} + \frac{\cos ax}{a}\left\{ \frac{U'}{a} - \frac{U'''}{a^3} + \frac{U^{V}}{a^5} - \dots \right\}$$

$$\int \frac{dx}{\cos x} = \log \operatorname{tg}\left(\frac{\pi}{4} + \frac{x}{2}\right) = \frac{1}{2}\log\frac{1 + \sin x}{1 - \sin x}.$$

$$\int \frac{dx}{\cos 2x} = \frac{1}{2}\log \operatorname{tg}\left(\frac{\pi}{4} + x\right).$$

$$\int \frac{dx}{1 + \cos x} = \operatorname{tg}\frac{x}{2} = \frac{\sin x}{1 + \cos x}.$$

$$\int \frac{dx}{1 - \cos x} = -\operatorname{cotg}\frac{x}{2}.$$

$$\int \frac{dx}{2 + \cos x} = \frac{2}{\sqrt{3}} \operatorname{arc\,tg}\left(\frac{1}{\sqrt{3}}\operatorname{tg}\frac{x}{2}\right) = \frac{1}{\sqrt{3}}\operatorname{arc\,cos}\frac{1 + 2\cos x}{2 + \cos x}.$$

$$\int \frac{dx}{\cos x + \cos \alpha} = \frac{1}{\sin \alpha}\log\frac{\cos\frac{x-\alpha}{2}}{\cos\frac{x+\alpha}{2}} = \frac{1}{\sin \alpha}\log\frac{1 + \cos(x-\alpha)}{\cos\alpha + \cos x}.$$

$$\int \frac{dx}{\cos x - \cos \alpha} = \frac{1}{\sin \alpha}\log\frac{\sin\frac{\alpha+x}{2}}{\sin\frac{\alpha-x}{2}}.$$

$$\int \frac{dx}{1 + \cos\alpha\cos x} = \frac{2}{\sin\alpha}\operatorname{arc\,tg}\left(\operatorname{tg}\frac{x}{2}\cdot\operatorname{tg}\frac{\alpha}{2}\right) = \frac{1}{\sin\alpha}\operatorname{arc\,tg}\frac{\sin\alpha\sin x}{\cos\alpha + \cos x}.$$

$$\int \frac{dx}{1 + e\cos x} = \frac{2}{\sqrt{1-e^2}}\operatorname{arc\,tg}\left\{\sqrt{\frac{1-e}{1+e}}\cdot\operatorname{tg}\frac{x}{2}\right\}.$$

$$\int \frac{dx}{1 + 2\cos x} = \frac{1}{\sqrt{3}}\log\frac{\sqrt{3} + \operatorname{tg}\frac{x}{2}}{\sqrt{3} - \operatorname{tg}\frac{x}{2}}$$

$$\int \frac{dx}{5 + 4\cos x} = \frac{2}{3}\operatorname{arc\,tg}\left(\frac{1}{3}\operatorname{tg}\frac{x}{2}\right).$$

$$\int \frac{dx}{4 + 5\cos x} = \frac{1}{3}\log\frac{\operatorname{tg}\frac{x}{2} + 3}{\operatorname{tg}\frac{x}{2} - 3}.$$

$$\int \frac{dx}{a^2 + b^2 + 2ab\cos x} = \frac{2}{a^2 - b^2}\operatorname{arc\,tg}\left\{\frac{a-b}{a+b}\operatorname{tg}\frac{x}{2}\right\}.$$

$$\int \frac{dx}{a + b\cos x} = \frac{2}{\sqrt{a^2-b^2}}\operatorname{arc\,tg}\left\{\sqrt{\frac{a-b}{a+b}}\operatorname{tg}\frac{x}{2}\right\} = \frac{1}{\sqrt{a^2-b^2}}\operatorname{arc\,tg}\frac{\sqrt{a^2-b^2}\sin x}{b + a\cos x} = \frac{2}{\sqrt{a^2-b^2}}\operatorname{arc\,tg}\frac{\sqrt{(a-b)(1-\cos x)}}{\sqrt{(a+b)(1+\cos x)}}.$$

$$= \frac{1}{\sqrt{a^2-b^2}}\operatorname{arc\,sin}\frac{\sqrt{a^2-b^2}\sin x}{a + b\cos x} = \frac{1}{\sqrt{a^2-b^2}}\operatorname{arc\,cos}\frac{b + a\cos x}{a + b\cos x}. \qquad a > b.$$

$$= \frac{1}{\sqrt{b^2-a^2}}\log\frac{b + a\cos x + \sqrt{b^2-a^2}\sin x}{a + b\cos x} = \frac{1}{\sqrt{b^2-a^2}}\log\frac{\sqrt{b+a} + \sqrt{b-a}\operatorname{tg}\frac{x}{2}}{\sqrt{b+a} - \sqrt{b-a}\operatorname{tg}\frac{x}{2}}$$

$$= \frac{1}{\sqrt{b^2 - a^2}} \log \frac{\sqrt{(b+a)(1+\cos x)} + \sqrt{(b-a)(1-\cos x)}}{\sqrt{(b+a)(1+\cos x)} - \sqrt{(b-a)(1-\cos x)}}. \qquad a < b.$$

$$= \frac{1}{a} \operatorname{tg} \frac{x}{2}. \qquad a = b.$$

$$\int \frac{\cos x}{x^2}\, dx = -\frac{\cos x}{x} - \int \frac{\sin x}{x}\, dx.$$

$$\int \frac{\cos x}{x^3}\, dx = -\frac{\cos x}{2x^2} + \frac{\sin x}{2x} - \frac{1}{2} \int \frac{\cos x}{x}\, dx.$$

$$\int \frac{\cos x}{x^4}\, dx = -\frac{\cos x}{3x^3} + \frac{\sin x}{6x^2} + \frac{\cos x}{6x} + \frac{1}{6} \int \frac{\sin x}{x}\, dx.$$

$$\int \frac{\cos x}{x^m}\, dx = -\frac{\cos x}{(m-1)\, x^{m-1}} - \frac{1}{m-1} \int \frac{\sin x}{x^{m-1}}\, dx.$$

$$\int \frac{\cos ax}{x^m}\, dx = \frac{ax \sin ax - (m-2) \cos ax}{(m-1)(m-2)\, x^{m-1}} - \frac{a^2}{(m-1)(m-2)} \int \frac{\cos ax}{x^{m-2}}\, dx.$$

$$\int \frac{\cos x}{2 + \cos x}\, dx = x - \frac{4}{\sqrt{3}} \operatorname{arc\,tg} \left(\frac{1}{\sqrt{3}} \operatorname{tg} \frac{x}{2} \right).$$

$$\int \frac{\cos x}{a + b \cos x}\, dx = \frac{x}{b} - \frac{a}{b} \int \frac{dx}{a + b \cos x}.$$

$$\int \frac{1 - a \cos x}{1 - 2a \cos x + a^2}\, dx = \frac{x}{2} + \operatorname{arc\,tg} \left\{ \frac{1+a}{1-a} \operatorname{tg} \frac{x}{2} \right\}.$$

$\cos^n U$

$$\int \cos^2 x\, dx = \frac{1}{2} (\sin x \cos x + x) = \frac{1}{2} \left(x + \frac{\sin 2x}{2} \right).$$

$$\int \cos^2 mx\, dx = \frac{x}{2} + \frac{\sin 2m\, x}{4m}.$$

$$\int (1 - \cos x)^2\, dx = \frac{3x}{2} - 2 \sin x + \frac{\sin 2x}{4}.$$

$$\int \frac{dx}{\cos^2 x} = \operatorname{tg} x.$$

$$\int \frac{dx}{\cos^2 nx} = \frac{1}{n} \operatorname{tg} nx.$$

$$\int \frac{dx}{\cos^2 (px + q)} = \frac{1}{p} \operatorname{tg} (px + q).$$

$$\int \frac{dx}{1 + \cos^2 x} = \frac{1}{\sqrt{2}} \operatorname{arc\,tg} \left(\frac{1}{\sqrt{2}} \operatorname{tg} x \right).$$

$$\int \frac{dx}{a + b \cos^2 x} = \frac{1}{\sqrt{a(a+b)}} \operatorname{arc\,tg} \frac{a \operatorname{tg} x}{\sqrt{a(a+b)}}. \qquad a(a+b) > 0.$$

$$= \frac{1}{2\sqrt{-a(a+b)}} \log \frac{a \operatorname{tg} x - \sqrt{-a(a+b)}}{a \operatorname{tg} x + \sqrt{-a(a+b)}}. \qquad a(a+b) < 0.$$

$$\int \frac{dx}{a^2 - b^2 \cos^2 x} = \frac{1}{2ab \sin \alpha} \log \frac{\sin(\alpha - x)}{\sin(\alpha + x)}. \qquad a = b \cos \alpha.$$

$$= \frac{1}{a^2 \sin^2 \beta} \operatorname{arc\,tg} \left(\frac{\operatorname{tg} x}{\sin \beta}\right). \qquad b = a \cos \beta.$$

$$\int \frac{dx}{(1 + e \cos x)^2} = \frac{2}{(1 - e)^{\frac{3}{2}}} \operatorname{arc\,tg} \left\{ \sqrt{\frac{1 - e}{1 + e}} \operatorname{tg} \frac{x}{2} \right\} - \frac{e}{1 - e^2} \frac{\sin x}{1 + e \cos x}.$$

$$\int \frac{dx}{(1 + e \cos mx)^2} = \frac{1}{m(1 - e^2)^{\frac{3}{2}}} \operatorname{arc\,tg} \left\{ \sqrt{\frac{1 - e}{1 + e}} \operatorname{tg} \frac{mx}{2} \right\} - \frac{e}{m(1 - e^2)} \frac{\operatorname{tg} \frac{mx}{2}}{1 + e + (1 - e) \operatorname{tg}^2 \frac{mx}{2}}.$$

$$\int \frac{dx}{(a + b \cos x)^2} = -\frac{b \sin x}{(a^2 - b^2)(a + b \cos x)} + \frac{a}{a^2 - b^2} \int \frac{dx}{a + b \cos x}.$$

$$= -\frac{1}{a^2 - b^2} \left\{ \frac{b \sin x}{a + b \cos x} - \frac{2a}{\sqrt{a^2 - b^2}} \operatorname{arc\,tg} \left(\sqrt{\frac{a - b}{a + b}} \operatorname{tg} \frac{x}{2} \right) \right\}. \qquad a > b.$$

$$= \frac{1}{b^2 - a^2} \left\{ \frac{b \sin x}{a + b \cos x} - \frac{a}{\sqrt{b^2 - a^2}} \log \frac{\sqrt{b - a} \operatorname{tg} \frac{x}{2} + \sqrt{b + a}}{\sqrt{b - a} \operatorname{tg} \frac{x}{2} - \sqrt{b + a}} \right\}. \qquad a < b.$$

$$\int \frac{dx}{(a + b \cos x) \cos x} = \frac{1}{a} \int \frac{dx}{\cos x} - \frac{b}{a} \int \frac{dx}{a + b \cos x}.$$

$$\int \frac{x}{\cos^2 x} dx = x \operatorname{tg} x + \log \cos x.$$

$$\int \frac{\cos x}{(1 + \cos x)^2} dx = \frac{1}{2} \operatorname{tg} \frac{x}{2} - \frac{1}{6} \operatorname{tg}^3 \frac{x}{2}.$$

$$\int \frac{\cos x}{(a + b \cos x)^2} dx = \frac{1}{a^2 - b^2} \left\{ \frac{a \sin x}{a + b \cos x} - b \int \frac{dx}{a + b \cos x} \right\}.$$

$$= -\frac{2b}{(a^2 - b^2)^{\frac{3}{2}}} \operatorname{arc\,tg} \left\{ \sqrt{\frac{a - b}{a + b}} \operatorname{tg} \frac{x}{2} \right\} + \frac{a}{a^2 - b^2} \frac{\sin x}{a + b \cos x}.$$

$$\int \frac{\alpha + \beta \cos x}{(a + b \cos x) \cos x} dx = \frac{\alpha}{a} \log \operatorname{tg} \left(\frac{\pi}{4} + \frac{x}{2} \right) + \frac{a\beta - b\alpha}{a} \int \frac{dx}{a + b \cos x}$$

$$\int \cos^3 x \, dx = \frac{1}{3} \sin x \cos^2 x + \frac{2}{3} \sin x = \frac{1}{12} \sin 3x + \frac{3}{4} \sin x.$$

$$\int x \cos^3 x \, dx = x \left(\frac{3}{4} \sin x + \frac{1}{12} \sin 3x \right) + \frac{3}{4} \cos x + \frac{1}{36} \cos 3x.$$

$$\int x^2 \cos^3 x \, dx = \frac{1}{12} \left(x^2 - \frac{2}{9} \right) \sin 3x + \frac{x \cos 3x}{18} + \frac{3}{4}(x^2 - 2) \sin x + \frac{3}{2} x \cos x.$$

$$\int \frac{dx}{\cos^3 x} = \frac{\sin x}{2 \cos^2 x} + \frac{1}{2} \log \operatorname{tg} \left(\frac{\pi}{4} + \frac{x}{2} \right).$$

$$\int \frac{dx}{(a + b \cos x)^3} = \frac{1}{(b^2 - a^2)^2} \left\{ \frac{(b^2 - a^2) b \sin x}{2\Phi^2} - \frac{3ab \sin x}{2\Phi} \right\} \frac{a^2 + \frac{1}{2} b^2}{(b^2 - a^2)^2} \int \frac{dx}{\Phi}.$$

$$= -b \sin x \frac{4a^2 - b^2 + 3ab \cos x}{2c^4 (a + b \cos x)} + \frac{2a^2 + b^2}{c^5} \operatorname{arc\,tg} \left(\frac{a - b}{c} \operatorname{tg} \frac{x}{2} \right). \qquad a^2 - b^2 = c^2$$

$$\int \frac{x}{\cos^3 x} dx = \frac{x \sin x - \cos x}{2 \cos^2 x} + \frac{1}{2} \int \frac{x}{\cos x} dx.$$

$$\int \frac{\cos^3 x}{1 - a^2 \cos^2 x} dx = \frac{1}{a^3 \sqrt{1 - a^2}} \text{arc tg} \frac{a \sin x}{\sqrt{1 - a^2}} - \frac{\sin x}{a^2}.$$

$$\int \cos^4 x \, dx = \frac{1}{32} \sin 4x + \frac{1}{4} \sin 2x + \frac{3}{8} x = \frac{1}{4} \cos^3 x \sin x + \frac{3}{8} (\cos x \sin x + x)$$

$$\int \frac{dx}{\cos^4 x} = \frac{\sin x}{3 \cos^3 x} + \frac{2}{3} \text{tg } x = \text{tg } x + \frac{1}{3} \text{tg}^3 x.$$

$$\int \frac{x}{\cos^4 x} dx = \frac{2 x \sin x - \cos x}{6 \cos^3 x} + \frac{2}{3} (x \text{ tg } x + \log \cos x).$$

$$\int \cos^n x \, dx = \frac{1}{n} \cos^{n-1} x \sin x + \frac{n-1}{n} \int \cos^{n-2} x \, dx.$$

$$\int \cos^{2n} x \, dx = \frac{1}{2^{2n-1}} \left\{ \frac{\sin 2nx}{2n} + 2n \frac{\sin (2n-2) x}{2n-2} + \frac{2n(2n-1)}{1.2} \frac{\sin (2n-4) x}{2n-4} + \cdots + \frac{2n(2n-1) \cdots (n+1)}{1.2 \cdots n} \frac{x}{2} \right\}.$$

$$= \frac{\sin x}{2n} \left\{ \cos^{2n-1} x + \frac{2n-1}{2n-2} \cos^{2n-3} x + \frac{(2n-1)(2n-3)}{(2n-2)(2n-4)} \cos^{2n-5} x + \cdots + \frac{1.3 \cdots (2n-1)}{2.4 \cdots (2n-2)} \cos x \right\} + \frac{1.3 \cdots (2n-1)}{2.4 \cdots 2n} x.$$

$$\int \cos^{2n+1} x \, dx = \frac{1}{2^{2n}} \left\{ \frac{\sin (2n+1) x}{2n+1} + (2n+1) \frac{\sin (2n-1) x}{2n-1} + \frac{(2n+1)(2n-1)}{1.2} \frac{\sin (2n-3) x}{2n-3} + \cdots \right.$$

$$\left. + \frac{(2n+1) \cdots (n+2)}{1.2 \cdots n} \sin x \right\}.$$

$$= \frac{\sin x}{2n+1} \left\{ \cos^{2n} x + \frac{2n}{2n-1} \cos^{2n-2} x + \cdots + \frac{2.4 \cdots 2n}{1.3 \cdots (2n-1)} \right\}.$$

$$\int (a + b \cos x)^n (\alpha + \beta \cos x) \, dx = \frac{\beta}{n+1} (a + b \cos x)^n \sin x + \int (a + b \cos x)^{n-1} \left[a\alpha + \frac{nb\beta}{n+1} + \left(b\alpha + \frac{na\beta}{n+1} \right) \cos x \right] dx.$$

$$\int \frac{dx}{\cos^n x} = \frac{\sin x}{(n-1) \cos^{n-1} x} + \frac{n-2}{n-1} \int \frac{dx}{\cos^{n-2} x}$$

$$\int \frac{dx}{\cos^{2n+2} x} = \text{tg } x + \frac{n_1}{3} \text{tg}^3 x + \frac{n_2}{5} \text{tg}^5 x + \cdots + \frac{1}{2n+1} \text{tg}^{2n+1} x.$$

$$\int \frac{dx}{(a + b \cos x)^n} = - \frac{b \sin x}{(n-1)(a^2 - b^2) \Phi^{n-1}} + \frac{(2n-3) a}{(n-1)(a^2 - b^2)} \int \frac{dx}{\Phi^{n-1}} - \frac{n-2}{(n-1)(a^2 - b^2)} \int \frac{dx}{\Phi^{n-2}}$$

$$\int \frac{x}{\cos^n x} dx = \frac{(n-2) x \sin x - \cos x}{(n-2)(n-1) \cos^{n-1} x} + \frac{n-2}{n-1} \int \frac{x}{\cos^{n-2} x} dx.$$

$$\int \frac{\cos x}{(a + b \cos x)^n} dx = - \frac{a}{b} \int \frac{dx}{\Phi^n} + \frac{1}{b} \int \frac{dx}{\Phi^{n-1}} = \frac{a \sin x}{(n-1)(a^2 - b^2) \Phi^{n-1}} - \frac{1}{(n-1)(a^2 - b^2)} \int \frac{(n-1) b - (n-2) a \cos x}{\Phi^{n-1}} dx.$$

$$\int \frac{\cos^2 x}{(a + b \cos x)^n} dx = \frac{a^2}{b^2} \int \frac{dx}{\Phi^n} - \frac{a}{b^2} \int \frac{dx}{\Phi^{n-1}} + \frac{1}{b} \int \frac{\cos x}{\Phi^{n-1}} dx.$$

$$\int \frac{a' + b' \cos x}{(a + b \cos x)^n} dx = \frac{(ab' - ba) \sin x}{(n-1)(a^2 - b^2)(a + b \cos x)^{n-1}} + \frac{1}{(n-1)(a^2 - b^2)} \int \frac{(n-1)(aa' - bb') + (n-2)(ab' - ba') \cos x}{(a + b \cos x)^{n-1}} dx.$$

$$\int \cos mx \cos nx \, dx = \frac{\sin (m+n) x}{2(m+n)} + \frac{\sin (m-n) x}{2(m-n)}.$$

$$\int \cos (ax+\alpha) \cos (ax+\beta) \, dx = \frac{\sin (2ax+\alpha+\beta)}{4a} + \frac{1}{2} x \cos (\alpha-\beta).$$

$$\int \cos (ax+\alpha) \cos (bx+\beta) \, dx = \frac{\sin \{ (a+b) x + \alpha + \beta \}}{2(a+b)} + \frac{\sin \{ (a-b) x + \alpha - \beta \}}{2(a-b)}.$$

$$\int \cos^n x \cos ax \, dx = \frac{\cos^n x \sin ax}{n+a} + \frac{n}{n+a} \int \cos^{n-1} x \cos (a-1) x \, dx.$$

$$= \frac{\cos^{n-1} x}{n^2-a^2} \left(n \sin x \cos ax - a \cos x \sin ax \right) + \frac{n(n-1)}{n^2-a^2} \int \cos^{n-2} x \cos ax \, dx.$$

$$\int \cos x \cos 2x \cos 3x \, dx = \frac{1}{4} \left\{ \frac{\sin 6x}{6} + \frac{\sin 4x}{4} + \frac{\sin 2x}{2} + x \right\}.$$

$$\int \frac{dx}{a+2b\cos x + c \cos 2x} = \frac{c}{m} \int \frac{dx}{2c \cos x + b - m} - \frac{c}{m} \int \frac{dx}{2c \cos x + b + m}. \qquad m = \sqrt{b^2 - 2c(a-c)}.$$

$$\int \frac{dx}{a + b\cos x + c\cos 2x} = -\frac{1}{2}\left(\frac{1}{n} - \frac{1}{m}\right) \frac{1}{\sqrt{\beta^2 - 4\alpha\gamma}} \operatorname{arc\,tg} \frac{2\sqrt{\beta^2 - 4\alpha\gamma} \sin x}{m+n+(m-n)\cos x}$$

$$+ \frac{1}{2}\left(\frac{1}{n} + \frac{1}{m}\right) \frac{1}{\alpha-\gamma} \operatorname{arc\,tg} \frac{2(\alpha-\gamma)\sin x}{m-n+(m+n)\cos x},$$

where $m^2 = a+b+c$, $\beta^2 - 4\alpha\gamma = \frac{1}{4}(m-n)^2 - c$, $n^2 = a-b+c$, $(\alpha-\gamma)^2 = \frac{1}{4}(m+n)^2 - 2c.$

$$\int \frac{\cos x}{\cos 2x} dx = \frac{1}{2\sqrt{2}} \log \frac{1+\sqrt{2}\sin x}{1-\sqrt{2}\sin x}.$$

$$\int \frac{\cos x}{\cos 3x} dx = \frac{1}{2\sqrt{3}} \log \frac{\cos\left(\frac{\pi}{3} - x\right)}{\cos\left(\frac{\pi}{3}+x\right)}.$$

$$\int \frac{\cos 2x}{\cos x} dx = 2 \sin x - \log \operatorname{tg} \left(\frac{\pi}{4} + \frac{x}{2}\right).$$

$$\int \frac{\cos 2x}{\cos^2 x} dx = 2x - \operatorname{tg} x.$$

$$\int \frac{\cos 2x}{\cos^3 x} dx = -\frac{\sin x}{2\cos^2 x} + \frac{3}{2} \log \operatorname{tg} \left(\frac{\pi}{4} + \frac{x}{2}\right).$$

$$\int \frac{\cos 2x}{\cos^4 x} dx = \frac{1}{2} \operatorname{tg} x \left(4 - \frac{1}{\cos^2 x}\right).$$

$$\int \frac{\cos 3x}{\cos x} dx = 2 \sin x \cos x - x.$$

$$\int \frac{\cos 3x}{\cos^2 x} dx = 4 \sin x - 3 \log \operatorname{tg} \left(\frac{\pi}{4} + \frac{x}{2}\right).$$

$$\int \frac{\cos 3x}{\cos^3 x} dx = 4x - 3 \operatorname{tg} x.$$

$$\int \frac{\cos 3x}{\cos^4 x}\,dx = -\frac{3\sin x}{2\cos^2 x} + \frac{5}{2}\log \operatorname{tg}\left(\frac{\pi}{4} + \frac{x}{2}\right).$$

$$\int \frac{\cos ax}{\cos^n x}\,dx = 2\int \frac{\cos(a-1)x}{\cos^{n-1} x}\,dx - \int \frac{\cos(a-2)x}{\cos^n x}\,dx.$$

Sin U $\cos^n$ U

$$\int \sin x \cos x\,dx = \frac{1}{2}\sin^2 x = -\frac{1}{4}\cos 2x.$$

$$\int \sin ax \cos bx\,dx = -\frac{\cos(a+b)x}{2(a+b)} - \frac{\cos(a-b)x}{2(a-b)}.$$

$$\int \sin x \cos^2 x\,dx = -\frac{1}{3}\cos^3 x = -\frac{1}{4}\left(\frac{1}{3}\cos 3x + \cos x\right).$$

$$\int \sin x \cos^3 x\,dx = -\frac{1}{4}\cos^4 x = -\frac{1}{8}\left(\frac{1}{4}\cos 4x + \cos 2x\right).$$

$$\int \sin x \cos^4 x\,dx = -\frac{1}{5}\cos^5 x = -\frac{1}{16}\left(\frac{1}{5}\cos 5x + \cos 3x + 2\cos x\right)$$

$$\int \sin x \cos^n x\,dx = -\frac{1}{n+1}\cos^{n+1} x.$$

$$\int \frac{dx}{\sin x \cos x} = \log \operatorname{tg} x.$$

$$\int \frac{dx}{\sin x \cos^2 x} = \frac{1}{\cos x} + \log \operatorname{tg}\frac{x}{2}.$$

$$\int \frac{dx}{\sin x \cos^3 x} = \frac{1}{2\cos^2 x} + \log \operatorname{tg} x.$$

$$\int \frac{dx}{\sin x \cos^4 x} = \frac{1}{3\cos^3 x} + \frac{1}{\cos x} + \log \operatorname{tg}\frac{x}{2}.$$

$$\int \frac{dx}{a\sin x + b\cos x} = \frac{1}{\sqrt{a^2+b^2}}\log \operatorname{tg}\frac{x+\alpha}{2}. \qquad \frac{b}{\sin\alpha} = \frac{a}{\cos\alpha} = \sqrt{a^2+b^2}.$$

$$= \frac{1}{\sqrt{a^2+b^2}}\log\frac{b\operatorname{tg}\frac{x}{2} - a + \sqrt{a^2+b^2}}{b\operatorname{tg}\frac{x}{2} - a - \sqrt{a^2+b^2}} = \frac{1}{\sqrt{a^2+b^2}}\log\frac{b\sin x - a\cos x + \sqrt{a^2+b^2}}{b\cos x + a\sin x}.$$

$$\int \frac{dx}{\sin x(a + b\cos x)} = \frac{b}{a^2-b^2}\log(a + b\cos x) + \frac{1}{b+a}\log\sin\frac{x}{2} + \frac{1}{b-a}\log\cos\frac{x}{2}.$$

$$= \frac{b}{a^2-b^2}\log(a + b\cos x) + \frac{\log(1-\cos x)}{2(a+b)} - \frac{\log(1+\cos x)}{2(a-b)}.$$

$$= \frac{b}{a^2-b^2}\log(a + b\cos x) + \frac{a}{a^2-b^2}\log\operatorname{tg}\frac{x}{2} + \frac{b}{a^2-b^2}\log\sin x.$$

$$= \frac{1}{a^2-b^2}\log\left\{\left(a\operatorname{cos\acute{e}c} x + b\operatorname{cotg} x\right)^b \operatorname{tg}^a\frac{x}{2}\right\}.$$

$$\int \frac{dx}{a\sin x + b\cos x + c} = \int \frac{dx}{c + k\cos(x-\alpha)}. \qquad b = k\cos\alpha, \qquad a = k\sin\alpha.$$

$$= \frac{2}{\sqrt{c^2 - b^2 - a^2}} \operatorname{arc\,tg} \frac{(c - b) \operatorname{tg} \frac{x}{2} + a}{\sqrt{c^2 - b^2 - a^2}}. \qquad c^2 > a^2 + b^2.$$

$$= \frac{1}{\sqrt{c^2 - b^2 - a^2}} \operatorname{arc\,cos} \frac{b^2 + a^2 + bc \cos x + ac \sin x}{(c + b \cos x + a \sin x)\sqrt{a^2 + b^2}}.$$

$$= \frac{1}{\sqrt{a^2 + b^2 - c^2}} \log \frac{(c - b) \operatorname{tg} \frac{x}{2} + a - \sqrt{a^2 + b^2 - c^2}}{(c - b) \operatorname{tg} \frac{x}{2} + a + \sqrt{a^2 + b^2 - c^2}}. \qquad c^2 < a^2 + b^2.$$

$$= \frac{1}{m} \log \frac{a^2 + b^2 + (bc - am) \cos x + (ac + mb) \sin x}{a \sin x + b \cos x + c}. \qquad m^2 = a^2 + b^2 - c^2.$$

$$\int \frac{\sin x}{\cos^2 x} dx = \frac{1}{\cos x} = \text{séc}\, x.$$

$$\int \frac{\sin nx}{\cos^2 nx} dx = \frac{1}{n} \text{séc}\, nx.$$

$$\int \frac{\sin (px + q)}{\cos^2 (px + q)} dx = \frac{1}{p} \text{séc}\, (px + q).$$

$$\int \frac{\sin x}{\cos^3 x} dx = \frac{1}{2} \operatorname{tg}^2 x.$$

$$\int \frac{\sin x}{\cos^n x} dx = \frac{1}{(n - 1) \cos^{n-1} x}.$$

$$\int \frac{\sin x}{a + b \cos x} dx = -\frac{1}{b} \log (a + b \cos x).$$

$$\int \frac{\sin x}{(a - b \cos x)^2} dx = -\frac{1}{b (a - b \cos x)}.$$

$$\int \frac{\cos^2 x}{\sin x} dx = \cos x + \log \operatorname{tg} \frac{x}{2}.$$

$$\int \frac{\cos^3 x}{\sin x} dx = \frac{1}{2} \cos^2 x + \log \sin x.$$

$$\int \frac{\cos^4 x}{\sin x} dx = \frac{1}{3} \cos^3 x + \cos x + \log \operatorname{tg} \frac{x}{2}.$$

$$\int \frac{\cos^n x}{\sin x} dx = \frac{1}{n - 1} \cos^{n-1} x + \int \frac{\cos^{n-2} x}{\sin x} dx.$$

$$\int \frac{\cos x}{a^2 + b^2 \sin x} dx = \frac{1}{b^2} \log (a^2 + b^2 \sin x).$$

$$\int \frac{\sin x \cos^2 x}{1 + a^2 \cos^2 x} dx = -\frac{\cos x}{a^2} + \frac{1}{a^3} \operatorname{arc\,tg} (a \cos x).$$

$$\int \frac{2 - \sin x}{2 + \cos x} dx = \log (2 + \cos x) + \frac{4}{\sqrt{3}} \operatorname{arc\,tg} \left(\frac{1}{\sqrt{3}} \operatorname{tg} \frac{x}{2}\right).$$

$$\int \frac{\alpha + \beta \cos x}{\sin x (a + b \cos x)} dx = \frac{b\alpha - a\beta}{a^2 - b^2} \log (a + b \cos x) - \frac{\alpha - \beta}{a - b} \log \cos \frac{x}{2} + \frac{\alpha + \beta}{a + b} \log \sin \frac{x}{2}$$

$$\int \frac{\sin x + \cos x + 1}{\sin x - \cos x + 1} dx = 2 \log \sin \frac{x}{2}.$$

$$\int \frac{m \sin x + n \cos x + k}{a \sin x + b \cos x} dx = \frac{nb + ma}{a^2 + b^2} \left\{ x - \int \frac{k}{a \sin x + b \cos x} dx \right\} + \frac{an - bm}{a^2 + b^2} \log (a \sin x + b \cos x).$$

$$\int \frac{A + B \cos x + C \sin x}{a + b \cos x + c \sin x} dx = A \int \frac{dz}{a + p \cos z} + (B \cos \alpha + C \sin \alpha) \int \frac{\cos z}{a + p \cos z} dz - (B \sin \alpha - C \cos \alpha) \int \frac{\sin z}{a + p \cos z} dz.$$

in which $b = p \cos \alpha, \quad c = p \sin \alpha, \quad x - \alpha = z.$

$\mathrm{Sin}^2\, U \cos^n U$

$$\int \sin^2 x \cos x \, dx = \frac{1}{3} \sin^3 x = -\frac{1}{4} \left(\frac{1}{3} \sin 3x - \sin x \right) = -\cos (\sin x).$$

$$\int \sin^2 x \cos^2 x \, dx = \frac{1}{4} \int \sin^2 2x \, dx = \frac{1}{8} x - \frac{1}{32} \sin 4x = \frac{1}{4} \sin^3 x \cos x - \frac{1}{8} \sin x \cos x + \frac{1}{8} x.$$

$$\int \sin^2 x \cos^3 x \, dx = -\frac{1}{16} \left\{ \frac{1}{5} \sin 5x + \frac{1}{3} \sin 3x - 2 \sin x \right\} = \frac{1}{3} \sin^3 x - \frac{1}{5} \sin^5 x.$$

$$\int \sin^2 x \cos^4 x \, dx = -\frac{1}{32} \left\{ \frac{1}{6} \sin 6x + \frac{1}{2} \sin 4x - \frac{1}{2} \sin 2x - 2x \right\}.$$

$$= \frac{\sin^3 x \cos^3 x}{6} + \frac{\sin^3 x \cos x}{8} - \frac{1}{16} (\sin x \cos x - x).$$

$$\int \sin^2 x \cos^n x \, dx = -\frac{\sin x \cos^{n+1} x}{n+2} + \frac{1}{n+2} \int \cos^n x \, dx.$$

$$\int \frac{dx}{\sin^2 x \cos x} = -\frac{1}{\sin x} + \log \operatorname{tg} \left(\frac{\pi}{4} + \frac{x}{2} \right).$$

$$\int \frac{dx}{\sin^2 x \cos^2 x} = \int \frac{dx}{\cos^2 x} + \int \frac{dx}{\sin^2 x} = \operatorname{tg} x - \operatorname{cotg} x = -2 \operatorname{cotg} 2x.$$

$$\int \frac{dx}{\sin^2 x \cos^3 x} = \frac{1}{2 \sin x \cos^2 x} - \frac{3}{2 \sin x} + \frac{3}{2} \log \operatorname{tg} \left(\frac{\pi}{4} + \frac{x}{2} \right).$$

$$\int \frac{dx}{\sin^2 x \cos^4 x} = \frac{1}{3 \sin x \cos^3 x} - \frac{8}{3} \operatorname{cotg} 2x = \frac{1}{3 \sin x \cos^3 x} + \frac{4}{3 \sin x \cos x} - \frac{8 \cos x}{3 \sin x}.$$

$$\int \frac{dx}{\sin^2 x \cos^n x} = \frac{1 - n \cos^2 x}{(n-1) \cos^{n-1} x \sin x} + \frac{n(n-2)}{n-1} \int \frac{dx}{\cos^{n-2} x}.$$

$$\int \frac{dx}{a \cos^2 x + b \sin^2 x} = \frac{1}{\sqrt{ab}} \operatorname{arc\,tg} \left(\sqrt{\frac{b}{a}} \operatorname{tg} x \right).$$

$$\int \frac{dx}{a^2 \cos^2 \alpha x + b^2 \sin^2 \alpha x} = \frac{1}{\alpha ab} \operatorname{arc\,tg} \left(\frac{b}{a} \operatorname{tg} \alpha x \right).$$

$$\int \frac{dx}{a^2 \cos^2 x - b^2 \sin^2 x} = \frac{1}{2ab} \log \frac{a \cos x + b \sin x}{a \cos x - b \sin x}.$$

$$\int \frac{dx}{a \cos^2 x + b \sin^2 x + c \sin x \cos x} = \int \frac{dz}{a + bz^2 + cz}. \qquad \operatorname{tg} x = z.$$

$$\int \frac{\sin^2 x}{\cos^4 x} dx = \frac{1}{3} \operatorname{tg}^3 x.$$

$$\int \frac{\sin^2 x}{(a + b \cos x)^n} dx = \frac{\sin x}{b\,(n-1)\,(a + b \cos x)^{n-1}} - \frac{1}{b\,(n-1)} \int \frac{\cos x}{(a + b \cos x)^{n-1}} dx.$$

$$\int \frac{\cos x}{\sin^2 x} dx = -\operatorname{coséc} x.$$

$$\int \frac{\cos nx}{\sin^2 nx} dx = -\frac{1}{n} \operatorname{coséc} nx = -\frac{1}{n \sin nx}.$$

$$\int \frac{\cos\ (px+q)}{\sin^2\ (px+q)} dx = -\frac{1}{p} \operatorname{coséc} (px+q).$$

$$\int \frac{\cos^2 x}{\sin^2 x} dx = -\operatorname{cotg} x - x.$$

$$\int \frac{\cos^3 x}{\sin^2 x} dx = -\frac{1 + \sin^2 x}{\sin x}.$$

$$\int \frac{\cos^4 x}{\sin^2 x} dx = -\frac{\cos^3 x}{\sin x} - \frac{3}{2} \sin x \cos x - \frac{3}{2} x.$$

$$\int \frac{\cos^n x}{\sin^2 x} dx = \frac{1}{n-2} \frac{\cos^{n-1} x}{\sin x} + \frac{n-1}{n-2} \int \frac{\cos^{n-2} x}{\sin^2 x} dx = -\frac{\cos^{n-1} x}{\sin x} - n \int \cos^n x\, dx.$$

$$\int \frac{\cos x}{1 + \sin^2 x} dx = \operatorname{arc\,tg} (\sin x).$$

$$\int \frac{x^2}{(x \sin x + \cos x)^2} dx = \frac{\sin x - x \cos x}{x \sin x + \cos x}.$$

$$\int \frac{x^2}{(x \cos x - \sin x)^2} dx = \frac{x \sin x + \cos x}{x \cos x - \sin x}.$$

$$\int \frac{x^2}{(\sin x - x \cos x)^2} dx = -\frac{x \sin x + \cos x}{\sin x - x \cos x}.$$

$$\int \frac{x^2}{\left\{(ax-b) \sin x + (a+bx) \cos x\right\}^2} dx = \frac{x \sin x + \cos x}{b \left\{(ax-b) \sin x + (a+bx) \cos x\right\}}.$$

$\text{Sin}^3\ \text{U} \cos^n \text{U}$

$$\int \sin^3 x \cos x\, dx = \frac{1}{8} \left\{ \frac{1}{4} \cos 4x - \cos 2x \right\} = \frac{1}{4} \sin^4 x.$$

$$\int \sin^3 x \cos^2 x\, dx = \frac{1}{16} \left\{ \frac{1}{5} \cos\ 5x - \frac{1}{3} \cos 3x - 2 \cos x \right\} = \frac{1}{5} \sin^4 x \cos x - \frac{1}{15} \sin^2 x \cos x - \frac{2}{15} \cos x$$

$$= -\frac{\cos^3 x}{5} \left(\sin^2 x + \frac{2}{3}\right) = -\frac{1}{3} \cos^3 x + \frac{1}{5} \cos^5 x.$$

$$\int \sin^3 x \cos^3 x\, dx = \frac{1}{32} \left\{ \frac{1}{6} \cos 6x - \frac{3}{2} \cos 2x \right\} = \frac{1}{6} \sin^4 x \cos^2 x + \frac{1}{12} \sin^4 x = -\frac{1}{4} \cos^4 x + \frac{1}{6} \cos^6 x.$$

$$\int \sin^3 x \cos^4 x\, dx = \frac{1}{64} \left\{ \frac{1}{7} \cos\ 7x + \frac{1}{5} \cos 5x - \cos 3x - 3 \cos x \right\} = -\frac{1}{5} \sin^5 x + \frac{1}{7} \sin^7 x.$$

$$= \frac{1}{7} \sin^4 x \cos^3 x - \frac{3}{35} \sin^4 x \cos x + \frac{1}{35} \sin^2 x \cos x + \frac{2}{35} \cos x.$$

$$\int \sin^3 x \cos^n x \, dx = -\frac{\cos^{n+1} x}{n+3}\left(\sin^2 x + \frac{2}{n+1}\right).$$

$$\int \frac{dx}{\sin^3 \cos x} = -\frac{1}{2 \sin^2 x} + \log \operatorname{tg} x.$$

$$\int \frac{dx}{\sin^3 x \cos^2 x} = \frac{1}{\sin^2 x \cos x} - \frac{3 \cos x}{2 \sin^2 x} + \frac{3}{2} \log \operatorname{tg} \frac{x}{2} = -\frac{1}{2 \cos x \sin^2 x} + \frac{3}{2 \cos x} + \frac{3}{2} \log \operatorname{tg} \frac{x}{2}.$$

$$\int \frac{dx}{\sin^3 x \cos^3 x} = \frac{1}{2 \sin^2 x \cos^2 x} - \frac{2}{\sin^2 x} + 2 \log \operatorname{tg} x = -\frac{2 \cos 2x}{\sin^2 2x} + 2 \log \operatorname{tg} x.$$

$$\int \frac{dx}{\sin^3 x \cos^4 x} = -\frac{1}{2 \cos^3 x \sin^2 x} + \frac{5}{6 \cos^3 x} + \frac{5}{2 \cos x} + \frac{5}{2} \log \operatorname{tg} \frac{x}{2}.$$

$$\int \frac{dx}{\sin^3 x \cos^n x} = \frac{2-(n+1)\cos^2 x}{2(n-1)\cos^{n-1} x \sin^2 x} + \frac{n+1}{2} \int \frac{dx}{\cos^{n-2} x \sin x}.$$

$$\int \frac{\sin^3 x}{\cos x} dx = \frac{1}{2} \cos^2 x - \log \cos x = \frac{1}{\cos^3 x} \left\{ \sin^2 x - \frac{2}{3} \right\}.$$

$$\int \frac{\sin^3 x}{\cos^2 x} dx = \cos x + \operatorname{séc} x.$$

$$\int \frac{\cos^2 x}{\sin^3 x} dx = -\frac{\cos x}{2 \sin^2 x} - \frac{1}{2} \log \operatorname{tg} \frac{x}{2}.$$

$$\int \frac{\cos^3 x}{\sin^3 x} dx = -\frac{1}{2} \operatorname{cotg}^2 x - \log \sin x.$$

$$\int \frac{\cos^4 x}{\sin^3 x} dx = \frac{1}{\sin^2 x}\left(\cos^5 x - \frac{3}{2} \cos x\right) - \frac{3}{2} \log \operatorname{tg} \frac{x}{2}.$$

$$\int \frac{\cos^n x}{\sin^3 x} dx = -\frac{\cos^{n+1} x}{2 \sin^2 x} - \frac{n-1}{2} \int \frac{\cos^n x}{\sin x} dx = \frac{\cos^{n-1} x}{(n-3) \sin^2 x} + \frac{n-1}{n-3} \int \frac{\cos^{n-2} x}{\sin^3 x} dx.$$

Sin⁴ U cosⁿ U

$$\int \operatorname{Sin}^4 x \cos x \, dx = \frac{1}{16} \left\{ \frac{1}{5} \sin 5x - \sin 3x + 2 \sin x \right\}.$$

$$\int \sin^4 x \cos^2 x \, dx = \frac{1}{32} \left\{ \frac{1}{6} \sin 6x - \frac{1}{2} \sin 4x - \frac{1}{2} \sin 2x + 2x \right\}.$$

$$\int \sin^4 x \cos^3 x \, dx = \frac{1}{64} \left\{ \frac{1}{7} \sin 7x - \frac{1}{5} \sin 5x - \sin 3x + 3 \sin x \right\} = \frac{1}{5} \sin^5 x - \frac{1}{7} \sin^7 x.$$

$$\int \sin^4 x \cos^4 x \, dx = \frac{1}{128} \left\{ \frac{1}{8} \sin 8x - \sin 4x + 3x \right\}.$$

$$\int \sin^4 x \cos^5 x \, dx = \int (z^4 - 2z^6 + z^8) \, dz. \qquad \sin x = z.$$

$$\int \sin^4 x \cos^n x \, dx = -\frac{\cos^{n+1} x}{n+4} \left\{ \sin^3 x + \frac{3}{n+2} \sin x \right\} + \frac{3}{(n+4)(n+2)} \int \cos^n x \, dx.$$

$$\int \frac{dx}{\sin^4 x \cos x} = -\frac{1}{3 \sin^3 x} - \frac{1}{\sin x} + \log \operatorname{tg}\left(\frac{\pi}{4} + \frac{x}{2}\right).$$

$$\int \frac{dx}{\sin^4 x \cos^2 x} = -\frac{1}{3\cos x \sin^3 x} - \frac{8}{3}\operatorname{cotg} 2x.$$

$$\int \frac{dx}{\sin^4 x \cos^3 x} = \frac{1}{2\cos^2 x \sin^3 x} - \frac{5}{6\sin^3 x} - \frac{5}{2\sin x} + \frac{5}{2}\log \operatorname{tg}\left(\frac{\pi}{4} + \frac{x}{2}\right).$$

$$\int \frac{dx}{\sin^4 x \cos^4 x} = -\frac{8}{3}\left\{\frac{1}{\sin^2 2x} + 2\right\}\operatorname{cotg} 2x.$$

$$\int \frac{dx}{\sin^4 x \cos^n x} = -\frac{1}{3\cos^{n-1} x \sin^3 x} + \frac{n+2}{3}\int \frac{dx}{\cos^n x \sin^2 x}.$$

$$\int \frac{dx}{(a^2\cos^2 x + b^2 \sin^2 x)^2} = \frac{(a^2+b^2)\,\alpha}{2a^3 b^3} - \frac{(a^2-b^2)\sin\alpha\cos\alpha}{4a^3 b^3}. \qquad \operatorname{tg}\alpha = \frac{b}{a}\operatorname{tg} x.$$

$$\int \frac{\sin^4 x}{\cos x}\,dx = -\frac{\sin^2 x + 3}{3}\sin x + \log \operatorname{tg}\left(\frac{\pi}{4} + \frac{x}{2}\right).$$

$$\int \frac{\cos^2 x}{\sin^4 x}\,dx = -\frac{\cos x}{3\sin^3 x} + \frac{1}{3}\operatorname{cotg} x.$$

$$\int \frac{\cos^3 x}{\sin^4 x}\,dx = -\frac{\cos^2 x}{3\sin^3 x} + \frac{2}{3\sin x}.$$

$$\int \frac{\cos^4 x}{\sin^4 x}\,dx = -\frac{1}{3}\operatorname{cotg}^3 x + \operatorname{cotg} x + x.$$

Sin^m U cos^n U

$$\int \sin^m x \cos x\,dx = \frac{1}{m+1}\sin^{m+1} x.$$

$$\int \sin^m x \cos^2 x\,dx = \frac{1}{m+2}\sin^{m+1} x \cos x + \frac{1}{m+2}\int \sin^m x\,dx.$$

$$\int \sin^m x \cos^3 x\,dx = \frac{1}{m+3}\sin^{m+1} x \cos^2 x + \frac{2}{m+3}\int \sin^m x \cos x\,dx = \frac{\sin^{m+1} x}{m+1}\left(\cos^2 x + \frac{2}{m+3}\right).$$

$$\int \sin^m x \cos^4 x\,dx = \frac{1}{m+4}\sin^{m+1} x \cos^3 x + \frac{3}{m+4}\int \sin^m x \cos^2 x\,dx.$$

$$\int \sin^m x \cos^5 x\,dx = \frac{1}{m+5}\sin^{m+1} x \cos^4 x + \frac{4}{m+5}\int \sin^m x \cos^3 x\,dx.$$

$$\int \sin^m x \cos^6 x\,dx = \frac{1}{m+6}\sin^{m+1} x \cos^5 x + \frac{5}{m+6}\int \sin^m x \cos^4 x\,dx.$$

$$\int \sin^m x \cos^n x\,dx = \int z^m (1-z^2)^{\frac{n-1}{2}}\,dz. \qquad z = \sin x.$$

$$= \int z^m (1+z^2)^{k-1}\,dz. \qquad z = \operatorname{tg} x \qquad m+n = -2k.$$

$$= \frac{\sin^{m+1} x \cos^{n-1} x}{m+1} + \frac{n-1}{m+1}\int \sin^{m+2} x \cos^{n-2} x\,dx.$$

$$= -\frac{\sin^{m-1} x \cos^{n+1} x}{n+1} + \frac{m-1}{n+1}\int \sin^{m-2} x \cos^{n+2} x\,dx.$$

$$= \frac{\sin^{m+1} x \cos^{n-1} x}{m+n} + \frac{n-1}{m+n} \int \sin^m x \cos^{n-2} x \, dx.$$

$$= -\frac{\sin^{m-1} x \cos^{n+1} x}{m+n} + \frac{m-1}{m+n} \int \sin^{m-2} x \cos^n x \, dx.$$

$$= \frac{\sin^{m-1} x \cos^{n-1} x}{m+n} \left\{ \sin^2 x - \frac{m-1}{m+n-2} \right\} + \frac{(m-1)(n-1)}{(m+n)(m+n-2)} \int \sin^{m-2} x \cos^{n-2} x \, dx.$$

$$= -\frac{\sin^{m+1} x \cos^{n+1} x}{n+1} + \frac{m+n+2}{n+1} \int \sin^m x \cos^{n+2} x \, dx.$$

$$= \frac{\sin^{m+1} x \cos^{n+1} x}{m+1} + \frac{m+n+2}{m+1} \int \sin^{m+2} x \cos^n x \, dx.$$

$$= \frac{\sin^{m+1} x}{m+n} \left\{ \cos^{n-1} x + \frac{n-1}{n+m-2} \cos^{n-3} x + \frac{(n-1)(n-3)}{(n+m-2)(n+m-4)} \cos^{n-5} x + \cdots \right\}$$

$$+ \frac{(n-1)(n-3)\cdots 3.1}{(n+m)(n+m-2)\cdots(n+2)} \int \sin^m x \, dx. \qquad \text{For } n \text{ even.}$$

$$= -\frac{\cos^{n+1} x}{m+n} \left\{ \sin^{m-1} x + \frac{m-1}{n+m-2} \sin^{m-3} + \frac{(m-1)(m-3)}{(n+m-2)(n+m-4)} \sin^{m-5} x + \cdots \right\}$$

$$+ \frac{(m-1)(m-3)\cdots 3.1}{(n+m)(n+m-2)\cdots(n+2)} \int \cos^n x \, dx. \qquad \text{For } m \text{ even.}$$

$$\int \frac{dx}{\sin^m x \cos x} = -\frac{1}{(m-1)\sin^{m-1} x} + \int \frac{dx}{\sin^{m-2} x \cos x}$$

$$\int \frac{dx}{\sin^m x \cos^2 x} = \frac{1}{\sin^{m-1} x \cos x} + m \int \frac{dx}{\sin^m x} = -\frac{1}{(m-1)\sin^{m-1} x \cos x} + \frac{m}{m-1} \int \frac{dx}{\sin^{m-2} x \cos^2 x}.$$

$$\int \frac{dx}{\sin^m x \cos^3 x} = \frac{1}{2\sin^{m-1} x \cos^2 x} + \frac{m+1}{2} \int \frac{dx}{\sin^m x \cos x} = -\frac{1}{(m-1)\sin^{m-1} x \cos^2 x} + \frac{m+1}{m-1} \int \frac{dx}{\sin^{m-2} x \cos^3 x}.$$

$$\int \frac{dx}{\sin^m x \cos^4 x} = \frac{1}{3\sin^{m-1} x \cos^3 x} + \frac{m+2}{3} \int \frac{dx}{\sin^m x \cos^2 x} = -\frac{1}{(m-1)\sin^{m-1} x \cos^3 x} + \frac{m+2}{m-1} \int \frac{dx}{\sin^{m-2} x \cos^4 x}.$$

$$\int \frac{dx}{\sin^m x \cos^m x} = 2^{m-1} \int \frac{dz}{\sin^m z}. \qquad z = 2x.$$

$$\int \frac{dx}{\sin^m x \cos^n x} = \int \frac{dx}{\sin^{m-2} x \cos^n x} + \int \frac{dx}{\sin^m x \cos^{n-2} x}.$$

$$= \frac{1}{(n-1)\sin^{m-1} x \cos^{n-1} x} + \frac{m+n-2}{n-1} \int \frac{dx}{\sin^m x \cos^{n-2} x}.$$

$$= -\frac{1}{(m-1)\sin^{m-1} x \cos^{n-1} x} + \frac{m+n-2}{m-1} \int \frac{dx}{\sin^{m-2} x \cos^n x}.$$

$$= \frac{(m-1)\sin^2 x - (n-1)\cos^2 x}{(n-1)(m-1)\cos^{n-1} x \sin^{m-1} x} + \frac{(n+m-2)(n+m-4)}{(n-1)(m-1)} \int \frac{dx}{\sin^{m-2} x \cos^{n-2} x}.$$

$$\int \frac{dx}{(a^2 \cos^2 x + b^2 \sin^2 x)^n} = \frac{1}{(ab)^{2n-1}} \int \left(a^2 \sin^2 z + b^2 \cos^2 z\right)^{n-1} dz. \qquad \operatorname{tg} x = \frac{a}{b} \operatorname{tg} z.$$

$$\int \frac{\sin^m x}{\cos x} dx = -\frac{\sin^{m-1} x}{m-1} + \int \frac{\sin^{m-2} x}{\cos x} dx.$$

$$\int \frac{\sin^{2n} x}{\cos x}\, dx = -\sin x - \frac{\sin^3 x}{3} - \frac{\sin^5 x}{5} - \cdots - \frac{\sin^{2n-1} x}{2n-1} + \int \frac{dx}{\cos x}.$$

$$\int \frac{\sin^{2n+1} x}{\cos x}\, dx = -\frac{\sin^2 x}{2} - \frac{\sin^4 x}{4} - \frac{\sin^6 x}{6} - \cdots - \frac{\sin^{2n} x}{2n} + \int \operatorname{tg} x\, dx.$$

$$\int \frac{\sin^m x}{\cos^2 x}\, dx = -\frac{\sin^{m-1} x}{(m-2)\cos x} + \frac{m-1}{m-2}\int \frac{\sin^{m-2} x}{\cos^2 x}\, dx = \frac{\sin^{m+1} x}{\cos x} - m\int \sin^m x\, dx.$$

$$\int \frac{\sin^m x}{\cos^3 x}\, dx = -\frac{\sin^{m-1} x}{(m-3)\cos^2 x} + \frac{m-1}{m-3}\int \frac{\sin^{m-2} x}{\cos^3 x}\, dx = \frac{\sin^{m+1} x}{2\cos^2 x} - \frac{m-1}{2}\int \frac{\sin^m x}{\cos x}\, dx.$$

$$\int \frac{\sin^m x}{\cos^4 x}\, dx = -\frac{\sin^{m-1} x}{(m-4)\cos^3 x} + \frac{m-1}{m-4}\int \frac{\sin^{m-2} x}{\cos^4 x}\, dx = \frac{\sin^{m+1} x}{3\cos^3 x} - \frac{m-2}{3}\int \frac{\sin^m x}{\cos^2 x}\, dx.$$

$$\int \frac{\sin^m x}{\cos^{m+2} x}\, dx = \frac{1}{m+1}\operatorname{tg}^{m+1} x.$$

$$\int \frac{\sin^m x}{\cos^n x}\, dx = -\int \frac{\cos^m z}{\sin^n z}\, dz. \qquad z = \frac{\pi}{2} - x.$$

$$= \frac{\sin^{m+1} x}{(n-1)\cos^{n-1} x} + \frac{n-m-2}{n-1}\int \frac{\sin^m x}{\cos^{n-2} x}\, dx.$$

$$= -\frac{\sin^{m-1} x}{(m-n)\cos^{n-1} x} + \frac{m-1}{m-n}\int \frac{\sin^{m-2} x}{\cos^n x}\, dx.$$

$$= \frac{\sin^{m-1} x}{(n-1)\cos^{n-1} x} - \frac{m-1}{n-1}\int \frac{\sin^{m-2} x}{\cos^{n-2} x}\, dx.$$

$$\int \frac{\cos x}{\sin^m x}\, dx = -\frac{1}{(m-1)\sin^{m-1} x}.$$

$$\int \frac{\cos^2 x}{\sin^m x}\, dx = -\frac{\cos x}{(m-2)\sin^{m-1} x} - \frac{1}{m-2}\int \frac{dx}{\sin^m x} = -\frac{\cos x}{(m-1)\sin^{m-1} x} - \frac{1}{m-1}\int \frac{dx}{\sin^{m-2} x}$$

$$\int \frac{\cos^3 x}{\sin^m x}\, dx = -\frac{\cos^2 x}{(m-3)\sin^{m-1} x} + \frac{2}{(m-1)(m-3)\sin^{m-1} x} = -\frac{\cos^2 x}{(m-1)\sin^{m-1} x} + \frac{2}{(m-1)(m-3)\sin^{m-3} x}.$$

$$\int \frac{\cos^4 x}{\sin^m x}\, dx = -\frac{1}{(m-4)\sin^{m-1} x}\left(\cos^3 x - \frac{3}{m-2}\cos x\right) + \frac{3}{(m-2)(m-4)}\int \frac{dx}{\sin^m x}.$$

$$= -\frac{\cos^3 x}{(m-1)\sin^{m-1} x} - \frac{3}{m-1}\int \frac{\cos^2 x}{\sin^{m-2} x}\, dx.$$

$$\int \frac{\cos^m x}{\sin^{m+2} x}\, dx = -\frac{1}{m+1}\operatorname{cotg}^{m+1} x.$$

$$\int \frac{\cos^n x}{\sin^m x}\, dx = -\frac{\cos^{n+1} x}{(m-1)\sin^{m-1} x} + \frac{m-n-2}{m-1}\int \frac{\cos^n x}{\sin^{m-2} x}\, dx.$$

$$= \frac{\cos^{n-1} x}{(n-m)\sin^{m-1} x} + \frac{n-1}{n-m}\int \frac{\cos^{n-2} x}{\sin^m x}\, dx.$$

$$= -\frac{\cos^{n-1} x}{(m-1)\sin^{m-1} x} - \frac{n-1}{m-1}\int \frac{\cos^{n-2} x}{\sin^{m-2} x}\, dx.$$

$$\int \sin mx \cos nx \, dx = -\frac{\cos(m+n)x}{2(m+n)} - \frac{\cos(m-n)x}{2(m-n)} = \frac{1}{n^2-m^2}\left\{ n \sin nx \cos mx + m \cos nx \cos mx \right\}.$$

$$\int \sin(ax+\alpha)\cos(ax+\beta)\,dx = -\frac{\cos(2ax+\alpha+\beta)}{4a} + \frac{x}{2}\sin(\alpha-\beta).$$

$$\int \sin(ax+\alpha)\cos(bx+\beta)\,dx = -\frac{\cos\{(a+b)x+\alpha+\beta\}}{2(a+b)} - \frac{\cos\{(a-b)x+\alpha-\beta\}}{2(a-b)}.$$

$$\int \cos x \sin 2x \sin 3x \, dx = \frac{1}{4}\left\{ -\frac{\sin 6x}{6} - \frac{\sin 4x}{4} + \frac{\sin 2x}{2} + x \right\}.$$

$$\int \sin 3x \cos^2 x \, dx = -\frac{1}{5}\cos x\left(2\sin x \sin 3x + 3\cos x \cos 3x\right) + \frac{2}{15}\cos 3x.$$

$$\int \sin 4x \cos^2 x \, dx = -\frac{1}{6}\cos x\left(\sin x \sin 4x + 2\cos x \cos 4x\right) + \frac{1}{24}\cos 4x.$$

$$\int \cos 3x \sin^2 x \, dx = \frac{1}{5}\sin x\left(2\cos x \cos 3x + 3\sin x \sin 3x\right) - \frac{2}{15}\sin 3x.$$

$$\int \cos 4x \sin^2 x \, dx = \frac{1}{6}\sin 4x \sin^2 x - \frac{1}{24}(2\sin 2x - \cos 4x).$$

$$\int \sin^{m-1} x \cos(m+1)x \, dx = \frac{1}{m}\cos mx \sin^m x.$$

$$\int \sin^m x \cos ax \, dx = \frac{\sin^m x \sin ax}{m+a} - \frac{m}{m+a}\int \sin^{m-1} x \sin(a-1)x\,dx.$$

$$= \frac{\sin^{m-1} x}{a^2-m^2}\left(m\cos x \cos ax + a \sin x \sin ax\right) - \frac{m(m-1)}{a^2-m^2}\int \sin^{m-2} x \cos ax \, dx.$$

$$\int \sin ax \cos^n x \, dx = -\frac{\cos^n x \cos ax}{n+a} + \frac{n}{n+a}\int \cos^{n-1} x \sin(a-1)x\,dx.$$

$$= \frac{\cos^{n-1} x}{n^2-a^2}\left(n \sin x \sin ax + a\cos x \cos ax\right) + \frac{n(n-1)}{n^2-a^2}\int \cos^{n-2} x \sin ax \, dx.$$

$$\int \frac{\sin x}{\cos 3x}\,dx = \frac{1}{3}\log\cos x - \frac{1}{6}\log\left(\cos^2 x - \frac{3}{4}\right).$$

$$\int \frac{\sin 2x}{\cos x}\,dx = -2\cos x.$$

$$\int \frac{\sin 2x}{\cos^2 x}\,dx = -2\log\cos x.$$

$$\int \frac{\sin 2x}{\cos^n x}\,dx = \frac{2}{(n-2)\cos^{n-2} x}.$$

$$\int \frac{\sin 2x}{\cos 3x}\,dx = \frac{1}{2\sqrt{3}}\log\left\{\operatorname{cotg}\left(15^0 - \frac{x}{2}\right)\operatorname{cotg}\left(15^0 + \frac{x}{2}\right)\right\}.$$

$$\int \frac{\sin 3x}{\cos x}\,dx = 2\sin^2 x + \log\cos x.$$

$$\int \frac{\sin 3x}{\cos^2 x}\, dx = -4 \cos x - \frac{1}{\cos x}$$

$$\int \frac{\sin 3x}{\cos^3 x}\, dx = -\frac{1}{2 \cos^2 x} - 4 \log \cos x.$$

$$\int \frac{\sin 3x}{\cos^4 x}\, dx = \frac{4}{\cos x} - \frac{1}{3 \cos^3 x}\cdot$$

$$\int \frac{\sin 3x}{\cos^n x}\, dx = \frac{4}{(n-3) \cos^{n-3} x} - \frac{1}{(n-1) \cos^{n-1} x}\cdot$$

$$\int \frac{\sin ax}{\cos^n x}\, dx = 2 \int \frac{\sin (a-1)\, x}{\cos^{n-1} x}\, dx - \int \frac{\sin (a-2)\, x}{\sin^n x}\, dx.$$

$$\int \frac{\sin 3x}{\cos 4x}\, dx = -\frac{(2+\sqrt{2})^{\frac{1}{2}}}{8} \log \frac{\cos x - \cos \frac{\pi}{8}}{\cos x + \cos \frac{\pi}{8}} - \frac{(2-\sqrt{2})^{\frac{1}{2}}}{8} \log \frac{\cos x - \cos \frac{3}{8}\pi}{\cos x + \cos \frac{3}{8}\pi}\cdot$$

$$\int \frac{\sin^2 x}{\cos 3x}\, dx = \frac{1}{3} \log \operatorname{tg} \left(\frac{\pi}{4} + \frac{3}{2} x\right)\cdot$$

$$\int \frac{\cos 2x}{\sin x}\, dx = 2 \cos x + \log \operatorname{tg} \frac{x}{2}\cdot$$

$$\int \frac{\cos 2x}{\sin^2 x}\, dx = -2x - \operatorname{cotg} x.$$

$$\int \frac{\cos 2x}{\sin^3 x}\, dx = -\frac{\cos x}{2 \sin^2 x} - \frac{3}{2} \log \operatorname{tg} \frac{x}{2}\cdot$$

$$\int \frac{\cos 2x}{\sin^4 x}\, dx = -\frac{\cos x}{3 \sin^3 x} + \frac{4}{3} \operatorname{cotg} x.$$

$$\int \frac{\cos 3x}{\sin x}\, dx = -2 \sin^2 x + \log \sin x.$$

$$\int \frac{\cos 3x}{\sin^2 x}\, dx = -4 \sin x - \frac{1}{\sin x}\cdot$$

$$\int \frac{\cos 3x}{\sin^3 x}\, dx = -\frac{1}{2 \sin^2 x} - 4 \log \sin x.$$

$$\int \frac{\cos 3x}{\sin^4 x}\, dx = -\frac{1}{3 \sin^3 x} + \frac{4}{\sin x}\cdot$$

$$\int \frac{\cos 3x}{\sin^m x}\, dx = -\frac{1}{(m-1) \sin^{m-1} x} + \frac{4}{(m-3) \sin^{m-3} x}\cdot$$

$$\int \frac{\cos mx}{\sin^n x}\, dx = -2 \int \frac{\sin (m-1)\, x}{\sin^{n-1} x}\, dx + \int \frac{\cos (m-2)\, x}{\sin^n x}\, dx.$$

$$\int \frac{\cos x}{\sin 2x}\, dx = \frac{1}{2} \log \operatorname{tg} \frac{x}{2}\cdot$$

$$\int \frac{\cos x}{\sin 3x}\, dx = \frac{1}{3} \log \sin x - \frac{1}{6} \log \left(\sin^2 x - \frac{3}{4}\right)\cdot$$

$$\int \frac{\cos^2 x}{\sin 3x}\, dx = \frac{1}{3} \log \operatorname{tg} \frac{3}{2} x.$$

$$\int \frac{\cos 2x}{\sin^2 x \cos^2 x}\, dx = -\frac{2}{\sin 2x}.$$

$$\int \frac{a + b \sin x + c \cos x}{\sin 2x}\, dx = \frac{a}{2} \log \operatorname{tg} x + \frac{b}{2} \log \operatorname{tg}\left(\frac{\pi}{4} + \frac{x}{2}\right) + \frac{c}{2} \log \operatorname{tg} \frac{x}{2}.$$

$$\int \frac{a + b \sin x + c \cos x}{\cos 2x}\, dx = \frac{a}{2} \log \operatorname{tg}\left(\frac{\pi}{4} + x\right) + \frac{c - b}{2\sqrt{2}} \log \operatorname{tg}\left(\frac{\pi}{8} + \frac{x}{2}\right) - \frac{c + b}{2\sqrt{2}} \log \operatorname{tg}\left(\frac{\pi}{8} - \frac{x}{2}\right).$$

Cir U

$$\int \operatorname{tg} x\, dx = -\log \cos x = \log \operatorname{s\acute{e}c} x.$$

$$\int \operatorname{tg} mx\, dx = -\frac{1}{m} \log \cos mx.$$

$$\int \operatorname{tg}(ax + b)\, dx = -\frac{1}{a} \log \cos(ax + b).$$

$$\int \operatorname{tg}^2 x\, dx = \operatorname{tg} x - x.$$

$$\int \operatorname{tg}^3 x\, dx = \frac{1}{2} \operatorname{tg}^2 x + \log \cos x.$$

$$\int \operatorname{tg}^4 x\, dx = \frac{1}{3} \operatorname{tg}^3 x - \operatorname{tg} x + x.$$

$$\int \operatorname{tg}^n x\, dx = \frac{1}{n - 1} \operatorname{tg}^{n-1} x - \int \operatorname{tg}^{n-2} x\, dx.$$

$$\int \operatorname{tg}^{2n} x\, dx = \frac{1}{2n - 1} \operatorname{tg}^{2n-1} x - \frac{1}{2n - 3} \operatorname{tg}^{2n-3} x + \frac{1}{2n - 5} \operatorname{tg}^{2n-5} x - \cdots + (-1)^{n-1} \operatorname{tg} x + (-1)^n x.$$

$$\int \operatorname{tg}^{2n+1} x\, dx = \frac{1}{2n} \operatorname{tg}^{2n} x - \frac{1}{2n - 2} \operatorname{tg}^{2n-2} x + \frac{1}{2n - 4} \operatorname{tg}^{2n-4} x - \cdots + (-1)^{n-1} \frac{1}{2} \operatorname{tg}^2 x + (-1)^{n-1} \log \cos x.$$

$$\int \frac{dx}{\operatorname{tg} x} = \log \sin x.$$

$$\int \frac{dx}{\operatorname{tg}^n x} = -\frac{1}{(n - 1) \operatorname{tg}^{n-1} x} - \int \frac{dx}{\operatorname{tg}^{n-2} x}.$$

$$\int \frac{dx}{1 + \operatorname{tg} x} = \frac{x}{2} + \frac{1}{2} \log(\sin x + \cos x).$$

$$\int \frac{dx}{a + b \operatorname{tg} x} = \frac{1}{a^2 + b^2} \left\{ ax + b \log(a \cos x + b \sin x) \right\}.$$

$$\int \frac{dx}{\{a + (ax + b) \operatorname{tg} x\}^2} = \frac{1}{a} \frac{\operatorname{tg} x}{a + (ax + b) \operatorname{tg} x}.$$

$$\int \operatorname{cotg} x\, dx = \log \sin x.$$

$$\int \operatorname{cotg} mx\, dx = \frac{1}{m} \log \sin mx.$$

$$\int \operatorname{cotg}(ax+b)\,dx = \frac{1}{a}\log\sin(ax+b).$$

$$\int \operatorname{cotg}^2 x\,dx = -\operatorname{cotg} x - x.$$

$$\int \operatorname{cotg}^3 x\,dx = -\frac{1}{2}\operatorname{cotg}^2 x - \log\sin x.$$

$$\int \operatorname{cotg}^n x\,dx = -\frac{1}{n-1}\operatorname{cotg}^{n-1} x - \int \operatorname{cotg}^{n-2} x\,dx.$$

$$\int \operatorname{cotg}^{2n} x\,dx = -\frac{\operatorname{cotg}^{2n-1} x}{2n-1} + \frac{\operatorname{cotg}^{2n-3} x}{2n-3} - \frac{\operatorname{cotg}^{2n-5} x}{2n-5} + \dots - (-1)^{n-1}\operatorname{cotg} x + (-1)^{n-1} x.$$

$$\int \operatorname{cotg}^{2n+1} x\,dx = -\frac{\operatorname{cotg}^{2n} x}{2n} + \frac{\operatorname{cotg}^{2n-2} x}{2n-2} - \frac{\operatorname{cotg}^{2n-4} x}{2n-4} + \dots + (-1)^n \frac{\operatorname{cotg}^2 x}{2} + (-1)^n \log\sin x.$$

$$\int \frac{dx}{\operatorname{cotg} x} = -\log\cos x.$$

$$\int \operatorname{séc} x\,dx = \log\operatorname{tg}\left(\frac{\pi}{4}+\frac{x}{2}\right)\cdot$$

$$\int \operatorname{séc}^2 x\,dx = \operatorname{tg} x.$$

$$\int \operatorname{séc}^4 dx = \operatorname{tg} x + \frac{1}{3}\operatorname{tg}^3 x.$$

$$\int \operatorname{cosé c} x\,dx = \log\operatorname{tg}\frac{x}{2}\cdot$$

$\mathrm{Cir}^{\frac{m}{n}}\,\mathrm{U}$

$$\int \sqrt{1+\cos x}\,dx = 2\sqrt{2}\sin\frac{x}{2} = 2\sqrt{1-\cos x}\cdot$$

$$\int \sqrt{1-\cos x}\,dx = -2\sqrt{2}\cos\frac{x}{2} = -2\sqrt{1+\cos x}\cdot$$

$$\int \sqrt{1-a^2\sin^2 x}\,\sin x\,dx = -\frac{1}{2}\sqrt{\Phi}\cos x - \frac{1-a^2}{2a}\log\left\{a\cos x + \sqrt{\Phi}\right\}\cdot$$

$$\int (1-a^2\sin^2 x)^{\frac{3}{2}}\sin x\,dx = -\frac{1}{4}\Phi^{\frac{3}{2}}\cos x + \frac{3}{4}(1-a^2)\int\sqrt{\Phi}\sin x\,dx.$$

$$\int \sqrt{1-a^2\sin^2 x}\,\cos x\,dx = \frac{1}{2}\sqrt{1-a^2\sin^2 x}\,\sin x + \frac{1}{2a}\operatorname{arc\,sin}(a\sin x).$$

$$\int \frac{dx}{\sqrt{\sin x\cos^3 x}} = 2\sqrt{\operatorname{tg} x}\cdot$$

$$\int \frac{dx}{\sqrt{\sin x}\,\cos^{\frac{3}{2}} x} = 2\sqrt{\operatorname{tg} x}\left(1+\frac{1}{5}\operatorname{tg}^2 x\right)$$

$$\int \frac{dx}{\sqrt{\sin 2x\cos^3 x}} = \sqrt{2\operatorname{tg} x}\left(1+\frac{1}{5}\operatorname{tg}^2 x\right)\cdot$$

$$\int \frac{dx}{\sqrt{a^2 + b^2 \operatorname{tg}^2 x}} = \frac{1}{\sqrt{a^2 - b^2}} \arcsin \left\{ \sqrt{1 - \frac{b^2}{a^2}} \sin x \right\}. \qquad a > b.$$

$$= \frac{1}{\sqrt{b^2 - a^2}} \log \left\{ \sqrt{b^2 - a^2} \sin x + \sqrt{a^2 \cos^2 x + b^2 \sin^2 x} \right\}. \qquad a < b.$$

$$\int \frac{dx}{\cos^2 x \sqrt{a + b \operatorname{tg} x}} = \frac{2}{b} \sqrt{a + b \operatorname{tg} x}.$$

$$\int \frac{dx}{\sqrt{1 - \operatorname{tg}^2 x} \cos^2 x} = \arcsin (\operatorname{tg} x).$$

$$\int \frac{\sin x}{\sqrt{c^2 \sin^2 x + 2ab \cos x - (a^2 + b^2)}} dx = \frac{1}{c} \arcsin \frac{ab - c^2 \cos x}{\sqrt{(c^2 - a^2)(c^2 - b^2)}}.$$

$$\int \frac{\cos x}{\sqrt{a^2 \sin^2 x + b^2 \cos^2 x}} dx = \frac{1}{\sqrt{a^2 - b^2}} \log \left\{ \sqrt{a^2 - b^2} \sin x + \sqrt{a^2 \sin^2 x + b^2 \cos^2 x} \right\}. \qquad a > b.$$

$$= \frac{1}{\sqrt{b^2 - a^2}} \arcsin \frac{\sqrt{b^2 - a^2} \sin x}{b}. \qquad a < b.$$

$$\int \frac{\operatorname{tg} x}{\sqrt{a^2 + b^2 \operatorname{tg}^2 x}} dx = \frac{1}{\sqrt{b^2 - a^2}} \arccos \left\{ \sqrt{1 - \frac{a^2}{b^2}} \cos x \right\}. \qquad a < b$$

$$= - \frac{1}{\sqrt{a^2 - b^2}} \log \left\{ \sqrt{a^2 - b^2} \cos x + \sqrt{a^2 \cos^2 x + b^2 \sin^2 x} \right\}. \qquad a > b.$$

$$\int \frac{\sin^3 x}{\sqrt{\cos x}} dx = \frac{2}{5} \left(\cos^2 x - 5 \right) \sqrt{\cos x}.$$

$$\int \frac{\cos^3 x}{\sqrt{\sin x}} dx = \frac{2}{5} \left(5 - \sin^2 x \right) \sqrt{\sin x}.$$

$$\int \frac{\cos^2 x \sin x}{\sqrt{1 + a^2 \sin^2 x}} dx = \frac{\sqrt{1 + a^2 \sin^2 x} \cos x}{2a^2} + \frac{1 + a^2}{2a^3} \arccos \frac{a \cos x}{\sqrt{1 + a^2}}.$$

$$\int \frac{x \cos (a^2 - x^2)}{\sqrt{\sin (a^2 - x^2)}} dx = - \sqrt{\sin (a^2 - x^2)}.$$

$$\int \frac{a - b \cos x}{\sin x \sqrt{c^2 \sin^2 x + 2ab \cos x - (a^2 + b^2)}} dx = \arcsin \frac{b - a \cos x}{\sqrt{c^2 - a^2} \sin x}.$$

$$\int \frac{\sqrt{a^2 + b^2 \sin^2 x}}{\sin x} dx = b \arccos \left\{ \frac{b \cos x}{\sqrt{a^2 + b^2}} \right\} - a \log \left(a \operatorname{cotg} x + \sqrt{b^2 + a^2 \operatorname{cos\acute{e}c}^2 x} \right).$$

$$\int \sqrt{1 - \frac{\sin^2 \alpha}{\cos^2 x}} dx = \arcsin \left(\frac{\sin x}{\cos \alpha} \right) - \sin \alpha \arcsin (\operatorname{tg} \alpha \operatorname{tg} x).$$

Arc Cir U

$$\int \arcsin x \, dx = x \arcsin x + \sqrt{1 - x^2}.$$

$$\int \left(\arcsin x \right)^n dx = x \Phi^n + n \Phi^{n-1} \sqrt{1 - x^2} - n (n - 1) \int \Phi^{n-2} dx.$$

$$= x \left\{ \Phi^n - n (n - 1) \Phi^{n-2} + \cdots \right\} + \sqrt{1 - x^2} \left\{ n \Phi^{n-1} - n (n - 1)(n - 2) \Phi^{n-3} + \cdots \right\}.$$

$$\int x \operatorname{arc\,sin} x \, dx = \frac{1}{4}(2x^2 - 1) \operatorname{arc\,sin} x + \frac{1}{4} x \sqrt{1 - x^2}.$$

$$\int x^m \operatorname{arc\,sin} x \, dx = \frac{x^{m+1}}{m+1} \operatorname{arc\,sin} x - \frac{1}{m+1} \int \frac{x^{m+1}}{\sqrt{1 - x^2}} dx.$$

$$\int \mathrm{U} \operatorname{arc\,sin} x \, dx = \mathrm{U}_1 \operatorname{arc\,sin} x - \int \frac{\mathrm{U}_1}{\sqrt{1 - x^2}} dx.$$

$$\int \operatorname{arc\,sin} \sqrt{\frac{x}{a + x}} \, dx = (a + x) \operatorname{arc\,sin} \sqrt{\frac{x}{a + x}} - \sqrt{ax} = (a + x) \operatorname{arc\,tg} \sqrt{\frac{x}{a}} - \sqrt{ax}.$$

$$\int x \operatorname{arc\,sin} \frac{1}{2} \sqrt{\frac{2a - x}{a}} \, dx = \frac{x^2}{2} \operatorname{arc\,sin} \frac{1}{2} \sqrt{\frac{2a - x}{a}} + \frac{a^2}{2} \operatorname{arc\,sin} \frac{x}{2a} - \frac{x}{8} \sqrt{4a^2 - x^2}.$$

$$\int \frac{dx}{\sqrt{1 - x^2} \operatorname{arc\,sin} x} = \log \operatorname{arc\,sin} x.$$

$$\int \frac{dx}{1 + \left(\operatorname{arc\,sin} \frac{x}{a}\right)^2} = \sqrt{a^2 - x^2} \operatorname{arc\,tg} \left(\operatorname{arc\,sin} \frac{x}{a}\right).$$

$$\int \frac{\operatorname{arc\,sin} x}{x^2} dx = -\frac{1}{x} \operatorname{arc\,sin} x + \log \frac{x}{1 + \sqrt{1 - x^2}}.$$

$$\int \frac{\operatorname{arc\,sin} x}{\sqrt{1 - x^2}} dx = \frac{1}{2} \left(\operatorname{arc\,sin} x\right)^2 = x - \sqrt{1 - x^2} \operatorname{arc\,sin} x.$$

$$\int \frac{\operatorname{arc\,sin} x}{(1 - x^2)^{\frac{3}{2}}} dx = \frac{x \operatorname{arc\,sin} x}{\sqrt{1 - x^2}} + \frac{1}{2} \log (1 - x^2).$$

$$\int \frac{x \operatorname{arc\,sin} x}{\sqrt{1 - x^2}} dx = x - \sqrt{1 - x^2} \operatorname{arc\,sin} x.$$

$$\int \frac{x \operatorname{arc\,sin} x}{(1 - x^2)^{\frac{3}{2}}} dx = \frac{1}{2} \log \frac{1 - x}{1 + x} + \frac{\operatorname{arc\,sin} x}{\sqrt{1 - x^2}}.$$

$$\int \frac{x^2 \operatorname{arc\,sin} x}{\sqrt{1 - x^2}} dx = \frac{1}{4} x^2 - \frac{1}{2} x \sqrt{1 - x^2} \operatorname{arc\,sin} x + \frac{1}{4} \left(\operatorname{arc\,sin} x\right)^2.$$

$$\int \frac{x^3 \operatorname{arc\,sin} x}{\sqrt{1 - x^2}} dx = \frac{2}{3} x + \frac{1}{9} x^3 - \frac{1}{3} (x^2 + 2) \sqrt{1 - x^2} \operatorname{arc\,sin} x.$$

$$\int \frac{\operatorname{arc\,sin} (1 - 2x)}{\sqrt{x - x^2}} dx = -\frac{1}{2} \left\{ \operatorname{arc\,sin} (1 - 2x) \right\}^2.$$

$$\int \frac{a - 2x}{\sqrt{ax - x^2}} \operatorname{arc\,sin} \sqrt{\frac{a - x}{a + x}} \, dx = 2 \sqrt{ax - x^2} \operatorname{arc\,sin} \sqrt{\frac{a - x}{a + x}} + a \sqrt{2} \log (a + x).$$

$$\int \operatorname{arc\,cos} x \, dx = x \operatorname{arc\,cos} x - \sqrt{1 - x^2}.$$

$$\int x^m \operatorname{arc\,cos} x \, dx = \frac{x^{m+1}}{m+1} \operatorname{arc\,cos} x + \frac{1}{m+1} \int \frac{x^{m+1}}{\sqrt{1 - x^2}} dx.$$

$$\int \mathrm{U} \operatorname{arc\,cos} x \, dx = \mathrm{U}_1 \operatorname{arc\,cos} x + \int \frac{\mathrm{U}_1}{\sqrt{1 - x^2}} dx.$$

$$\int \frac{dx}{\sqrt{1 - x^2} \operatorname{arc\,cos} x} = -\log \operatorname{arc\,cos} x.$$

$$\int \frac{\text{arc cos } x}{\sqrt{1-x^2}}\, dx = -\frac{1}{2}\left(\text{arc cos } x\right)^2.$$

$$\int \text{arc tg } x\, dx = x \text{ arc tg } x - \log \sqrt{1+x^2}.$$

$$\int x \text{ arc tg } x\, dx = -\frac{1}{2}x + \frac{1}{2}(x^2+1) \text{ arc tg } x.$$

$$\int x^2 \text{ arc tg } x\, dx = \frac{1}{3}x^3 \text{ arc tg } x - \frac{1}{6}x^2 + \frac{1}{3}\log \sqrt{1+x^2}.$$

$$\int x^m \text{ arc tg } x\, dx = \frac{x^{m+1}}{m+1} \text{ arc tg } x - \frac{1}{m+1}\int \frac{x^{m+1}}{1+x^2}\, dx.$$

$$\int \text{U arc tg } x\, dx = \text{U}_1 \text{ arc tg } x - \int \frac{\text{U}_1}{1+x^2}\, dx.$$

$$\int \frac{dx}{(1+x^2) \text{ arc tg } x} = \log \text{ arc tg } x.$$

$$\int \frac{\text{arc tg } x}{x^2}\, dx = -\frac{1}{x} \text{ arc tg } x + \log \frac{x}{\sqrt{1+x^2}}.$$

$$\int \frac{\text{arc tg } x}{1+x^2}\, dx = \frac{1}{2}\left(\text{arc tg } x\right)^2.$$

$$\int \frac{\text{arc tg } x}{(1+x^2)^{\frac{3}{2}}}\, dx = \frac{1 + x \text{ arc tg } x}{\sqrt{1+x^2}}.$$

$$\int \frac{x \text{ arc tg } x}{1+x^2}\, dx = \frac{1}{2} \text{ arc tg } x \log(1+x^2) - \frac{1}{2}\int \frac{\log(1+x^2)}{1+x^2}\, dx.$$

$$\int \frac{x^2 \text{ arc tg } x}{1+x^2}\, dx = -\log \sqrt{1+x^2} + \left(x - \frac{1}{2} \text{ arc tg } x\right) \text{ arc tg } x.$$

$$\int \frac{x^3 \text{ arc tg } x}{1+x^2}\, dx = -\frac{1}{2}x + \frac{1}{2}(1+x^2) \text{ arc tg } x - \int \frac{x \text{ arc tg } x}{1+x^2}\, dx.$$

$$\int \frac{x^4 \text{ arc tg } x}{1+x^2}\, dx = -\frac{1}{6}x^2 + \frac{2}{3}\log(1+x^2) + \left(\frac{x^3}{3} - x\right) \text{ arc tg } x + \frac{1}{2}\left(\text{arc tg } x\right)^2.$$

$$\int \frac{x(3-x^2) \text{ arc tg } x}{(1-x^2)^{\frac{3}{2}}}\, dx = \frac{1+x^2}{\sqrt{1-x^2}} \text{ arc tg } x - \text{arc sin } x.$$

$$\int \text{arc cotg } x\, dx = x \text{ arc cotg } x + \log \sqrt{1+x^2}.$$

$$\int x^m \text{ arc cotg } x\, dx = \frac{x^{m+1}}{m+1} \text{ arc cotg } x + \frac{1}{m+1}\int \frac{x^{m+1}}{1+x^2}\, dx.$$

$$\int \text{U arc cotg } x\, dx = \text{U}_1 \text{ arc cotg } x + \int \frac{\text{U}_1}{1+x^2}\, dx.$$

$$\int \frac{dx}{(1+x^2) \text{ arc cotg } x} = -\log \text{ arc cotg } x.$$

$$\int \frac{\text{arc cotg } x}{1+x^2}\, dx = -\frac{1}{2}\left(\text{arc cotg } x\right)^2.$$

$$\int \text{arc séc } x\, dx = x \text{ arc séc } x - \log\left(x + \sqrt{x^2 - 1}\right).$$

$$\int x^m \text{ arc séc } x\, dx = \frac{x^{m+1}}{m+1} \text{ arc séc } x - \frac{1}{m+1} \int \frac{x^m}{\sqrt{x^2-1}} dx.$$

$$\int \text{U arc séc } x\, dx = \text{U}_1 \text{ arc séc } x - \int \frac{\text{U}_1}{x\sqrt{x^2-1}} dx.$$

$$\int \frac{dx}{x\sqrt{x^2-1} \text{ arc séc } x} = \log \text{ arc séc } x.$$

$$\int \frac{\text{arc séc } x}{x\sqrt{x^2-1}} dx = \frac{1}{2}\left(\text{arc séc } x\right)^2.$$

$$\int \frac{x}{\sqrt{1-x^2}} \text{ arc séc} \sqrt{\frac{1+x^2}{1-x^2}}\, dx = -\sqrt{1-x^2} \text{ arc séc} \sqrt{\frac{1+x^2}{1-x^2}} + \sqrt{2} \text{ arc tg } x.$$

$$\int \text{arc coséc } x\, dx = x \text{ arc coséc } x + \log\left(x + \sqrt{x^2-1}\right).$$

$$\int x^m \text{ arc coséc } x\, dx = \frac{x^{m+1}}{m+1} \text{ arc coséc } x + \frac{1}{m+1} \int \frac{x^m}{\sqrt{x^2-1}} dx.$$

$$\int \text{U arc coséc } x\, dx = \text{U}_1 \text{ arc coséc } x + \int \frac{\text{U}_1}{x\sqrt{x^2-1}} dx.$$

$$\int \frac{dx}{x\sqrt{x^2-1} \text{ arc coséc } x} = -\log \text{ arc coséc } x.$$

$$\int \frac{\text{arc coséc } x}{x\sqrt{x^2-1}} dx = -\frac{1}{2}\left(\text{arc coséc } x\right)^2.$$

Log U

$$\int \log x\, dx = x(\log x - 1).$$

$$\int x \log x\, dx = \frac{x^2}{2}\left(\log x - \frac{1}{2}\right).$$

$$\int x^m \log x\, dx = \frac{x^{m+1}}{m+1}\left(\log x - \frac{1}{m+1}\right).$$

$$\int \left(a + bx\right) \log x\, dx = \frac{(a+bx)^2}{2b} \log x - \frac{a^2}{2b} \log x - ax - \frac{1}{4} bx^2.$$

$$\int \left(a + bx\right)^2 \log x\, dx = \frac{(a+bx)^3}{3b} \log x - \frac{a^3}{3b} \log x - a^2 x - \frac{abx^2}{2} - \frac{b^2 x^3}{9}.$$

$$\int \left(a + bx\right)^m \log x\, dx = \frac{(a+bx)^{m+1}}{(m+1)b} \log x - \frac{1}{(m+1)b} \int \frac{(a+bx)^{m+1}}{x} dx.$$

$$\int \text{U} \log x\, dx = \text{U}_1 \log x - \int \frac{\text{U}_1}{x} dx.$$

$$\int x^m \log(a + bx)\, dx = \frac{x^{m+1}}{m+1} \log(a+bx) - \frac{b}{m+1} \int \frac{x^{m+1}}{a+bx} dx.$$

$$\int x^3 \log(1+x^2)\,dx = \frac{x^4}{4}\log(1+x^2) - \frac{1}{8}x^4 + \frac{1}{4}x^2 - \frac{1}{2}\log(1+x^2).$$

$$\int x^2 \log(x^2-1)\,dx = \frac{1}{3}x^3\log(x^2-1) - \frac{2}{3}x - \frac{2}{9}x^3 - \frac{1}{3}\log\frac{x-1}{x+1}.$$

$$\int x \log(1-x^4)\,dx = \frac{1}{2}(1+x^2)\log(1+x^2) - \frac{1}{2}(1-x^2)\log(1-x^2) - x^2.$$

$$\int V \log U\,dx = V_1 \log U - \int \frac{U' V_1}{U}\,dx.$$

$$\int \log\left\{\sqrt{x-a}+\sqrt{x-b}\right\}dx = \frac{1}{2}\left(2x-a-b\right)\log\left\{\sqrt{x-a}+\sqrt{x-b}\right\} - \frac{1}{2}\sqrt{(x-a)(x-b)}.$$

$$\int \frac{dx}{\log x} = \log\log x + \frac{\log x}{1} + \frac{1}{2}\frac{(\log x)^2}{1.2} + \frac{1}{3}\frac{(\log x)^3}{1.2.3} + \frac{1}{4}\frac{(\log x)^4}{1.2.3.4} + \cdots$$

$$\int \frac{dx}{\log\frac{1}{x}} = \log\log x - \frac{\log x}{1} + \frac{1}{2}\frac{(\log x)^2}{1.2} - \frac{1}{3}\frac{(\log x)^3}{1.2.3} + \frac{1}{4}\frac{(\log x)^4}{1.2.3.4} - \cdots$$

$$\int \frac{dx}{x\log x} = \log\log x.$$

$$\int \frac{\log x}{x}\,dx = \frac{1}{2}\left(\log x\right)^2.$$

$$\int \frac{\log x}{a+bx}\,dx = \frac{1}{b}\log x\log(a+bx) - \frac{1}{b}\int\frac{\log(a+bx)}{x}\,dx.$$

$$= \frac{1}{b}\log x\log\frac{a+bx}{a} - \frac{x}{a} + \frac{bx^2}{(2a)^2} - \frac{b^2x^3}{(3a)^2\,a} + \frac{b^3x^4}{(4a)^2\,a^2} - \cdots$$

$$= \frac{1}{b}\log x\log(a+bx) - \frac{1}{2b}\left(\log bx\right)^2 + \frac{a}{b^2x} - \frac{a^2}{2^2b^3x^2} + \frac{a^3}{3^2b^4x^3} - \cdots$$

$$\int \frac{\log x}{(1-x)^2}\,dx = \frac{x\log x}{1-x} + \log(1-x).$$

$$\int \frac{\log x}{(a+bx)^2}\,dx = -\frac{\log x}{b(a+bx)} + \frac{1}{ab}\log\frac{x}{a+bx}.$$

$$\int \frac{\log x}{(1-x)^{n+1}}\,dx = \frac{\log x}{n(1-x)^n} - \frac{1}{n}\int\frac{dx}{x(1-x)^n}.$$

$$\int \frac{\log x}{(a+bx)^m}\,dx = -\frac{\log x}{(m-1)\,b\,(a+bx)^{m-1}} + \frac{1}{(m-1)\,b}\int\frac{dx}{x\,(a+bx)^{m-1}}.$$

$$= -\frac{\log x}{(m-1)\,b\,\Phi^{m-1}} + \frac{1}{(m-1)ab}\left\{\frac{1}{(m-2)\Phi^{m-2}} + \frac{1}{(m-3)\,a\,\Phi^{m-3}} + \frac{1}{(m-4)\,a^2\Phi^{m-4}} + \cdots + \frac{1}{2a^{m-4}\Phi^2} + \frac{1}{a^{m-3}\Phi}\right\}$$

$$+ \frac{1}{(m-1)\,a^{m-1}b}\log\frac{x}{\Phi}.$$

$$\int \frac{\log x}{(a^2+b^2x^2)^{\frac{3}{2}}}\,dx = \frac{1}{a^2b}\left\{\frac{bx\log x}{\sqrt{a^2+b^2x^2}} - \log\frac{a+bx}{a-bx}\right\}.$$

$$\int \frac{x\log x}{\sqrt{a^2+x^2}}\,dx = \left(\log x - 1\right)\sqrt{a^2+x^2} - \frac{a}{2}\log\frac{\sqrt{a^2+x^2}-a}{\sqrt{a^2+x^2}+a}.$$

$$\int \frac{\log(a+bx)}{x}\,dx = \log a \log x + \frac{bx}{a} - \frac{b^2x^2}{2^2a^2} + \frac{b^3x^3}{3^2a^3} - \cdots$$

$$= \frac{1}{2}\left(\log bx\right)^2 - \frac{a}{bx} + \frac{a^2}{2^2b^2x^2} - \frac{a^3}{3^2b^3x^3} + \cdots$$

$$\int \frac{\log(1-x)}{x\sqrt{x}}\,dx = -\frac{2}{\sqrt{x}}\log(1-x) + 2\log\frac{1-\sqrt{x}}{1+\sqrt{x}}.$$

$$\int \frac{\log\log x}{x}\,dx = \left\{\log\log x - 1\right\}\log x.$$

$$\int \frac{x}{(1+x^2)\log(1+x^2)}\,dx = \frac{1}{2}\log\left\{\log(1+x^2)\right\}.$$

$$\int \frac{x^m}{\log x}\,dx = \log\log x + (m+1)\log x + \frac{(m+1)^2(\log x)^2}{1.2^2} + \frac{(m+1)^3(\log x)^3}{1.2.3^2} + \frac{(m+1)^4(\log x)^4}{1.2.3.4^2} + \cdots$$

$$\int \frac{1}{x}\log\frac{1}{1-x}\,dx = x + \frac{1}{4}x^2 + \frac{1}{9}x^3 + \frac{1}{16}x^4 + \cdots$$

$$\int \frac{1}{x\sqrt{x}}\log\frac{1}{1-x}\,dx = \frac{2\log(1-x)}{\sqrt{x}} + 2\log\frac{1+\sqrt{x}}{1-\sqrt{x}}.$$

$\left(\text{Log U}\right)^n$

$$\int \left(\log x\right)^2 dx = x\left\{\left(\log x\right)^2 - 2\log x + 2\right\}.$$

$$\int x\left(\log x\right)^2 dx = \frac{x^2}{2}\left\{\left(\log x\right)^2 - \log x + \frac{1}{2}\right\}.$$

$$\int x^2\left(\log x\right)^2 dx = \frac{x^3}{3}\left\{\left(\log x\right)^2 - \frac{2}{3}\log x + \frac{2}{9}\right\}.$$

$$\int x^3\left(\log x\right)^2 dx = \frac{x^4}{4}\left\{\left(\log x\right)^2 - \frac{1}{2}\log x + \frac{1}{8}\right\}.$$

$$\int x^m\left(\log x\right)^2 dx = x^{m+1}\left\{\frac{(\log x)^2}{m+1} - \frac{2\log x}{(m+1)^2} + \frac{2}{(m+1)^3}\right\}.$$

$$\int \frac{(\log x)^2}{x}\,dx = \frac{1}{3}\left(\log x\right)^3$$

$$\int \frac{x^m}{(\log x)^2}\,dx = -\frac{x^{m+1}}{\log x} + (m+1)\int \frac{x^m}{\log x}\,dx.$$

$$\int x^m\left(\log x\right)^3 dx = \frac{x^{m+1}}{m+1}\left\{\left(\log x\right)^3 - \frac{3}{m+1}\left(\log x\right)^2 + \frac{6}{(m+1)^2}\log x - \frac{6}{(m+1)^3}\right\}.$$

$$\int \frac{x^m}{(\log x)^3}\,dx = -\frac{x^{m+1}}{2(\log x)^2} - \frac{(m+1)x^{m+1}}{2\log x} + \frac{(m+1)^2}{2}\int \frac{x^m}{\log x}\,dx.$$

$$\int \left(\log x\right)^n dx = \int z^n e^z\,dz. \qquad \log x = z.$$

$$= x\left(\log x\right)^n - n\int \left(\log x\right)^{n-1} dx.$$

$$= x \left\{ \left(\log x\right)^n - n \left(\log x\right)^{n-1} + n\,(n-1)\left(\log x\right)^{n-2} - \ldots \right\}.$$

$$\int x^m \left(\log x\right)^n dx = \int z^n\, e^{(m+1)z}\, dz. \qquad \log x = z.$$

$$= \frac{1}{(m+1)^{n+1}} \int \left(\log z\right)^n dz. \qquad x^{m+1} = z.$$

$$= \frac{x^{m+1}\,(\log x)^n}{m+1} - \frac{n}{m+1} \int x^m \left(\log x\right)^{n-1} dx.$$

$$= \frac{x^{m+1}}{m+1} \left\{ \left(\log x\right)^n - \frac{n}{m+1}\left(\log x\right)^{n-1} + \frac{n\,(n-1)}{(m+1)^2}\left(\log x\right)^{n-2} - \frac{n\,(n-1)\,(n-2)}{(m+1)^3}\left(\log x\right)^{n-3} + \ldots \pm \frac{n\,(n-1)\ldots 2.1}{(m+1)^n} \right\}.$$

$$\int U \left(\log x\right)^n dx = U_1 \left(\log x\right)^n - n \int \frac{U_1\,(\log x)^{n-1}}{x}\, dx.$$

$$= U_1 \left(\log x\right)^n - n\, U_2 \left(\log x\right)^{n-1} + n\,(n-1)\, U_3 \left(\log x\right)^{n-2} - \ldots$$

$$\int \left\{ \log\left(x + \sqrt{1+x^2}\right) \right\}^n dx = \left(\log \Phi\right)^{n-1} \left\{ x \log \Phi - n \sqrt{1+x^2} \right\} + n\,(n-1) \int \left(\log \Phi\right)^{n-2} dx.$$

$$\int \frac{dx}{x\,(\log x)^n} = -\frac{1}{(n-1)\,(\log x)^{n-1}}.$$

$$\int \frac{(\log x)^n}{x}\, dx = \frac{1}{n+1}\left(\log x\right)^{n+1}.$$

$$\int \frac{x^m}{(\log x)^n}\, dx = -\frac{x^{m+1}}{(n-1)\,(\log x)^{n-1}} + \frac{m+1}{n-1} \int \frac{x^m}{(\log x)^{n-1}}\, dx.$$

$$= -x^{m+1} \left\{ \frac{1}{(n-1)\,(\log x)^{n-1}} + \frac{m+1}{(n-1)\,(n-2)\,(\log x)^{n-2}} + \frac{(m+1)^2}{(n-1)\,(n-2)\,(n-3)\,(\log x)^{n-3}} + \ldots + \frac{(m+1)^{n-2}}{(n-1)\ldots 2.1 \log x} \right\}$$

$$+ \frac{(m+1)^{n-1}}{(n-1)\ldots 2.1} \int \frac{x^m}{\log x}\, dx.$$

$$\int \frac{U}{(\log x)^n}\, dx = -\frac{Ux}{(n-1)\,(\log x)^{n-1}} - \frac{U'x}{(n-1)\,(n-2)\,(\log x)^{n-2}} - \frac{U''x}{(n-1)\,(n-2)\,(n-3)\,(\log x)^{n-3}} - \ldots$$

e^U

$$\int e^x\, dx = e^x.$$

$$\int e^{mx}\, dx = \frac{1}{m}\, e^{mx}.$$

$$\int e^{mx+n}\, dx = \frac{1}{m}\, e^{mx+n}.$$

$$\int e^{m \arcsin x}\, dx = \frac{x + m\sqrt{1-x^2}}{1+m^2}\, e^{m \arcsin x}.$$

$$\int x e^x\, dx = (x-1)\, e^x.$$

$$\int x e^{ax}\, dx = \frac{ax-1}{a^2}\, e^{ax}.$$

$$\int x^2 e^x\,dx = (x^2 - 2x + 2)\,e^x.$$

$$\int x^2 e^{ax}\,dx = \frac{a^2x^2 - 2ax + 2}{a^3}\,e^{ax}.$$

$$\int x^3 e^x\,dx = (x^3 - 3x^2 + 6x - 6)\,e^x.$$

$$\int x^4 e^x\,dx = (x^4 - 4x^3 + 12x^2 - 24x + 24)\,e^x.$$

$$\int x^m e^{ax}\,dx = \frac{x^m e^{ax}}{a} - \frac{m}{a}\int x^{m-1} e^{ax}\,dx = \frac{e^{ax}}{a}\left\{ x^m - \frac{m}{a}x^{m-1} + \frac{m\,(m-1)}{a^2}x^{m-2} - \dots \right\}.$$

$$\int \mathrm{U}\,e^{ax}\,dx = \frac{1}{a}\,\mathrm{U}\,e^{ax} - \frac{1}{a}\int \mathrm{U}'\,e^{ax}\,dx = \frac{e^{ax}}{a}\left\{ \mathrm{U} - \frac{\mathrm{U}'}{a} + \frac{\mathrm{U}''}{a^2} - \dots + \frac{(-1)^n}{a^n}\,\mathrm{U}^{(n)} \right\}.$$

$$\int \frac{dx}{e^{ax}} = -\frac{1}{ae^{ax}}.$$

$$\int \frac{dx}{x^m e^x} = -\frac{1}{(m-1)\,x^{m-1}\,e^x} - \frac{1}{m-1}\int \frac{dx}{x^{m-1}\,e^x}.$$

$$\int \frac{dx}{1+e^x} = \log\frac{e^x}{1+e^x} = x - \log(1+e^x).$$

$$\int \frac{dx}{a+be^x} = \frac{x}{a} - \frac{1}{a}\log(a+be^x).$$

$$\int \frac{dx}{1-e^{-x}} = \log\,(e^x - 1).$$

$$\int \frac{dx}{e^x + e^{-x}} = \text{arc tg}\; e^x.$$

$$\int \frac{dx}{e^{ax} + e^{-ax}} = \frac{1}{a}\,\text{arc tg}\; e^{ax}.$$

$$\int \frac{dx}{a^2e^x + b^2e^{-x}} = \frac{1}{ab}\,\text{arc tg}\left(\frac{a}{b}\,e^x\right).$$

$$\int \frac{e^x}{x}\,dx = \int \frac{dz}{\log z}. \qquad e^x = z.$$

$$= \log x + \frac{x}{1} + \frac{1}{2}\,\frac{x^2}{1.2} + \frac{1}{3}\,\frac{x^3}{1.2.3} + \frac{1}{4}\,\frac{x^4}{1.2.3.4} + \cdot$$

$$\int \frac{e^{ax}}{x}\,dx = \log x + \frac{ax}{1} + \frac{1}{2}\,\frac{(ax)^2}{1.2} + \frac{1}{3}\,\frac{(ax)^3}{1.2.3} + \dots$$

$$\int \frac{e^x}{x^2}\,dx = -\frac{e^x}{x} + \int \frac{e^x}{x}\,dx.$$

$$\int \frac{e^x}{x^3}\,dx = -\frac{1}{2}\,e^x\left(\frac{1}{x^2} + \frac{1}{x}\right) + \frac{1}{2}\int \frac{e^x}{x}\,dx.$$

$$\int \frac{e^{ax}}{x^3}\,dx = -\frac{e^{ax}}{2x^2}\left(1 + ax\right) + \frac{a^2}{2}\int \frac{e^{ax}}{x}\,dx.$$

$$\int \frac{e^x}{x^4}\,dx = -\frac{e^x}{6x^3}\left(x^2 + x + 2\right) + \frac{1}{6}\int \frac{e^x}{x}\,dx.$$

$$\int \frac{e^{ax}}{x^m}\, dx = -\frac{e^{ax}}{(m-1)\, x^{m-1}} + \frac{a}{m-1} \int \frac{e^{ax}}{x^{m-1}}\, dx = \frac{e^{ax}}{a} \left\{ \frac{1}{x^m} + \frac{m}{ax^{m+1}} + \frac{m\,(m+1)}{a^2\, x^{m+2}} + \cdots \right\}.$$

$$\int \frac{e^{nx}}{(x-a)^m}\, dx = -\frac{e^{nx}}{(m-1)\,(x-a)^{m-1}} + \frac{n}{m-1} \int \frac{e^{ax}}{(x-a)^{m-1}}\, dx.$$

$$\int \frac{e^{a \operatorname{arc\,tg} x}}{(1+x^2)^{\frac{3}{2}}}\, dx = \frac{(a+x)\; e^{a \operatorname{arc\,tg} x}}{(1+a^2)\sqrt{1+x^2}}.$$

$$\int \frac{xe^x}{(1+x)^2}\, dx = \frac{e^x}{1+x}.$$

$$\int \frac{x\; e^{\operatorname{arc\,sin} x}}{\sqrt{1-x^2}}\, dx = \frac{1}{2}\,(x + \sqrt{1-x^2})\; e^{\operatorname{arc\,sin} x}.$$

$$\int \frac{x\; e^{a \operatorname{arc\,tg} x}}{(1+x^2)^{\frac{3}{2}}}\, dx = \frac{(ax-1)\; e^{a \operatorname{arc\,tg} x}}{(1+a^2)\sqrt{1+x^2}}.$$

$$\int \frac{(x^2+1)\, e^x}{(x+1)^2}\, dx = \frac{x-1}{x+1}\, e^x.$$

$$\int \frac{(2-x^2)\, e^x}{\sqrt{1+x}\,(1-x)^{\frac{3}{2}}}\, dx = e^x \sqrt{\frac{1+x}{1-x}}.$$

$$\int \frac{(ax-1)\,(bx-1)\; e^{(a+b)\,x}}{x^2}\, dx = ab\; e^{(a+b)\,x} \left\{ \frac{1}{a+b} - \frac{1}{abx} \right\}.$$

$$\int \frac{x}{e^{2x}}\, dx = -\frac{x}{2\, e^{2x}} - \frac{1}{4\, e^{2x}}.$$

$$\int \frac{x^3}{e^x}\, dx = -\frac{1}{e^x}\,(x^3 + 3x^2 + 6x + 6).$$

$$\int \frac{x}{(e^x-1)^2}\, dx = -\frac{1}{2}\, \frac{e^x+1}{e^x-1}.$$

$$\int \frac{e^x}{1+e^{2x}}\, dx = \operatorname{arc\,tg} e^x.$$

$$\int \frac{e^x-1}{e^x+1}\, dx = 2 \log \left(e^{\frac{x}{2}} + e^{-\frac{x}{2}} \right).$$

$$\int \frac{a+e^x}{b+e^x}\, dx = \frac{ax}{b} - \frac{a-b}{b} \log\,(b+e^x).$$

$$\int \frac{a + be^x + ce^{2x}}{m+e^x}\, dx = \frac{ax}{m} + ce^x - \frac{a - bm + cm^2}{m} \log\,(m+e^x).$$

a^U

$$\int a^x\, dx = \frac{a^x}{\log a}.$$

$$\int a^{mx}\, dx = \frac{a^{mx}}{m \log a}.$$

$$\int x a^x \, dx = \frac{x a^x}{\log a} - \frac{a^x}{(\log a)^2}.$$

$$\int x^2 a^x \, dx = \frac{a^x}{\log a} \left\{ x^2 - \frac{2x}{\log a} + \frac{2}{(\log a)^2} \right\}.$$

$$\int x^3 a^x \, dx = a^x \left\{ \frac{x^3}{\log a} - \frac{3x^2}{(\log a)^2} + \frac{6x}{(\log a)^3} - \frac{6}{(\log a)^4} \right\}.$$

$$\int x^m a^x \, dx = \frac{a^x}{\log a} \left\{ x^m - \frac{m x^{m-1}}{\log a} + \frac{m(m-1) x^{m-2}}{(\log a)^2} - \frac{m(m-1)(m-2) x^{m-3}}{(\log a)^3} + \cdots \pm \frac{m(m-1)\cdots 2.1}{(\log a)^m} \right\}.$$

$$\int x^m a^{nx} \, dx = \int \left\{ 1 + \frac{nx \log x}{1} + \frac{n^2 x^2 (\log x)^2}{1.2} + \frac{n^3 x^3 (\log x)^3}{1.2.3} + \cdots \right\} x^m \, dx.$$

$$\int \mathrm{U} a^x \, dx = \frac{\mathrm{U} a^x}{\log a} - \frac{1}{\log a} \int \mathrm{U}' a^x \, dx = \mathrm{U}_1 a^x - \log a \int \mathrm{U}_1 a^x \, dx.$$

$$= \frac{\mathrm{U} a^x}{\log a} - \frac{\mathrm{U}' a^x}{(\log a)^2} + \frac{\mathrm{U}'' a^x}{(\log a)^3} - \cdots = \mathrm{U}_1 a^x - \mathrm{U}_2 a^x \log a + \mathrm{U}_3 a^x \left(\log a\right)^2 - \cdots$$

$$\int \frac{a^x}{x} \, dx = \log x + \frac{x \log a}{1} + \frac{1}{2} \frac{x^2 (\log a)^2}{1.2} + \frac{1}{3} \frac{x^3 (\log a)^3}{1.2.3} + \frac{1}{4} \frac{x^4 (\log a)^4}{1.2.3.4} + \cdots$$

$$\int \frac{a^x}{x^2} \, dx = -\frac{a^x}{x} + \log a \int \frac{a^x}{x} \, dx.$$

$$\int \frac{a^x}{x^3} \, dx = -\frac{a^x}{2x^2} - \frac{a^x \log a}{2x} + \frac{(\log a)^2}{2} \int \frac{a^x}{x} \, dx.$$

$$\int \frac{a^x}{x^m} \, dx = -\frac{a^x}{(m-1) x^{m-1}} - \frac{a^x \log a}{(m-1)(m-2) x^{m-2}} - \frac{a^x (\log a)^2}{(m-1)(m-2)(m-3) x^{m-3}} - \cdots - \frac{a^x (\log a)^{m-2}}{(m-1)\cdots 2.1\, x}$$

$$+ \frac{(\log a)^{m-1}}{(m-1)\cdots 2.1} \int \frac{a^x}{x} \, dx.$$

$$\int \frac{a^x}{\sqrt{x}} \, dx = \frac{a^x}{\sqrt{x}} \left\{ \frac{1}{\log a} + \frac{1}{2x (\log a)^2} + \frac{1.3}{2^2 x^2 (\log a)^3} + \frac{1.3.5}{2^3 x^3 (\log a)^4} + \cdots \right\}.$$

$$= \frac{a^x}{\sqrt{x}} \left\{ \frac{2x}{1} - \frac{2^2 x^2 \log a}{1.3} + \frac{2^3 x^3 (\log a)^2}{1.3.5} - \frac{2^4 x^4 (\log a)^3}{1.3.5.7} + \cdots \right\}.$$

$$\int \frac{a^x}{1-x} \, dx = a^x \left\{ \frac{1}{(1-x) \log a} - \frac{1}{(1-x)^2 (\log a)^2} + \frac{1.2}{(1-x)^3 (\log a)^3} - \frac{1.2.3}{(1-x)^4 (\log a)^4} + \cdots \right\}.$$

e^{U} cir V

$$\int e^x \sin x \, dx = \frac{1}{2} e^x (\sin x - \cos x).$$

$$\int e^{ax} \sin x \, dx = \frac{e^{ax} (a \sin x - \cos x)}{1 + a^2}.$$

$$\int e^{ax} \sin bx \, dx = \frac{e^{ax} (a \sin bx - b \cos bx)}{a^2 + b^2}.$$

$$\int x e^x \sin x \, dx = \frac{1}{2} x e^x (\sin x - \cos x) + \frac{1}{2} e^x \cos x.$$

$$\int x^m e^x \sin x\, dx = \frac{1}{2} x^m e^x (\sin x - \cos x) - \frac{m}{2}\int x^{m-1} e^x \sin x\, dx + \frac{m}{2}\int x^{m-1} e^x \cos x\, dx.$$

$$\int x^m e^{ax} \sin bx\, dx = x^m e^{ax} \frac{a \sin bx - b \cos bx}{a^2 + b^2} - \frac{m}{a^2 + b^2}\int x^{m-1} e^{ax} (a \sin bx - b \cos bx)\, dx.$$

$$= e^{ax}\left[\frac{1}{\rho} x^m \sin (bx - \alpha) - \frac{m}{\rho^2} x^{m-1} \sin (bx - 2\alpha) + \cdots \pm \frac{m(m-1)\cdots 1}{\rho^{m+1}} \sin \left\{ bx - (m+1)\alpha \right\}\right].$$

where $a + b\sqrt{-1} = \rho (\cos \alpha + \sqrt{-1} \sin \alpha)$.

$$\int e^{ax} \sin^2 x\, dx = \frac{e^{ax} \sin x (a \sin x - 2 \cos x)}{4 + a^2} + \frac{2e^{ax}}{a(4 + a^2)}.$$

$$= \frac{e^{ax}}{a(4 + a^2)}\left(a^2 \sin^2 x - a \sin 2x + 2\right).$$

$$\int e^{ax} \sin^3 x\, dx = \frac{e^{ax} \sin^2 x (a \sin x - 3 \cos x)}{a^2 + 9} + \frac{6 e^{ax} (a \sin x - \cos x)}{(a^2 + 9)(a^2 + 1)}.$$

$$\int e^{ax} \sin^n x\, dx = \frac{e^{ax} \sin^{n-1} x (a \sin x - n \cos x)}{n^2 + a^2} + \frac{n(n-1)}{n^2 + a^2}\int e^{ax} \sin^{n-2} x\, dx.$$

$$\int e^{ax} \sin^n bx\, dx = \frac{e^{ax} \sin^{n-1} bx (a \sin bx - nb \cos bx)}{a^2 + n^2 b^2} + \frac{n(n-1) b^2}{a^2 + n^2 b^2}\int e^{ax} \sin^{n-2} bx\, dx.$$

$$\int \frac{\sin x}{e^x}\, dx = -\frac{\sin x + \cos x}{2e^x}.$$

$$\int \frac{x \sin x}{e^x}\, dx = -\frac{x(\sin x + \cos x)}{2e^x} - \frac{\cos x}{2e^x}.$$

$$\int \frac{e^{ax}}{\sin^n x}\, dx = -\frac{e^{ax}\left\{ a \sin x + (n-2) \cos x \right\}}{(n-1)(n-2) \sin^{n-1} x} + \frac{a^2 + (n-2)^2}{(n-1)(n-2)}\int \frac{e^{ax}}{\sin^{n-2} x}\, dx.$$

$$\int e^x \cos x\, dx = \frac{1}{2} e^x (\sin x + \cos x).$$

$$\int e^{ax} \cos x\, dx = \frac{e^{ax} (\sin x + a \cos x)}{1 + a^2}.$$

$$\int e^{ax} \cos bx\, dx = \frac{e^{ax} (a \cos bx + b \sin bx)}{a^2 + b^2}.$$

$$\int x e^x \cos x\, dx = \frac{1}{2} x e^x \left(\sin x + \cos x\right) - \frac{1}{2} e^x \sin x.$$

$$\int x^m e^x \cos x\, dx = \frac{1}{2} x^m e^x \left(\sin x + \cos x\right) - \frac{m}{2}\int x^{m-1} e^x \sin x\, dx - \frac{m}{2}\int x^{m-1} e^x \cos x\, dx.$$

$$\int x^2 e^{ax} \cos bx\, dx = \frac{e^{ax}}{c^3}\left\{ c^2x^2 \cos (bx - \beta) - 2cx \cos (bx - 2\beta) + 2 \cos (bx - 3\beta) \right\}.$$

where $a = c \cos \beta$, $b = c \sin \beta$.

$$\int x^m e^{ax} \cos bx\, dx = x^m e^{ax} \frac{a \cos bx + b \sin bx}{a^2 + b^2} - \frac{m}{a^2 + b^2}\int x^{m-1} e^{ax} (a \cos bx + b \sin bx)\, dx.$$

$$= e^{ax}\left[\frac{1}{\rho} x^m \cos (bx - \alpha) - \frac{m}{\rho^2} x^{m-1} \cos (bx - 2\alpha) + \cdots \pm \frac{m(m-1)\cdots 1}{\rho^{m+1}} \cos \left\{ bx - (m+1)\alpha \right\}\right].$$

where $a + b\sqrt{-1} = \rho (\cos \alpha + \sqrt{-1} \sin \alpha)$.

$$\int e^{ax} \cos^2 x\, dx = \frac{e^{ax} \cos x\,(2 \sin x + a \cos x)}{4 + a^2} + \frac{2e^{ax}}{a\,(4 + a^2)}.$$

$$= \frac{e^{ax}}{a\,(a^2 + 4)}\,(a^2 \cos^2 x + a \sin 2x + 2).$$

$$\int e^{ax} \cos^3 x\, dx = \frac{e^{ax} \cos^2 x\,(3 \sin x + a \cos x)}{9 + a^2} + \frac{6\, e^{ax}\,(\sin x + a \cos x)}{(1 + a^2)(9 + a^2)}.$$

$$\int e^{ax} \cos^n x\, dx = \frac{e^{ax} \cos^{n-1} x\,(a \cos x + n \sin x)}{a^2 + n^2} + \frac{n\,(n-1)}{a^2 + n^2} \int e^{ax} \cos^{n-2} x\, dx.$$

$$\int e^{ax} \cos^n bx\, dx = \frac{e^{ax} \cos^{n-1} bx\,(a \cos bx + nb \sin bx)}{a^2 + n^2 b^2} + \frac{n\,(n-1)\, b^2}{a^2 + n^2 b^2} \int e^{ax} \cos^{n-2} bx\, dx.$$

$$\int \frac{\cos x}{e^x}\, dx = \frac{\sin x - \cos x}{2e^x}.$$

$$\int \frac{\cos x}{e^{ax}}\, dx = \frac{\sin x - a \cos x}{(1 + a^2)\, e^{ax}}.$$

$$\int \frac{x \cos x}{e^x}\, dx = \frac{x\,(\sin x - \cos x)}{2e^x} + \frac{\sin x}{2e^x}.$$

$$\int \frac{e^{ax}}{\cos^n x}\, dx = -\frac{e^{ax} \left\{ a \cos x - (n-2) \sin x \right\}}{(n-1)\,(n-2) \cos^{n-1} x} + \frac{a^2 + (n-2)^2}{(n-1)\,(n-2)} \int \frac{e^{ax}}{\cos^{n-2} x}\, dx.$$

$$\int e^{ax} \sin x \cos x\, dx = \frac{e^{ax}\,(a \sin 2x - 2 \cos 2x)}{2\,(a^2 + 4)}.$$

$$\int e^{ax} \sin x \cos bx\, dx = \frac{e^{ax}}{c} \left\{ (a \sin x - \cos x) \cos (bx - \beta) - b \sin x \sin (bx - \beta) \right\}.$$

where $\quad 1 + a^2 - b^2 = c \cos \beta, \qquad 2ab = c \sin \beta.$

$$\int e^{ax} \cos^m x \sin^n x\, dx = \frac{e^{ax} \cos^{m-1} x \sin^n x \left\{ a \cos x + (m+n) \sin x \right\}}{(m+n)^2 + a^2}.$$

$$- \frac{na}{(m+n)^2 + a^2} \int e^{ax} \cos^{m-1} x \sin^{n-1} x\, dx + \frac{(m-1)\,(m+n)}{(m+n)^2 + a^2} \int e^{ax} \cos^{m-2} x \sin^n x\, dx.$$

$$= \frac{e^{ax} \cos^m x \sin^{n-1} x \left\{ a \sin x - (m+n) \cos x \right\}}{(m+n)^2 + a^2}$$

$$+ \frac{ma}{(m+n)^2 + a^2} \int e^{ax} \cos^{m-1} x \sin^{n-1} x\, dx + \frac{(n-1)\,(m+n)}{(m+n)^2 + a^2} \int e^{ax} \cos^m x \sin^{n-2} x\, dx.$$

$$= \frac{e^{ax} \cos^{m-1} x \sin^{n-1} x\,(a \sin x \cos x + m \sin^2 x - n \cos^2 x)}{(m+n)^2 + a^2}$$

$$+ \frac{m\,(m-1)}{(m+n)^2 + a^2} \int e^{ax} \cos^{m-2} x \sin^n x\, dx + \frac{n\,(n-1)}{(m+n)^2 + a^2} \int e^{ax} \cos^m x \sin^{n-2} x\, dx.$$

$$= \frac{e^{ax} \cos^{m-1} x \sin^{n-1} x\,(a \cos x \sin x + m \sin^2 x - \cos^2 x)}{(m+n)^2 + a^2}$$

$$+ \frac{m\,(m-1)}{(m+n)^2 + a^2} \int e^{ax} \cos^{m-2} x \sin^{n-2} x\, dx + \frac{(n-m)\,(n+m-1)}{(m+n)^2 + a^2} \int e^{ax} \cos^m x \sin^{n-2} x\, dx$$

$$\int e^{ax} \operatorname{tg}^2 x\, dx = \frac{e^{ax}}{a} \left(a \operatorname{tg} x - 1 \right) - a \int e^{ax} \operatorname{tg} x\, dx.$$

$$\int e^{ax} \operatorname{tg}^3 x\, dx = \frac{e^{ax}}{2}\left(\operatorname{tg}^2 x - a \operatorname{tg} x + 1\right) + \frac{1}{2}\left(a^2 - 2\right)\int e^{ax} \operatorname{tg} x\, dx.$$

$$\int e^{ax} \operatorname{tg}^4 x\, dx = \frac{e^{ax}}{6a}\left\{ 2a \operatorname{tg}^3 x - a^2 \operatorname{tg}^2 x + a(a^2 - 6)\operatorname{tg} x - a^2 + 6 \right\} - \frac{a(a^2 - 8)}{6}\int e^{ax} \operatorname{tg} x\, dx.$$

$$\int e^{ax} \operatorname{tg}^n x\, dx = \frac{e^{ax} \operatorname{tg}^{n-1} x}{n-1} - \frac{a}{n-1}\int e^{ax} \operatorname{tg}^{n-1} x\, dx - \int e^{ax} \operatorname{tg}^{n-2} x\, dx.$$

Diverse Functions

$$\int \operatorname{séc} x \operatorname{tg} x\, dx = \operatorname{séc} x.$$

$$\int \operatorname{cosé c} x \operatorname{cotg} x\, dx = -\operatorname{cosé c} x.$$

$$\int \log(a + \cos x)\, dx = x \log(a + \cos x) + \int \frac{x \sin x}{a + \cos x}.$$

$$\int \cos x \log \sin x\, dx = \sin x(\log \sin x - 1).$$

$$\int \sin x \log \cos x\, dx = \cos x(1 - \log \cos x).$$

$$\int \sin \log x\, dx = \frac{x}{2}\left(\sin \log x - \cos \log x\right).$$

$$\int \frac{\cos(\log x)}{x}\, dx = \sin(\log x).$$

$$\int e^{\cos x} \sin x\, dx = -e^{\cos x}.$$

$$\int \frac{x^m}{\sqrt{\log \frac{1}{x}}}\, dx = \frac{x^{m+1}}{(m+1)\sqrt{\log \frac{1}{x}}}\left\{ 1 + \frac{1}{(2m+2)\log x} + \frac{1.3}{\{(2m+2)\log x\}^2} + \frac{1.3.5}{\{(2m+2)\log x\}^3} + \cdots \right\}$$

$$\int \sqrt{1 + e^{ax}}\, dx = \frac{2}{a}\sqrt{1 + e^{ax}} + \frac{1}{a}\log \frac{\sqrt{1 + e^{ax}} + 1}{\sqrt{1 + e^{ax}} - 1}.$$

$$\int \frac{dx}{\sqrt{e^x + a^2}} = \frac{2}{a}\log\left\{ \sqrt{e^x + a^2} - a \right\} - \frac{x}{a}.$$

$$\int \frac{dx}{\sqrt{e^{2x} - 1}} = \operatorname{arc\,cos} e^{-x}.$$

$$\int \frac{e^{mx}}{\sqrt{a + be^{mx}}}\, dx = \frac{2}{bm}\sqrt{a + be^{mx}}.$$

$$\int x^{nx}\, dx = x\left\{ 1 - \frac{nx}{2^2} + \frac{n^2x^2}{3^3} - \frac{n^3x^3}{4^4} + \cdots \right\} + \frac{nx^2 \log x}{1}\left\{ \frac{1}{2} - \frac{nx}{3^2} + \frac{n^2x^2}{4^3} - \frac{n^3x^3}{5^4} + \cdots \right\}$$

$$+ \frac{n^2x^3(\log x)^2}{1.2}\left\{ \frac{1}{3} - \frac{nx}{4^2} + \frac{n^2x^2}{5^3} - \frac{n^3x^3}{6^4} + \cdots \right\} + \frac{n^3x^4(\log x)^3}{1.2.3}\left\{ \frac{1}{4} - \frac{nx}{5^2} + \frac{n^2x^2}{6^3} - \frac{n^3x^3}{7^4} + \cdots \right\} + \cdots$$

$$\int x^{m+nx}\, dx = \int x^m\left\{ 1 + nx \log x + \frac{1}{2}\left(nx \log x\right)^2 + \cdots \right\} dx.$$